Recent Advances in Biostatistics

False Discovery Rates, Survival Analysis, and Related Topics

SERIES IN BIOSTATISTICS

Series Editor: Heping Zhang *(Yale University School of Medicine, USA)*

Vol. 1 Development of Modern Statistics and Related Topics
— In Celebration of Prof. Yaoting Zhang's 70th Birthday
edited by Heping Zhang & Jian Huang

Vol. 2 Contemporary Multivariate Analysis and Design of Experiments
— In Celebration of Prof. Kai-Tai Fang's 65th Birthday
edited by Jianqing Fan & Gang Li

Vol. 3 Advances in Statistical Modeling and Inference
— Essays in Honor of Kjell A. Doksum
edited by Vijay Nair

Vol. 4 Recent Advances in Biostatistics:
False Discovery Rates, Survival Analysis, and Related Topics
edited by Manish Bhattacharjee, Sunil K. Dhar & Sundarraman Subramanian

Series in Biostatistics – Vol. 4

Recent Advances in Biostatistics

False Discovery Rates, Survival Analysis, and Related Topics

Editors

Manish Bhattacharjee
Sunil K Dhar
Sundarraman Subramanian

New Jersey Institute of Technology, USA

World Scientific

NEW JERSEY • LONDON • SINGAPORE • BEIJING • SHANGHAI • HONG KONG • TAIPEI • CHENNAI

Published by

World Scientific Publishing Co. Pte. Ltd.
5 Toh Tuck Link, Singapore 596224
USA office: 27 Warren Street, Suite 401-402, Hackensack, NJ 07601
UK office: 57 Shelton Street, Covent Garden, London WC2H 9HE

British Library Cataloguing-in-Publication Data
A catalogue record for this book is available from the British Library.

Series in Biostatistics — Vol. 4
RECENT ADVANCES IN BIOSTATISTICS
False Discovery Rates, Survival Analysis, and Related Topics

ISBN-13 978-981-4329-79-8
ISBN-10 981-4329-79-7

Printed by FuIsland Offset Printing (S) Pte Ltd. Singapore

To Our Parents

Contents

Foreword

The present volume, a title in the *Series in Biostatistics*, published by World Scientific Publishing, is a noteworthy collection of some recent research in several themes of contemporary interest in biostatistics. Its contents cover a broad spectrum ranging from review articles to applications of cutting edge statistical methodology, as well as development of new methods and results.

The articles in this volume are based on research presented in a recent conference at the New Jersey Institute of Technology, which over the years has organized an annual conference series entitled "Frontiers in Applied and Computational Mathematics" with focus on several areas in mathematical sciences such as applied mathematics, mathematical biology, and statistics. As expected, the representation of statistical sciences in these conferences has grown significantly over time. Within this broad spectrum, one burgeoning field of research activity with vast scope of interdisciplinary applications is biostatistics, which embraces important areas such as survival analysis, clinical trials, bioinformatics/genomics, and false discovery rates.

It is indeed a pleasure to welcome this volume to the statistics and biostatistics literature and to write this foreword at the request of Manish Bhattacharjee, Sunil K. Dhar and Sundarraman Subramanian, who have ably edited this collection. As editors, they have invested significant effort in screening the articles for their relevance and timeliness, and guiding the whole process through to publication with due care given to established practices for reporting of scientific research that includes getting thorough peer reviews from knowledgeable experts. It should provide inspiration for further fruitful interaction between methodology and applications in a diverse interdisciplinary field.

Pranab K. Sen
University of North Carolina, Chapel Hill
December 1, 2010

Preface

Every book has a story behind its origin, and this one is no exception. The idea for this volume was conceived in the summer of 2009 during a conference at the New Jersey Institute of Technology, hosted by the Department of Mathematical Sciences, which organizes an annual meeting on various focus areas and themes within the broad umbrella of Mathematical Sciences. These annual "Frontiers in Applied and Computational Mathematics" (FACM) conferences bring together researchers to share their recent research and exchange ideas on contemporary developments and trends that provide a glimpse into where their specialties are headed. One of the focus areas in these meetings is biostatistics that continues to grow in importance both as an area of meaningful applications of statistics to public health as well as a clearinghouse for posing problems with novel statistical challenges which in turn foster new statistical methods to address them.

Our thinking in putting together this edited volume and share the articles herein with a wider audience has been shaped by the belief that such a collection would be useful and of interest to advanced graduate students, researchers, and practitioners of biostatistics. World Scientific Publishing Co., well known for its commitment to scholarly publishing, has been a willing partner in this endeavor. The papers for the biostatistics sessions in the 2009 conference appeared to us to be a particularly well balanced mix of methodology and applications, representing traditional areas of continuing interest such as survival analysis as well as topics of more recent vintage such as false discovery rates and multiple testing methods that are currently still undergoing active development at an accelerated pace.

The present volume is not a conference proceedings in the sense in which it is usually understood. While it primarily consists of papers given at the conference, it would be more accurate to say that the articles included in this volume are *based* on presentations in the biostatistics sessions in the FACM 2009 meeting. In several instances, the articles as they appear here have undergone substantial changes and modifications in scope and coverage in the interest of timeliness, relevance and contemporary research importance. Most of the articles are based on presentations by invited speakers who are eminent specialists and recognized experts in their respective fields, and a few have been chosen from among the contributed papers for their relevance and interest. Each article has gone through a careful peer review and corresponding revision(s) based on constructive suggestions by the reviewers. We gratefully acknowledge the refereeing services provided by a distinguished panel of reviewers, who come from reputed institutions such as University of Michigan; University of North Carolina, Chapel Hill; North Carolina State University, Raleigh; University of Wisconsin, Madison; Computational Biology Center at the Memorial Sloan Kettering Cancer Center, New York; University of Aachen, Germany; and Tilburg University, Netherlands — to name a few.

We thank our colleague and the Chair of our home department, Professor D. S. Ahluwalia, for encouraging us to plan and organize Statistics and Biostatistics sessions in the annual FACM meetings over the last several years, and for providing corresponding funding support; without which these sessions and hence the present volume would not have been possible.

We owe a special debt of gratitude to the eminent researcher and academician Dr. Pranab K. Sen, Professor of Biostatistics, Statistics and Operations Research at the University of North Carolina, Chapel Hill, and the recipient of the 2010 distinguished S.S. Wilks award of the American Statistical Association, for his unwavering support of our efforts. He has been an inspiration to us and we thank him for his advice and counsel that have been invaluable.

It has been a pleasure to work with Ms. Ling Xiao of the Singapore office of World Scientific throughout the process of editing this volume, and we would like to thank her for her professionalism. We also thank Ms. Jessie Tan for her technical assistance in preparing final camera ready copies, which she promptly provided to several authors when such help was asked for. Finally and most importantly, we would like to thank all the authors for patiently working with us in a timely manner throughout the editorial process to ensure that the articles appearing here meet the accepted standards of peer review and scientific publishing.

Manish Bhattacharjee
Sunil K. Dhar
Sundarraman Subramanian
New Jersey Institute of Technology
December 1, 2010

Overview

This volume has 15 chapters authored by leading researchers on a broad range of topics in biostatistics, and is divided into three main parts, consisting of chapters on false discovery rates and multiple testing, survival analysis, and other related topics such as clinical trials, genomics and bioinformatics. Almost all the articles include an application of the methods proposed and discussed, to real life medical data sets or/and appropriate simulation studies; features that will be appreciated by readers focused on applications. What follows is a brief overview of each part and the articles therein.

The theme of Part I, which consists of three articles, is on recent research developments in "False Discovery Rates" and associated multiple testing methods, an area of active current research. Each of the three contributions here breaks new grounds by proposing and investigating new methods and results which should be of significant interest to the community of researchers in multiple testing methods.

Liu and Sarkar, in their chapter, propose an adaptive method as an alternative to the method proposed in 2006 by Benjamini, Krieger and Yekutieli (BKY), for controlling the false discovery rate, by using a different estimate of the number of true null hypotheses. The new method that the authors propose also controls the false discovery rate (FDR) under independence as does the BKY method. Using simulation studies, the authors show that their proposed method (i) can control the FDR under positive dependence of the p-values, and is more powerful than the BKY method under positive but not very high correlations among the test statistics, and (ii) appears to outperform the BKY method even under high correlations, if the fraction of true null hypotheses is large. A comparative illustration of the methods using a benchmark breast cancer data is provided.

Control of the familywise error rate (FWER) in multiple testing is one of the two main approaches that deal with the multiplicity problem, which refers to the fact of sharply increasing probability of falsely rejecting at least one true null hypothesis as the number of hypotheses tested increases; the other approach being control of the false discovery rate (FDR). In their article on multiple testing procedures, Guo, Sarkar and Peddada present new adaptive test procedures which are shown to control the FWER in finite samples under several well-known types of positive dependence. Using a simulation study, the proposed adaptive tests are also shown to be more powerful than the corresponding non-adaptive procedures.

In multiple testing problems, when at least one of the hypotheses uses a categorical data endpoint, it is possible to further increase the power of procedures that control FWER (Tarone, 1990) or FDR (Gilbert, 2005), by exploiting the discreteness of the test statistic's distribution to essentially reduce the effective number of hypotheses considered for the multiplicity adjustment. Heyse introduces a modified fully discrete FDR sequential procedure using the exact conditional distribution of potential study outcomes, for which he demonstrates FDR control and power gains using simulation. The author includes a discussion of potential uses of the method and reviews an application of the proposed FDR procedure in a genetic data analysis setting.

Survival Analysis is an area of traditional interest in biostatistics, which remains a focus of continuing new developments. The eight articles in Part II focus on several facets of this active area of research and cover a broad range of topics that include new estimators of survival and hazard functions, modeling and analysis of recurrent event data with dependent censoring, failure time mixture models, Bayesian approaches to modeling survival data, and new Gibbs sampling algorithms for Bayesian regression.

The focus of the article by Luo and Tsai, is on estimating conditional cumulative hazard and survival functions for censored time-to-event data, in which their first contribution is the extension of asymptotic results to the entire support of covariates. This is achieved via a smart exploitation of Müller–Wang boundary kernels. The authors also obtain improved rates for remainder terms in asymptotic representations for estimators. The new methodological contributions and results reported in this article will be appreciated by readers with a preference for theoretical rigor that ultimately justify their use in applied data driven contexts.

Lee, Scheike and Sun, in their article, propose a mixture model for the event-free survival probability in a population that is a mixture of both susceptible and non-susceptible or cured subjects. This has been a continuing research focus, with several proposals on how to model each component of the mixture that the event-free survival probability represents. The authors propose failure time regression models with change points both in the latency hazard function model and logistic regression model for the cure fraction. The authors conduct a simulation study to check the performance of the proposed estimation and testing methods, and illustrate their use through an application to the melanoma survival study.

To model survival data, Demarqui, Dey, Loschi, and Colosimo consider a full Bayesian version of the piecewise exponential model (PEM) with random time grid, with a joint non-informative improper prior distribution, and show that the clustering structure of the model can be adapted to accommodate improper prior distributions in the framework of the PEM. They discuss the model properties, provide comparison with existing approaches and present an illustration of their method using a real data set.

Statistical analysis of recurrent time to event, an important topic in many applied fields, is of continuing interest to biostatisticians in the context of survival analysis methods. In a paper, which many readers will find to be of current interest, authors Amorim, Cai and Zeng, consider proportional rate models for such data under dependent censoring, which often occurs in recurrent event data, leading to complications in their analysis that are largely absent for independent censoring models. They review two methods for analyzing recurrent event data with dependent censoring and perform a comparison study between them. The methods attempt to surmount the difficulty brought about by the dependence through additional modeling requirements leading to complete or conditional independence between the censoring and the recurrent event process. Based on simulation results, the authors conclude that the approaches are effective for handling dependent censoring when the source of informative censoring is correctly specified.

Farias and Branco's article is an example of the increasing popularity and use of computationally intensive Bayesian approaches such as Markov Chain Monte Carlo (MCMC) methods in many areas of statistics and biostatistics. They propose new Gibbs sampling algorithms for Bayesian regression models for binary data using latent variables, and skew link functions which are more flexible compared to symmetric ones for modeling binary data. Specifically, they propose and investigate two new simulation algorithms for a Bayesian skew-probit regression model, which are then compared for two different measures of efficiency from which the authors find that one of the proposed algorithms leads to around 160% improvement in the effective sample size measure of efficiency. An application to an actual medical data set illustrating the methods developed is included.

A robust M-estimator proposed by Lim, Sen and Peddada and investigation of its asymptotic properties is the subject of the next article. The article is motivated by applications of statistical methods in toxicology, where researchers are interested in developing nonlinear statistical models to describe the relationship between a response variable and a set of covariates. More specifically, the ordinary least squares and maximum likelihood estimators typically would not perform well when the response distribution is different from the assumed one or when the error variance changes with changes in the covariate levels or both. In such cases robust M-estimation methods would be preferable. The authors illustrate the use of their proposed methodology with real data from a study conducted by the US National Toxicology Program.

Subramanian and Zhang investigate inverse probability of censoring weighted estimation of survival functions from homogeneous left and right censored data. The equivalence between the Kaplan-Meier and inverse censoring weighted estimators that holds under right censoring breaks down in the simplified double censoring scheme in which left censoring is always observed. Interesting extensions to the non-homogeneous case are also discussed. The authors provide an illustration of the proposed inverse censoring weighted estimator using an AIDS clinical trials data set, as well as numerical comparisons among the different estimators using simulation.

Competing risks analysis, highly relevant in medical research, is the subject of a review article by Kim. The inter-relationship between competing events manifests in dependent censoring, wherein an event of interest is censored by competing events that vie for "first event" honors. Much research has focused on cumulative incidence functions, which are useful for decision making regarding optimal treatment. In her review article on analysis and design of competing risks data in clinical research, Kim reviews a number of issues related to cumulative incidence functions, including semi-parametric estimation and model selection, with practical illustrations.

Part III consists of four articles in the areas of genomics/bioinformatics, medical imaging and diagnosis, and clinical trials, respectively. The use of information theoretic ideas in comparative genomic studies is the subject of the first article here. The next two articles that follow are on the application of data analytic methods in clinical settings in the context of medical imaging and diagnosis. The last article in Part III considers a new sequential screening design of treatments in the context of clinical trials.

A main goal of comparative genomic studies and bioinformatics is to unravel the molecular evolution of proteins. With the advent of high-throughput genome sequencing and the massive amount of data generated by sequencings, the possibilities of exploiting novel approaches to enhance our understanding of protein structure evolution have multiplied. Attributes pertinent to function diverge slowly during protein evolution and despite sequence diversity across species, structural conservation has been observed to be a salient feature in protein evolution. In their article, authors Fatakia, Costanzi and Chow utilize graph theoretic ideas and concepts from information theory to discuss a method they have recently proposed, to study the super-family of human G protein-coupled receptors (GPCRs), evolved from some common ancestral gene, to identify statistically related positions that share high mutual information and close spatial proximity with each other in amino acid sequences of the protein family. Such information is important when attempting to infer protein structure. The usefulness of these studies derives from the fact that the GPCR super-family is one of the most diverse and the largest super-families of membrane proteins in humans, involved in a variety of essential physiological functions and is a vital target for pharmaceutical intervention.

Medical imaging and diagnosis are getting increasing attention from biostatistics researchers and practitioners in recent years. Statistical modeling and analysis of Positron Emission Tomography (PET) as a diagnostic imaging tool, used in many areas of biomedical research, such as oncology, pharmacology, and psychiatry, is the subject of a review article by Chang and Ogden, who introduce the readers to PET data acquisition and some commonly used kinetic models for their analysis. The authors indicate how PET can be used to obtain the distribution of an appropriate neuro-receptor in the brain as an aid for diagnosis of MDD, a mental illness associated with a pervasive low mood, and loss of interest in usual activities.

In the context of medical decision making, Rindskopf discusses the use of latent class analysis, which hypothesizes the existence of unobserved categorical variable(s) to explain the relationships among a set of observed categorical variables, as an aid to medical diagnosis. In a medical context, the observed variables are signs, symptoms, or test results, usually dichotomized (positive or negative); while the latent variable is the true status of the disease, also often dichotomous (presence or absence of disease). He provides an overview of latent class analysis with an illustration with medical data sets, and discusses the advantages of such analysis over traditional methods of estimating sensitivity and specificity.

Leu, Cheung and Levin propose a class of sequential procedures in multi-arm randomized clinical trials within a large Phase III setting when only limited resources are available, to introduce a novel way that selects a subset of treatments which offer clinically meaningful improvements over the control group, if such a subset exists, preserving the type I error rates, and declares a null result otherwise. The comparative subset selection trials can be implemented on a monitored flexible calendar time schedule. While the method suggested in this article may not be the current industry practice for screening of promising treatment regimens via clinical trials, according to the authors, FDA has shown some interest in the type of methodologies covered here, and should therefore be of potential interest.

Manish Bhattacharjee
Sunil K. Dhar
Sundarraman Subramanian
New Jersey Institute of Technology
December 1, 2010

Part I

False Discovery Rates

Chapter 1

A New Adaptive Method to Control the False Discovery Rate

Fang Liu* and Sanat K. Sarkar†

Temple University, Statistic Department
1810 North 13th Street, Philadelphia, PA, 19122-6083, USA
**fangliu@temple.edu*
†sanat@temple.edu

Benjamini, Krieger and Yekutieli (BKY, 2006) have given an adaptive method of controlling the false discovery rate (FDR) by incorporating an estimate of n_0, the number of true null hypotheses, into the FDR controlling method of Benjamini and Hochberg (BH, 1995). The BKY method improves the BH method in terms of the FDR control and power. Benjamini, Krieger and Yekutieli have proved that their method controls the FDR when the p-values are independent and provided numerical evidence showing that the control over the FDR continues to hold when the p-values have some type of positive dependence. In this paper, we propose an alternative adaptive method via a different estimate of n_0. Like the BKY method, this new method controls the FDR under the independence, and can maintain a control over the FDR, as shown numerically, under the same type of positive dependence of the p-values. More importantly, as our simulations indicate, the proposed method can often outperform the BKY method in terms of the FDR control and power, particularly when the correlation between the test statistics is moderately low or the proportion of true null hypotheses is very high. When applied to a real microarray data, the new method is seen to pick up a few more significant genes than the BKY method.

1.1. Introduction

Multiple hypothesis testing plays a pivotal role in analyzing data from modern scientific investigations, such as DNA microarray, functional magnetic resonance imaging (fMRI), and many other biomedical studies. For instance, identification of differentially expressed genes across various experimental conditions in a microarray study or active voxels in an fMRI study is carried out through multiple testing. Since these investigations typically require tens and thousands of hypotheses to be tested simultaneously, the traditional multiple testing methods, like those designed to control the probability of at least one false rejection, the familywise error rate (FWER), become too conservative to use in these investigations. Benjamini and Hochberg (1995) introduced the false discovery rate (FDR), the expected proportion of false rejections among all rejections, which is less conservative than the FWER and has become the most popular measure of type I error rate in modern multiple testing.

Benjamini and Hochhberg (1995) gave a method, referred to as the BH method, for controlling the FDR. The FDR of this method at level α is equal to $n_0\alpha/n$, where n_0 is the number of true null hypotheses, when the underlying test statistics are independent, and less than or equal to $n_0\alpha/n$ when these statistics are positively dependent in a certain sense [Benjamini and Hochberg (1995), Benjamini and Yekutieli (2001) and Sarkar (2002)]. Since n_0 is unknown, by estimating it and modifying the BH method using this estimate can potentially make the BH method less conservative and thus more powerful. A number of such adaptive BH methods have been proposed in the literature, among which the one in Benjamini, Krieger and Yekutieli (2006) has received much attention and will be our main focus in this paper.

We consider estimating n_0 using a different estimate than the one considered in Benjamini, Krieger and Yekutieli (2006) before modifying the BH method. Like the BKY method, this new adaptive version of the BH method is proved to control the FDR when the p-values are independent and numerically shown to control the FDR under normal distributional setting with equal positive correlation. Moreover, as our simulations indicate, it outperforms the BKY method, in the sense of providing better FDR control and power, when the correlation between the test statistics is moderately low or the proportion of true null hypotheses is quite large.

This paper is organized as follows. We start with a background in the next section for our proposed method providing notations, the definition of the FDR, and some basic formulas. Section 1.3 revisits some FDR controlling methods, especially adaptive FDR controlling methods. The new estimate of n_0 is proposed in Sec. 1.4. Our proposed alternative version of adaptive BH method based on this new n_0 estimate is developed in Sec. 1.5. The results of a simulation study conducted to investigate the FDR controlling property and power performance of our proposed method relative to the BKY method are also presented in Sec. 1.5. Both BKY and the new adaptive FDR methods are applied to a real microarray data; the comparative results are presented in Sec. 1.6. The paper concludes with some final remarks made in Sec. 1.7.

1.2. Notation, Definition and Formulas

Consider testing n null hypotheses $H_1, \ldots, H_n$ simultaneously against certain alternatives using their respective p-values $p_1, \ldots, p_n$. A multiple testing of these hypotheses is typically carried out using a stepwise or single-step procedure. Let $p_{1:n} \leq \cdots \leq p_{n:n}$ be the ordered versions of these p-values, with $H_{1:n}, \ldots, H_{n:n}$ being their corresponding null hypotheses. Then, given a non-decreasing set of critical constants $0 < \alpha_1 \leq \cdots \leq \alpha_n < 1$, a *step-up procedure* rejects the set $\{H_{i:n},\ i \leq i^*_{SU}\}$ and accepts the rest, where $i^*_{SU} = \max\{1 \leq i \leq n : p_{i:n} \leq \alpha_i\}$, if the maximum exists, otherwise accepts all the null hypotheses. A *step-down procedure*, on the other hand, rejects the set of null hypotheses $\{H_{i:n},\ i \leq i^*_{SD}\}$ and accepts the rest, where $i^*_{SD} = \max\{1 \leq i \leq n : p_{j:n} \leq \alpha_j\ \forall\ j \leq i\}$, if the maximum exists, otherwise accepts all the null hypotheses. When the constants are same in a step-up or step-down procedure, it reduces to what is defined as a *single-step procedure.*

Let R denote the total number of rejections and V denote the number of those that are false, the type I errors, while testing n null hypotheses using a multiple testing method. Then, the FDR of this method is defined by

$$\text{FDR} = E\left(\text{FDP}\right), \text{ where } \text{FDP} = \frac{V}{\max\{R, 1\}} \tag{1.1}$$

is the false discovery proportion. Different formulas for the FDR of a stepwise procedure - step-up, step-down or single-step - have been considered in different papers [see, for example, Benjamin and Yekutielli (2001), Sarkar (2002, 2006)]. However, we will present an alternative expression for the FDR, given recently in Sarkar (2008b), that provides better insight and will be of use in the present paper.

For any multiple testing method,

$$\text{FDP} = \frac{V}{\max\{R, 1\}} = \sum_{i \in I_0} \sum_{r=1}^{n} \frac{1}{r} I\left(H_i \text{ is rejected},\ R = r\right), \tag{1.2}$$

where I_0 is the set of indices of true null hypotheses. For a step-up procedure, this expectation can be written more explicitly as follows, with P_i denoting the random variable corresponding to the observed p-value p_i.

Formula 1.1. *For a step-up procedure of testing the n null hypotheses $H_1, \ldots, H_n$ using the critical values $\alpha_1 \leq \cdots \leq \alpha_n$, the FDR is given by*

$$\mathrm{FDR} = \sum_{i \in I_0} E\left[\frac{I\left(P_i \leq \alpha_{R^{(-i)}_{SU,n-1}(\alpha_2,\ldots,\alpha_n)+1}\right)}{R^{(-i)}_{SU,n-1}(\alpha_2, \ldots, \alpha_n) + 1}\right],$$

where $R^{(-i)}_{SU,n-1}(\alpha_2, \ldots, \alpha_n)$ is the number of rejections in testing the $n-1$ null hypotheses other than H_i using the step-up procedure based on their p-values and the critical constants $\alpha_2 \leq \cdots \leq \alpha_n$.

By taking $\alpha_i = c$, for all $i = 1, \ldots, n$, in the above formula, one gets the following formula for a single-step procedure that rejects H_i if $p_i \leq c$:

$$\mathrm{FDR} = \sum_{i \in I_0} E\left[\frac{I(P_i \leq c)}{R^{(-i)}_{n-1}(c) + 1}\right],$$

where $R^{(-i)}_{n-1}(c)$ is the number of rejections in testing the $n-1$ null hypotheses other than H_i using the single-step procedure based on the p-values other than p_i.

Formula 1.2. *For a step-down procedure of testing the n null hypotheses $H_1, \ldots, H_n$ using the critical constants $\alpha_1 \leq \cdots \leq \alpha_n$, the FDR satisfies the following inequality:*

$$\mathrm{FDR} \leq \sum_{i \in I_0} E\left[\frac{I\left(P_i \leq \alpha_{R^{(-i)}_{SD,n-1}(\alpha_1,\ldots,\alpha_{n-1})+1}\right)}{R^{(-i)}_{SD,n-1}(\alpha_1, \ldots, \alpha_{n-1}) + 1}\right],$$

where $R^{(-i)}_{SD,n-1}(\alpha_1, \ldots, \alpha_{n-1})$ is the number of rejections in testing the $n-1$ null hypotheses other than H_i using the step-down procedure based on their p-values and the critical constants $\alpha_1 \leq \cdots \leq \alpha_{n-1}$.

1.3. A Review of FDR Controlling Methods

A number of FDR controlling methods have been proposed in the literature, among which the BH method has received the most attention. In this section, we will briefly review this and some of its adaptive versions.

1.3.1. *The BH method*

The BH method is a step-up procedure with the critical values $\alpha_i = i\alpha/n$, $i = 1, \ldots, n$; that is, it rejects the null hypotheses $H_{1:n}, \ldots, H_{r:n}$ and accepts the rest, where

$$r = \max\left\{1 \leq i \leq n : p_{i:n} \leq \frac{i}{n}\alpha\right\}, \tag{1.3}$$

provided this maximum exists; otherwise, accepts all the null hypotheses.

These critical values are the same ones as Simes (1986) originally considered while testing the global null hypotheses $H_0 = \bigcap_{i=1}^n H_i$. Simes also proposed to use them in a step-up manner for multiple testing of the H_i's upon rejection of the global null hypothesis. However, as an FWER controlling method at level α, it works only in a weak sense, that is, when all the null hypotheses are true, with the p-values being either independent (Simes, 1986) or positively dependent in a certain sense [Sarkar and Chang (1997), Sarkar (1998, 2008a)], but it does not work in a strong sense, that is, under any configuration of true and false null hypotheses, even when the p-values are independent (Hommel, 1988). Benjamini and Hochberg (1995) showed that this step-up procedure can be used to control the FDR in a strong sense, at least when the p-values are independent. In particular, they proved that FDR $\leq n_0\alpha/n$ for this method when the p-values are independent with each having $U(0,1)$ distribution under the corresponding null hypothesis.

Later, it was proved that the FDR of the BH method is actually equal to $n_0\alpha/n$ under the independence of the p-values [Benjamini and Yekutieli (2001), Finner and Roters (2001), Sarkar (2002, 2008b), Storey, Taylor and Siegmund (2004), of course, assuming that a null p-value is distributed as $U(0,1)$], and is less than or equal to $n_0\alpha/n$ under the following type of positive dependence among the p-values:

$$\mathrm{E}\left\{\psi\left(P_1, \ldots, P_n\right) \mid P_i = u\right\} \text{ is non-decreasing in } u \text{ for each } i \in I_0, \tag{1.4}$$

for any (coordinatewise) non-decreasing function ψ [Benjamini and Yekutieli (2001), Sarkar (2002, 2008b)]. This is referred to as the positive regression dependence on subset (PRDS) condition, which is satisfied by a number of multivariate distributions arising in many multiple testing situations, among which the multivariate normal with non-negative correlations is the most common. Other commonly arising multivariate distributions for which the BH method works are multivariate t with the associated multivariate normal with non-negative correlations (when $\alpha \leq 1/2$), absolute valued multivariate t with the associated normals being independent and some type of multivariate F [Benjamini and Yekutieli (2001), Sarkar (2002, 2004)].

Sarkar (2002) proved that the step-down analog of the BH method, that is, the method that rejects the null hypotheses $H_{1:n}, \ldots, H_{r:n}$ and accepts the rest, where

$$r = \max\left\{1 \leq i \leq n : p_{j:n} \leq \frac{j}{n}\alpha \text{ for all } j = 1, \ldots, i\right\}, \qquad (1.5)$$

provided this maximum exists, otherwise, accepts all the null hypotheses, also controls the FDR under the independence or the same type of positive dependence as above for the p-values.

The positive dependence condition required for the FDR control of the BH method or its step-down analog can be slightly relaxed from Eq. (1.4) to the following:

$$\mathrm{E}\left\{\psi\left(P_1, \ldots, P_n\right) \mid P_i \leq u\right\} \text{ is non-decreasing in } u \text{ for each } i \in I_0, \qquad (1.6)$$

for any (coordinatewise) non-decreasing function ψ [Finner, Dickhaus and Roters (2009) and Sarkar (2008b)].

If n_0 were known, the step-up procedure with the critical values $\alpha_i = i\alpha/n_0$ for $i = 1, \ldots, n$, would control the FDR precisely at the desired level α, when the p-values are independent. This has been the rationale for considering an adaptive version of the BH method that looks for a way to estimate n_0 with $\hat{n}_0$ from the available data and modifies the BH critical values to $\hat{\alpha}_i = i\alpha/\hat{n}_0$ for $i = 1, \ldots, n$. We will briefly review a number of such adaptive BH methods in the following subsections.

1.3.2. *The adaptive BH method of Benjamini & Hochberg*

Benjamini and Hochberg (2000) introduced this adaptive BH method for independent p-values based on an estimate of n_0 developed using the so called the lowest slope (LSL) method.

When all the null hypotheses are true and the test statistics are independent, the p-values should be iid as $U(0,1)$ with the expectations of the ordered p-values as $E(P_{i:n}) = i/(n+1)$, $i = 1, \ldots, n$. Therefore, the plot of $p_{i:n}$ versus i should exhibit a linear relationship, along the line with the slope $S = 1/(n+1)$ and passing through the origin and the point $(n+1, 1)$ (assuming $p_{n+1:n} = 1$).

When $n_0 \leq n$, the p-values corresponding to the false null hypotheses tend to be small, so they concentrate on the left side of the above plot.

The relationship over the right side of the plot remains approximately linear with the slope $\beta = 1/(n_0+1)$. Therefore, using a suitable set of the largest p-values, a straight line through the point $(n+1, 1)$ can be fitted with slope $\hat{\beta}$ and n_0 can be estimated as $\hat{n}_0 = 1/\hat{\beta}$. Benjamini and Hochberg (2000) suggested estimating n_0 using the LSL method and the corresponding adaptive BH method as follows:

(1) Apply the original BH method. If none is rejected, accept all hypotheses and stop; otherwise, continue.
(2) Calculate the slopes $S_i = 1 - p_{i:n}/(n+1-i)$.
(3) Start with $i = 1$, proceed as long as $S_i \geq S_{i-1}$ and stop when the first time $S_j < S_{j-1}$. Let $\hat{n}_0^{BH} = \min[n, 1/S_j + 1]$.
(4) Apply the BH method with $\alpha_i = i\alpha/\hat{n}_0^{BH}$.

Though there is no theoretical proof that this version of adaptive BH method guarantees an FDR control, simulation studies indicate that it does.

1.3.3. *The adaptive BH method of Storey, Taylor and Siegmund*

Storey, Taylor and Siegmund (2004) used the following estimate of n_0:

$$\hat{n}_0^{STS}(\lambda) = \frac{n - R(\lambda) + 1}{(1-\lambda)}, \quad \text{where } R(\lambda) = \#\{P_i \leq \lambda\}, \tag{1.7}$$

for some $\lambda \in [0, 1)$, and considered the adaptive method with the critical values $\alpha_i = \min\{i\alpha/\hat{n}_0^{STS}, \lambda\}$, $i = 1, \ldots, n$. It controls the FDR under the independence of the p-values [Benjamini, Krieger and Yekutieli (2006), Storey, Taylor and Siegmund (2004), Sarkar (2004, 2008b)], as well as under certain form of weak dependence asymptotically as $n \to \infty$ [Storey, Taylor and Siegmund (2004)].

This adaptive BH method is closely connected to Storey's (2002) estimation based approach to controlling the FDR. Storey (2002) derived a class of point estimates of the FDR for a single-step test that rejects H_i if $p_i \leq t$, for some fixed threshold t, under the following model:

Mixture Model. *Let P_i denote the random p-value corresponding to p_i and $H_i = 0$ or 1 according as the associated null hypothesis is true or false. Let (P_i, H_i), $i = 1, \ldots, n$, be independently and identically distributed with* $\Pr(P_i \leq u \mid H_i) = (1 - H_i)u + H_i F_1(u)$, $u \in (0, 1)$, *for some continuous cdf $F_1(u)$, and* $\Pr(H_i = 0) = \pi_0 = 1 - \Pr(H_i = 1)$.

Having proved that the FDR of the above single-step test for this mixture model is given by

$$\mathrm{FDR}(t) = \frac{\pi_0 t}{F(t)} \Pr\{R(t) > 0\}, \tag{1.8}$$

where

$$F(t) = \Pr(P_i \leq t) = \pi_0 t + (1-\pi_0)F_1(t), \tag{1.9}$$

[see also Liu and Sarkar (2009)], Storey (2002) proposed the following class of point estimates of the FDR(t):

$$\widehat{\mathrm{FDR}}_\lambda(t) = \frac{\hat{\pi}_0(\lambda)t}{\hat{F}(t)}, \quad \lambda \in [0,1) \tag{1.10}$$

where

$$\hat{F}(t) = \frac{1}{n}\max\{R(t),1\} \quad \text{and} \quad \hat{\pi}_0(\lambda) = \frac{\hat{n}_0}{n} = \frac{n - R(\lambda)}{n(1-\lambda)}. \tag{1.11}$$

This estimate of n_0 was originally suggested by Schweder and Spjotvoll (1982) in a different context. Storey (2002) showed that $E(\widehat{\mathrm{FDR}}_\lambda(t)) \geq FDR(t)$, that is, $\widehat{\mathrm{FDR}}_\lambda(t)$ is conservatively biased as an estimate of FDR(t), which he argued is desirable, because by controlling it one can control the true FDR(t). He suggested using

$$t_\alpha = \sup\left\{0 \leq t \leq 1 : \widehat{\mathrm{FDR}}_\lambda(t) \leq \alpha\right\} \tag{1.12}$$

to threshold the p-values, that is, to use it as the cut-off point below which a p-value should be declared significant at a level α. He pointed out that if one approximates t_α by $p_{\hat{l}_\alpha(\lambda):n}$, that is, rejects the null hypotheses $H_{1:n}, \ldots, H_{\hat{l}_\alpha(\lambda):n}$, where

$$\hat{l}_\alpha(\lambda) = \max\{1 \leq i \leq n : \widehat{\mathrm{FDR}}_\lambda(p_{i:n}) \leq \alpha\}, \tag{1.13}$$

then one gets the BH method when $\lambda = 0$. For $\lambda \neq 0$, thresholding the p-values at $p_{\hat{l}_\alpha(\lambda):n}$ is same as using an adaptive BH method. Unfortunately, however, the FDR of such an adaptive BH method is not less than or equal to α, even under independence, unless the $\hat{n}_0$ in Eq. (1.11) is modified, which Storey, Taylor and Siegmund (2004) did.

1.3.4. *The adaptive BH method of Benjamini, Krieger and Yekutieli*

Unlike Storey (2002) or Storey, Taylor and Siegmund (2004) where n_0 is estimated based on the number of significant p-values observed in a single-step test with an arbitrary critical value λ, Benjamini, Krieger and Yekutieli (2006) considered estimating n_0 from the BH method at level $\alpha/(1+\alpha)$. Their adaptive version of the BH method, the BKY method, runs as follows:

(1) Apply the BH method at level $q = \frac{\alpha}{1+\alpha}$. Let r_1 be the number of rejections. If $r_1 = 0$, accept all the null hypotheses and stop; if $r_1 = n$, reject all the null hypotheses and stop; otherwise continue to the next step.
(2) Estimate n_0 as

$$\hat{n}_0^{BKY} = \frac{n - r_1}{1 - q} = (n - r_1)(1 + \alpha). \tag{1.14}$$

(3) Apply the BH method with the critical values $\alpha_i = i\alpha/\hat{n}_0^{BKY}$, $i = 1, \ldots, n$.

As Benjamini, Krieger and Yekutieli (2006) have proved, the BKY method controls the FDR at α under independence of the p-values. While it is less powerful than the adaptive method proposed in Storey, Taylor and Siegmund (2004) when the p-values are independent, simulation studies have shown that with the test statistics generated from multivariate normals with common positive correlations it can also control the FDR [Benjamini, Krieger and Yekutieli (2006) and Romano, Shaikh and Wolf (2008)]

Benjamini, Krieger and Yekutieli (2006) also extended the BKY method to a multiple-stage procedure (MST) by repeating the two-stage procedure as long as more hypotheses are rejected, which is stated as follows:

(1) Let $r = \max\{i :\ \text{for all } j \leq i, \text{ there exists } l \geq j \text{ so that } p_{l:n} \leq \alpha l/[n+1-j(1-\alpha)]\}$.
(2) If such an r exists, reject $p_{1:n}, \ldots, p_{r:n}$; otherwise reject no hypotheses.

This multiple-stage procedure is a combination of step-up and step-down procedures. They offered no analytical proof of its FDR control. Benjamini, Krieger and Yekutieli (2006) also mentioned that a multiple-stage step-down procedure (MSD) can be developed by choosing $l = j$ in MST. They provided numerical results showing that the MST method can also control the FDR, the theoretical justification of which is given later in Gavrilov, Benjamini and Sarkar (2009) to be reviewed in the following section.

1.3.5. *The adaptive method of Gavrilov, Benjamini and Sarkar*

As mentioned above, Gavrilov, Benjamini and Sarkar (2009) re-examined the multiple-stage step-down procedure, the MSD procedure, mentioned in Benjamini, Krieger and Yekutieli (2006) and proved that this multiple-stage step-down procedure can control the FDR under the independence of the p-values. The following is the MSD method:

Find $k = \max\{1 \leq i \leq n : p_{j:n} \leq j\alpha/(n+1-j(1-\alpha))$ for all $j = 1, \dots, i\}$ and reject $H_{1:n}, \dots, H_{k:n}$ if k exists; otherwise reject no hypotheses.

Although it has been referred to as a multiple-stage step-down procedure by Benjamini, Krieger and Yekutieli (2006), it is actually, as Sarkar (2008b) argued, an adaptive version of the step-down analog of the BH method considered in Sarkar (2002). To see this, first note that, under the same setup involving the mixture model and a constant rejection threshold t for each p-value as in Storey (2002) or Storey, Taylor and Siegmund (2004), one can consider estimating n_0 based on the number of significant p-values compared to the t itself, rather than a different arbitrary constant λ. In other words, by considering the Storey, Taylor and Siegmund (2004) type estimate of $n_0 = n\pi_0$ with $\lambda = t$ and using this estimate in $\widehat{\text{FDR}}_\lambda(t)$, Storey's original estimate of the FDR(t), one can develop the following alternative estimate of FDR(t):

$$\widehat{\text{FDR}}^*(t) = \frac{[n - R(t) + 1]t}{(1-t)\max\{R(t), 1\}}.$$

A step-down procedure developed through this estimate, that is, the one that rejects $H_{1:n}, \dots, H_{r:n}$, where

$$\begin{aligned} r &= \max\left\{1 \leq i \leq n : \widehat{\text{FDR}}^*(p_{j:n}) \leq \alpha \text{ for all } j = 1, \dots, i\right\} \\ &= \max\left\{1 \leq i \leq n : \frac{p_{j:n}}{1 - p_{j:n}} \leq \frac{j\alpha}{n - j + 1} \text{ for all } j = 1, \dots, i\right\}, \quad (1.15) \end{aligned}$$

which is the same as the MSD, is an adaptive version of the step-down analog of the BH method.

There are some other methods to estimate n_0 in the literature, such as parametric beta-uniform mixture model by Pounds and Morris (2003), the Spacing LOESS Histogram (SPLOSH) method by Pounds and Cheng (2004), the nonparametric MLE method by Langaas and Lindqvist (2005), the moment generating function approach by Broberg (2005), and the resampling strategy by Lu and Perkins (2007). These other n_0 estimates could also be used while developing adaptive versions of the BH method or its step-down analog. However, whether or not any of these can control the FDR theoretically, at least when the p-values are independent, is an important open problem.

1.4. A New Estimate of n_0

We present in this section the new estimate of n_0 and the results of a simulation study comparing this estimate to $\hat{n}_0^{STS}$ and $\hat{n}_0^{BKY}$ before we use it to propose our version of adaptive BH method in the next section.

1.4.1. *The estimate*

Our estimate of n_0 is developed somewhat along the line of that in the BKY method. However, instead of deriving it from the number of significant p-values in the original BH method at level $q = \alpha/(1+\alpha)$, which is being done in the BKY method, we consider deriving it from the number of significant p-values in the step-down analog of the BH method at the same level q but using a formula that is similar to that in Storey, Taylor and Siegmund (2004). More specifically, our proposed estimate of n_0 is given by:

$$\hat{n}_0^{NEW}(\gamma) = (n - k + 1)/(1 - \gamma_{k+1}), \tag{1.16}$$

where k is the number of rejections in the step-down version of the BH method with the critical values $\gamma_i = i\gamma/n$, for $i = 1, \dots, n$, where $\gamma = \alpha/(1 + \alpha)$ and $\gamma_{n+1} \in [\gamma, (1 + \gamma)/2)$. The choice of γ_{n+1} in this particular interval is dictated by our main result proved in the section that for such γ_{n+1} the FDR of the corresponding adaptive BH method can be controlled at α, at least when the p-values are independent.

The results presented in the following section favoring $\hat{n}_0^{NEW}$ as an estimate of n_0 over $\hat{n}_0^{BKY}$ provide some rationale for our choice of this new estimate.

1.4.2. *Simulation study*

We ran a simulation study to investigate numerically how $\hat{n}_0^{NEW}$ performs compared to $\hat{n}_0^{STS}$ (with $\lambda = 0.5$) and $\hat{n}_0^{BKY}$ as an estimate of n_0. We generated n dependent random variables $X_i \sim N(\mu_i, 1)$, $i = 1, \ldots, n$, with a common non-negative correlation ρ, and determined their p-values for testing $\mu_i = 0$ against $\mu_i > 0$. We repeated this 10,000 times by setting n at 5000, ρ at $0, 0.25, 0.5, 0.75$, the proportion of the true null hypotheses π_0 at $0, 0.25, 0.5, 0.75$ and 1, the value of μ_i for each false null hypothesis at 1, and the value of α at 0.05. Each time, we calculated the values of the three estimates. From these 10,000 values, we constructed the boxplot and calculated the estimated mean and variance for each estimate. We present these boxplots in Fig. 1.1 and the estimated means and variances in Table 1.1 only for $\pi_0 = 0.5$, as they provide very similar comparative pictures for other values of π_0.

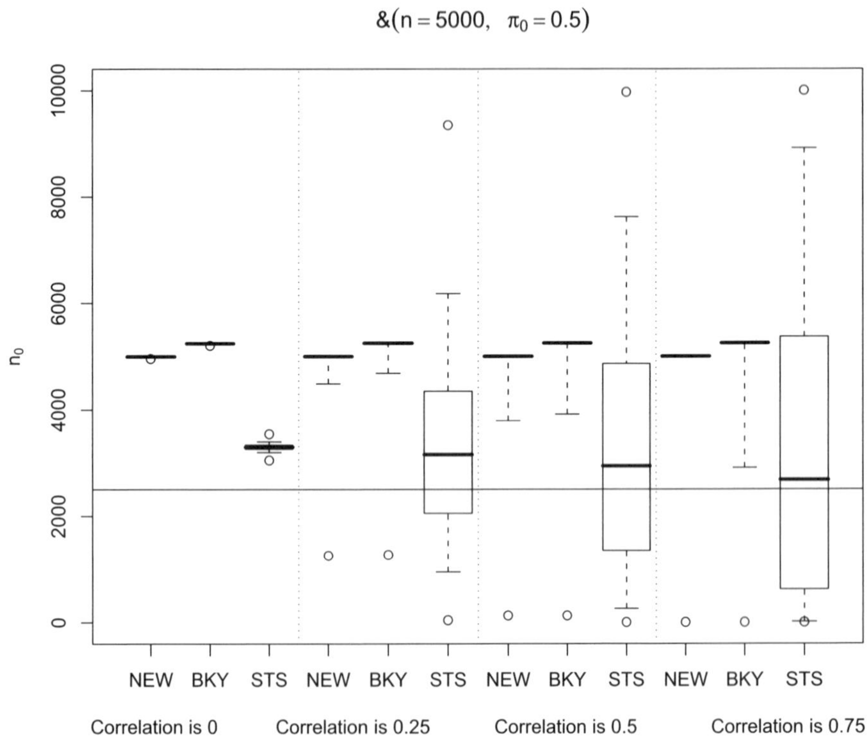

Fig. 1.1. The simulated distribution of $\hat{n}_0^{NEW}$, $\hat{n}_0^{STS}$ and $\hat{n}_0^{BKY}$ for the cases of $n = 5000$, $n_0 = 2500$ and $\rho = 0, 0.25, 0.5$ and 0.75. Each box displays the median and quartiles as usual. The whiskers extend to the 5th and 95th percentiles. The circles are the extreme values, i.e. the 0.01th and 99.99th percentiles.

Table 1.1. The estimated mean and variance of $\hat{n}_0^{NEW}$, $\hat{n}_0^{STS}$ and $\hat{n}_0^{BKY}$ for the cases of $n = 5000$ and $n_0 = 2500$.

	mean			variance		
	NEW	BKY	STS	NEW	BKY	STS
$\rho = 0$	4996	5242	3296	33.87	55.82	3782
$\rho = 0.25$	4927	5166	3284	69133	81661	2539704
$\rho = 0.5$	4862	5088	3280	279038	325556	5257668
$\rho = 0.75$	4881	5008	3276	448641	712831	8263566

As seen from Fig. 1.1 and Table 1.1, $\hat{n}_0^{NEW}$ is a better estimate of n_0 than $\hat{n}_0^{BKY}$. Looking at $\hat{n}_0^{STS}$ and comparing it to the other two, one notices that although it is more centrally located at the true n_0, it is more variable, and the variability increases quite dramatically with increasing ρ. The variabilities of both $\hat{n}_0^{NEW}$ and $\hat{n}_0^{BKY}$, on the other hand, remain relatively more stable with increasing ρ.

The above findings seem to suggest that the adaptive BH method based on our estimate $\hat{n}_0^{NEW}$ may perform well compared to that based on $\hat{n}_0^{BKY}$ in some situations. Moreover, both these adaptive BH methods seem to behave similarly in terms of the FDR control and power compared to that based on $\hat{n}_0^{STS}$. For instance, like the BKY method, the adaptive BH method based on $\hat{n}_0^{NEW}$ which controls the FDR under independence, which we will prove in the next section, can also control the FDR under positive dependence, which we will verify also in the next section.

1.5. New Adaptive Method to Control the FDR

In this section, we present our adaptive version of the BH method based on the estimate $\hat{n}_0^{NEW}$ of n_0. We will prove that the FDR of this adaptive BH method is controlled under independence of the p-values and numerically show that this control continues to hold even when the p-values are positively dependent under normal distributional setting with equal positive correlation. The performance of this adaptive procedure is examined by comparing it to the BKY procedure.

1.5.1. *The new adaptive BH method*

The following is our proposed adaptive BH method:

Procedure 1.1.

(1) Observe $R_{SD}(\gamma_1, \ldots, \gamma_n)$, the number of rejections in a step-down method with the critical values $\gamma_i = i\gamma/n$, $i = 1, \ldots, n$, with $\gamma = \alpha/(1+\alpha)$, and calculate

$$\hat{n}_0^{NEW} = \frac{n - R_{SD}(\gamma_1, \ldots, \gamma_n) + 1}{1 - \gamma_{R_{SD}(\gamma_1, \ldots, \gamma_n)+1}}, \tag{1.17}$$

with an arbitrary $\gamma_{n+1} \in [\gamma, (1+\gamma)/2)$.

(2) Apply the step up procedure with the critical values $\alpha_i = i\alpha/\hat{n}_0^{NEW}$, $i = 1, \ldots, n$, for testing the null hypotheses.

Theorem 1.1. *Procedure 1.1 controls the FDR at α when the p-values are independent.*

The following two lemmas will facilitate our proof of this theorem. These lemmas will be proved later in this section.

Lemma 1.1. *Let $U \sim U(0,1)$. Then, for any non-increasing function $\phi(U) > 0$ and a constant $c > 0$, we have*

$$E\left\{\frac{I\left(U \leq c\phi(U)\right)}{\phi(U)}\right\} \leq c. \tag{1.18}$$

Lemma 1.2. *Let $R^{(-i)}_{n-1,SD}(c_1, \ldots, c_{n-1})$ be the number of rejections in a step-down method based on the $n-1$ p-values other than p_i, where $i \in I_0$, and a set of critical values $0 < c_1 \leq \cdots \leq c_{n-1} < 1$. Then, under independence of the p-values, we have*

$$\sum_{i \in I_0} E\left\{\frac{1 - c_{R^{(-i)}_{SD,n-1}(c_1,\ldots,c_{n-1})+1}}{n - R^{(-i)}_{SD,n-1}(c_1, \ldots, c_{n-1})}\right\} \leq 1 - Pr\left\{P_{1:n} \leq c_1, \ldots, P_{n:n} \leq c_n\right\}, \tag{1.19}$$

for an arbitrary, fixed $c_n \in [c_{n-1}, 1)$.

Proof. [Proof of Theorem 1.1] Using Formula 1.1, we first note that

$$\mathrm{FDR} = \sum_{i \in I_0} E\left[\frac{I\left(P_i \le \alpha_{R^{(-i)}_{SU,n-1}(\alpha_2,\ldots,\alpha_n)+1}\right)}{R^{(-i)}_{SU,n-1}(\alpha_2,\ldots,\alpha_n)+1}\right]$$
$$= \sum_{i \in I_0} E\left[\frac{I\left(P_i \le \left\{R^{(-i)}_{SU,n-1}(\alpha_2,\ldots,\alpha_n)+1\right\}\alpha_1\right)}{R^{(-i)}_{SU,n-1}(\alpha_2,\ldots,\alpha_n)+1}\right], \quad (1.20)$$

with

$$\alpha_i = \frac{i\alpha\left(1-\gamma_{R_{SD}(\gamma_1,\ldots,\gamma_n)+1}\right)}{n - R_{SD}(\gamma_1,\ldots,\gamma_n)+1}$$
$$= \begin{cases} \frac{i\alpha[n-\gamma(R_{SD}(\gamma_1,\ldots,\gamma_n)+1)]}{n[n-R_{SD}(\gamma_1,\ldots,\gamma_n)+1]} & \text{when } R_{SD}(\gamma_1,\ldots,\gamma_n) = 0,\ldots,n-1, \\ i\alpha\left(1-\gamma_{n+1}\right) & \text{when } R_{SD}(\gamma_1,\ldots,\gamma_n) = n, \end{cases} \quad (1.21)$$

$i = 1, \ldots, n$.

Now, notice that $R_{SD}(\gamma_1,\ldots,\gamma_n)$, with fixed $(\gamma_1,\ldots,\gamma_n)$, is a decreasing function of each of the p-values, and as a function of $R_{SD}(\gamma_1,\ldots,\gamma_n)$, α_i is an increasing function if $\gamma \le n/(n+2)$ and $\gamma_{n+1} \le (1+\gamma)/2$. But, $\gamma \le n/(n+2)$ means that $\alpha \le n/2$, which is obviously true, since $n \ge 2$. Thus, as long as $\gamma_{n+1} \le (1+\gamma)/2$, each α_i is a (componentwise) decreasing function of $\mathbf{P} = (P_1,\ldots,P_n)$. So, by letting $P_i \to 0$ in α_1 we see that

$$\alpha_1 \le \frac{\alpha\left(1-\gamma_{R^{(-i)}_{SD,n-1}(\gamma_2,\ldots,\gamma_n)+2}\right)}{n - R^{(-i)}_{SD,n-1}(\gamma_2,\ldots,\gamma_n)}, \quad (1.22)$$

since $R_{SD}(\gamma_1,\ldots,\gamma_n) \to R^{(-i)}_{SD,n-1}(\gamma_2,\ldots,\gamma_n)+1$ as $P_i \to 0$. Let us define $g(\mathbf{P}) = R^{(-i)}_{SU,n-1}(\alpha_2,\ldots,\alpha_n)+1$ and $h(\mathbf{P}^{(-i)})$ equal the right-hand side of (21), with $\mathbf{P}^{(-i)} = (P_1,\ldots,P_n)\setminus\{P_i\}$. Then, we have

$$\mathrm{FDR} \le \sum_{i\in I_0} E\left[\frac{I\left(P_i \le g(\mathbf{P})\,h\left(\mathbf{P}^{(-i)}\right)\right)}{g(\mathbf{P})}\right]$$
$$= \sum_{i\in I_0} E\left[E\left\{\frac{I\left(P_i \le g(\mathbf{P})\,h\left(\mathbf{P}^{(-i)}\right)\right)}{g(\mathbf{P})} \mid \mathbf{P}^{(-i)}\right\}\right]$$
$$\le \sum_{i\in I_0} E\left\{h\left(\mathbf{P}^{(-i)}\right)\right\}$$
$$\le \alpha\left[1 - Pr\left\{P_{1:n} \le \gamma_2,\ldots,P_{n:n} \le \gamma_{n+1}\right\}\right]$$
$$\le \alpha, \quad (1.23)$$

with the second and third inequalities following from Lemmas 1.1 and 1.2 respectively. Thus, the theorem is proved. □

We will now give proofs of Lemmas 1.1 and 1.2.

Proof. [Proof of Lemma 1.1] Consider the function $\psi(u) = u - c\phi(u)$. Since this is non-decreasing, there exists a constant c^* such that $\{\psi(u) \leq 0\} \subseteq \{u \leq c^*\}$ and $\psi(c^*) \leq 0$, that is, $c^* \leq c\phi(c^*)$. Since $\phi(u) \geq \phi(c^*)$ when $u \leq c^*$, we have

$$E\left\{\frac{I\left(U \leq c\phi(U)\right)}{\phi(U)}\right\} \leq E\left\{\frac{I\left(U \leq c^*\right)}{\phi(c^*)}\right\} = \frac{c^*}{\phi(c^*)} \leq \frac{c\phi(c^*)}{\phi(c^*)} = c.$$

Thus, the lemma is proved. □

Proof. [Proof of Lemma 1.2]

$$\begin{aligned}
&\sum_{i\in I_0} E\left\{\frac{1 - c_{R^{(-i)}_{SD,n-1}(c_1,\ldots,c_{n-1})+1}}{n - R^{(-i)}_{SD,n-1}(c_1,\ldots,c_{n-1})}\right\} \\
&= \sum_{i\in I_0}\sum_{r=0}^{n-1} \frac{1-c_{r+1}}{n-r} Pr\left\{R^{(-i)}_{SD,n-1}(c_1,\ldots,c_{n-1}) = r\right\} \\
&= \sum_{i\in I_0}\sum_{r=0}^{n-1} \frac{1}{n-r} Pr\left\{P^{(-i)}_{1:n-1} \leq c_1,\ldots,P^{(-i)}_{r:n-1} \leq c_r, P^{(-i)}_{r+1:n-1} > c_{r+1}, P_i > c_{r+1}\right\} \\
&\leq \sum_{i=1}^{n}\sum_{r=0}^{n-1} \frac{1}{n-r} Pr\left\{P^{(-i)}_{1:n-1} \leq c_1,\ldots,P^{(-i)}_{r:n-1} \leq c_r, P^{(-i)}_{r+1:n-1} > c_{r+1}, P_i > c_{r+1}\right\} \\
&= \sum_{r=0}^{n-1} Pr\left\{P_{1:n} \leq c_1,\ldots,P_{r:n} \leq c_r, P_{r+1:n} > c_{r+1}\right\} \\
&= 1 - Pr\left\{P_{1:n} \leq c_1,\ldots,P_{n:n} \leq c_n\right\},
\end{aligned} \tag{1.24}$$

where $P^{(-i)}_{1:n-1} \leq \cdots \leq P^{(-i)}_{n-1:n-1}$ are the ordered components of $\mathbf{P}^{(-i)}$. The third equality in (23) follows from results on ordered random variables given in Sarkar (2002). Thus, the lemma is proved. □

1.5.2. *Simulation study*

A simulation study was performed to compare the FDR control and power of our proposed method with those of the BKY method. The study consisted of two parts, the first part was designed for small number of hypotheses, while the second part was designed for relatively large number of hypotheses as seen in most applications of the FDR.

In the first part of the study, we generated n dependent random variables $X_i \sim N(\mu_i, 1)$, $i = 1, \ldots, n$, with a common non-negative correlation ρ, and applied both the BKY and our proposed methods to test $\mu_i = 0$ against $\mu_i > 0$, simultaneously for $i = 1, \ldots, n$ at a level α. We repeated this 10,000 times by setting n at $4, 8, 16, 64, 128, 256$ and 512, the value of ρ at $0, 0.1, 0.25$ and 0.5, the proportion of the true null hypotheses π_0 at $0, 0.25, 0.5, 0.75$ and 1, α at 0.05, and μ_i at 1 for each false null hypothesis, to simulate the FDR and average power (the expected proportion of alternative μ_i's that are correctly identified) for both methods.

Figure 1.2 compares the FDR control and Table 1.2 lists the ratios of power of both methods to the 'Oracle' method when $n = 32, 128$ and 512. The 'Oracle' method is the BH method based on the critical values $\alpha_i = i\alpha/n_0$, which controls the FDR at the exact level α under the independence of the test statistics. Obviously, it is not implementable in practice as n_0 is unknown, but it serves as a benchmark against which other methods can be compared. As seen in Fig. 1.2, our proposed method, which is known to control the FDR at the desired level $\alpha = 0.05$ under independence, can continue to maintain a control over the FDR even under positive dependence, like the BKY method, although ours is often less conservative. Also in terms of power, as seen from Table 1.2, our method appears to be more powerful than the BKY method in most of the cases considered, especially when the correlation is not very high.

The second part of the study was conducted by setting $n = 5000$. The simulated FDR and power were also based on 10,000 iterations. The comparison between simulated FDR of the two methods is presented in Fig. 1.3. Again, there is evidence that our method can continue to control the FDR under positive dependence, at least when the p-values are equally correlated. The power comparisons in this case are displayed in Figs. 1.4 and 1.5. Figure 1.4 indicates that the proposed method is more powerful than the BKY when the correlation between the test statistics is moderately low. Figure 1.5 compares the power of the two methods under the condition of high proportion of true null, $\pi_0 \geq 0.9$, which is often the case in modern multiple testing situations. The proposed method seems to be more powerful then the BKY method in such situations.

In conclusion, the simulation study seems to indicate that the new proposed method can control the FDR under positive dependence of the p-values. It is more powerful than the BKY method under positive but not very high correlations between the test statistics. When there is a large proportion of true null hypotheses, the new method appears to perform better than the BKY method even in the case of high correlations.

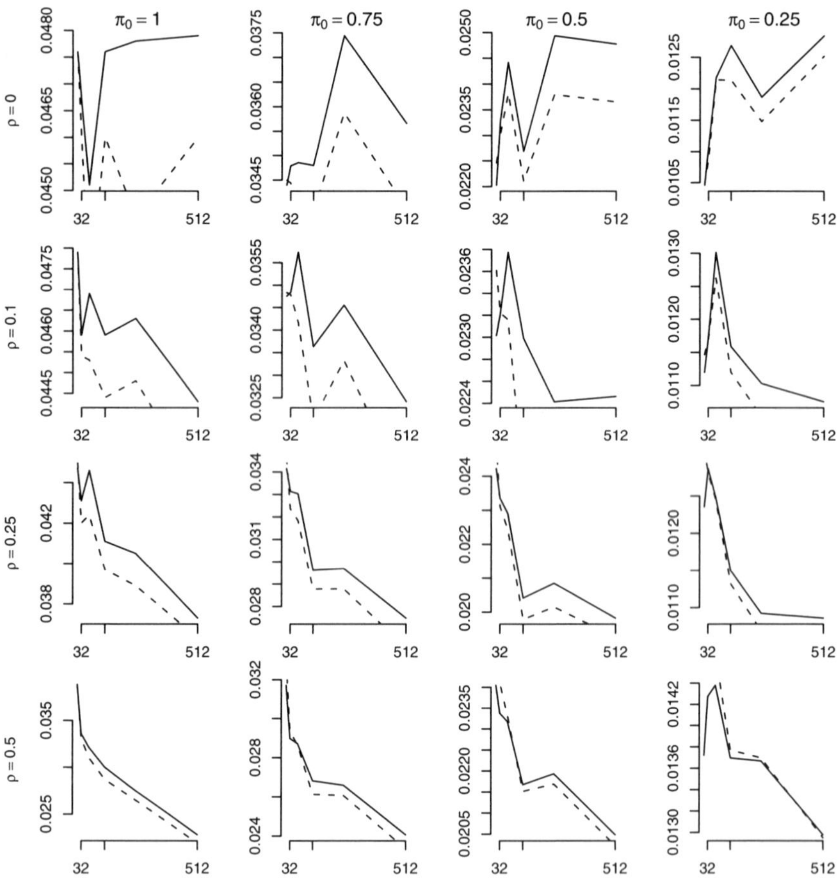

Fig. 1.2. Estimated FDR values for $n = 16, 32, \ldots, 512$ and $\rho = 0, 0.1, 0.25, 0.5$. Legend: NEW — solid line; BKY — dashed line.

1.6. An Application to Breast Cancer Data

We applied both the new adaptive BH and the BKY methods to the breast cancer data of Hedenfalk *et al.* (2001) available at http://www.nejm.org/general/content/supplemental/hedenfalk/index.html; see also Storey and Tibshirani (2003) from http://genomine.org/qvalue-/results.html. The results are presented in this section.

The data consists of 3,226 genes on 7 BRCA1 arrays, 8 BRCA2 arrays and 7 sporadic tumors. The goal of the study is to establish differences in gene expression patterns between these tumor groups. Here we analyzed

Table 1.2. Estimated power for $n = 32, 128, 512$ and $\rho = 0, 0.1, 0.25, 0.5$.

		$\pi_0 = 0.25$			$\pi_0 = 0.5$		
n		32	128	512	32	128	512
$\rho = 0$	new/oracle	0.1881	0.1160	0.0737	0.4661	0.3935	0.3111
	BKY/oracle	0.1867	0.1105	0.0682	0.4618	0.3746	0.2925
$\rho = 0.1$	new/oracle	0.2281	0.1769	0.1479	0.4956	0.4142	0.3550
	BKY/oracle	0.2293	0.1714	0.1403	0.4937	0.3979	0.3354
$\rho = 0.25$	new/oracle	0.2846	0.2530	0.2373	0.5291	0.4949	0.4582
	BKY/oracle	0.2908	0.2495	0.2305	0.5333	0.4829	0.4423
$\rho = 0.5$	new/oracle	0.3618	0.3391	0.3288	0.5942	0.5729	0.5551
	BKY/oracle	0.3755	0.3438	0.3305	0.6096	0.5725	0.5478

	$\pi_0 = 0.75$		
n	32	128	512
$\rho = 0$	0.7553	0.7188	0.7088
	0.7460	0.6914	0.6692
$\rho = 0.1$	0.7868	0.7060	0.6408
	0.7796	0.6794	0.6024
$\rho = 0.25$	0.7542	0.7502	0.6979
	0.7467	0.7294	0.6650
$\rho = 0.5$	0.7808	0.7895	0.7618
	0.7876	0.7781	0.7432

this data with permutation t-test to compare BRCA1 and BRCA2. The data were entered into R and all analyses were done using R. As Storey and Tibshirani (2003) did, if any gene had one or more measurement ($\log_2$ expression value) exceeding 20, then this gene was eliminated. This left $n = 3170$ genes for permutation t-test.

We tested each gene for differential expression between BRCA1 and BRCA2 by using a two-sample t-test. The p-values were calculated using a permutation method as in Storey and Tibshirani (2003). We did $B = 100$ permutations for each gene and got a set of null statistics $t_1^{0b}, \ldots, t_n^{0b}$, $b = 1, \ldots, B$. The p-value of the permutation t-test for gene i was calculated by

$$p_i = \sum_{b=1}^{B} \frac{\#\{j : |t_j^{0b}| \geq |t_i|, j = 1, \ldots, n\}}{nB}$$

The new adaptive method identifies 94 significant genes at the 0.05 level of false discovery rate, whereas, the BKY method gets 93 significant genes.

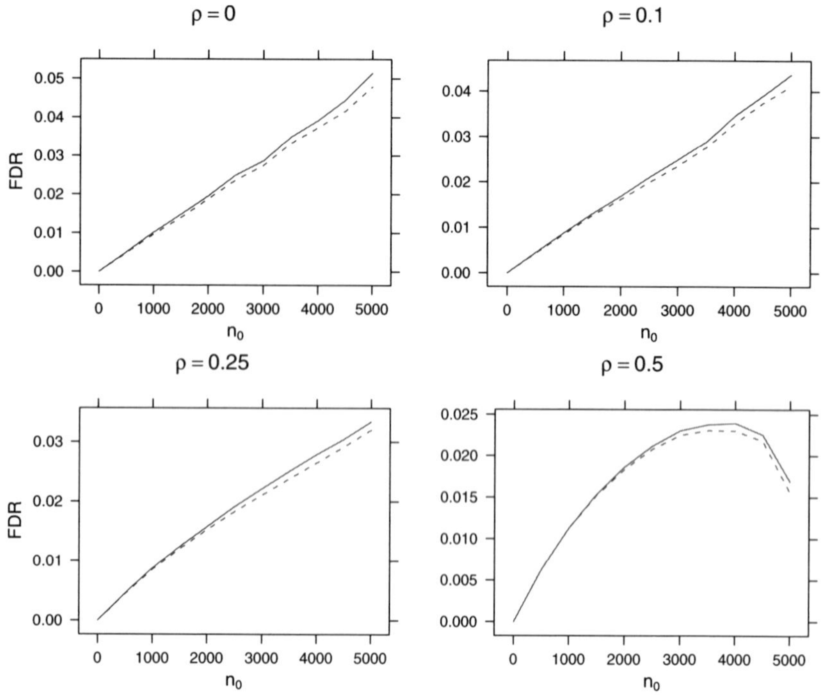

Fig. 1.3. Estimated FDR values for $n = 5000$ and $\rho = 0, 0.1, 0.25, 0.5$. Legend: NEW — solid line; BKY — dashed line.

This additional significant gene picked up by our method is intercellular adhesion molecule 2 (clone 471918).

1.7. Concluding Remarks

Adaptive BH methods other than those reviewed here have been proposed in the literature; see, for instance, Sarkar (2008b). Among these, the BKY method has received much attention since there is numerical evidence that it can continue to control the FDR under some form of positive dependence among the test statistics. The new adaptive BH method we propose in this article competes well with the BKY method. Like the BKY method, it controls the FDR with independent p-values and, as can be seen numerically, continue to maintain the control with the same type of positively dependent p-values as in the BKY method. More importantly, it can perform better than the BKY method in some instances, especially when the proportion of true null hypotheses is very large, which happens in many applications.

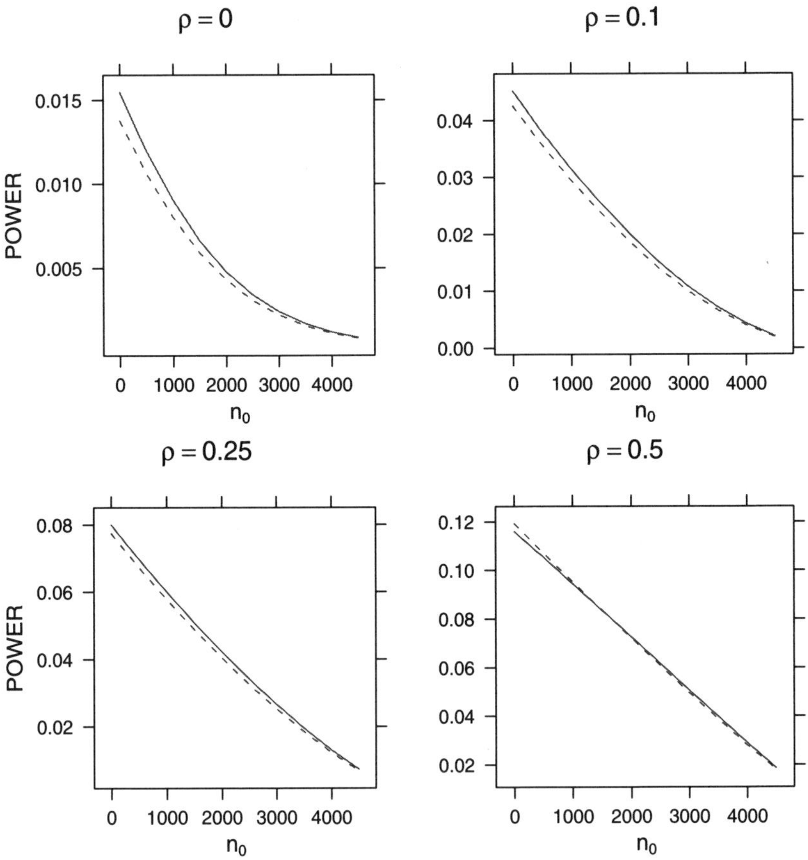

Fig. 1.4. Estimated power for $n = 5000$ and $\rho = 0, 0.1, 0.25, 0.5$. Legend: NEW — solid line; BKY — dashed line.

We have considered using $\lambda = 0.5$ in the STS procedure, since this is what Storey, Taylor and Siegmund (2004) have suggested, even though it may not control the FDR under positive dependence, and $\alpha/(1+\alpha)$ for γ in our procedure, since this is what Benjamini, Krieger and Yekutieli (2006) have also considered for the q in their procedure. All these procedures can be proven to control the FDR under independence if other different values are chosen for λ, q and γ. But, the BKY as well as our procedures may not continue to control the FDR under positive dependence with these other values of γ and q.

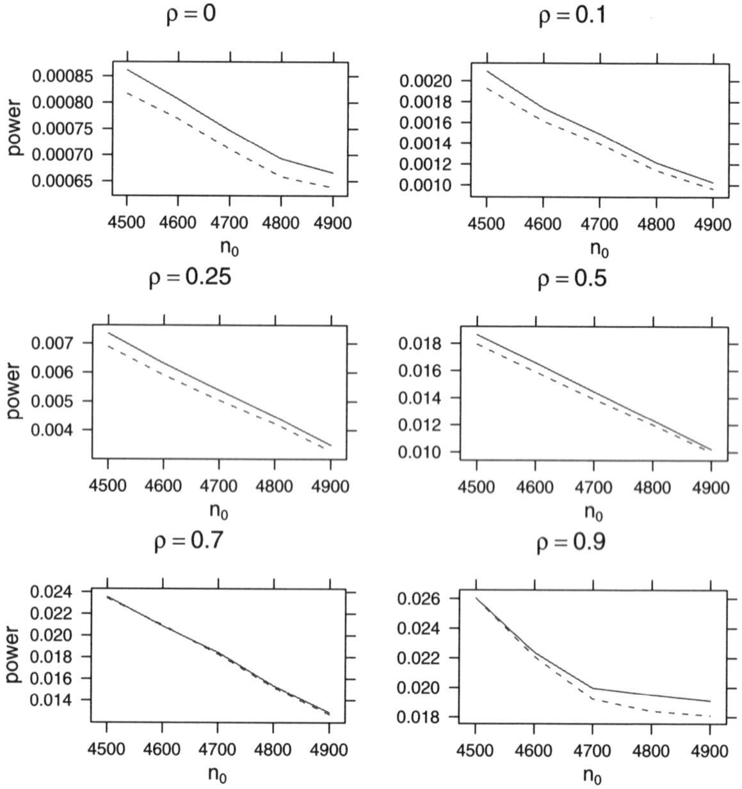

Fig. 1.5. Estimated power for $n = 5000$, $\pi_0 = 0.9, 0.92, 0.94, 0.96, 0.98$ and $\rho = 0, 0.1, 0.25, 0.5, 0.75$ and 0.9. Legend: NEW — solid line; BKY — dashed line.

Acknowledgment

We thank the referee who made some useful comments. The research is supported by NSF Grant DMS-0603868.

References

Benjamini, Y. and Hochberg, Y. (1995). Controlling the false discovery rate: a practical and powerful approach to multiple testing, *J. R. Stat. Soc. B.* **57**, 289–300.

Benjamini, Y. and Hochberg, Y. (2000). On the adaptive control of the false discovery rate in multiple testing with independent statistics, *Journal of educational and Behavioral Statistics* **25**, 60–83.

Benjamini, Y., Krieger, A. and Yekutieli, D. (2006). Adaptive linear step-up procedures that control the false discovery rate, *Biometrica* **93**, 491–507.

Benjamini, Y. and Yekutieli, D. (2001). The control of the false discovery rate in multiple testing under dependency. *Ann. Stat.* **29**, 1165–1188.

Broberg, P. (2005). A comparative review of estimates of the proportion unchanged genes and the false discovery rate. *BMC Bioinformatics* **6**, 199.

Finner, H., Dickhaus,T. and Roters, M. (2009). On the false discovery rate and an asymptotically optimal rejection curve. *Ann. Statist.* **37**(2), 596–618.

Finner, H. and Roters, M. (2001). On the false discovery rate and expected type I errors. *Biometrical Journal* **43**, 985–1005.

Gavrilov, Y., Benjamini, Y. and Sakar, S. K. (2009). An adaptive step-down procedure with proven FDR control. *Ann. Statist.* **37**, 2, 619–629.

Hommel, G. (1988). A stepwise rejective multiple test procedure based on a modified Bonferroni test. *Biometrika* **75**, 383–386.

Langaas, M. and Lindqvist, B. H. (2005). Estimating the proportion of true null hypotheses, with application to DNA microarray data. *J. Roy. Stat. Soc. B (Methodological)* **67**, 555–572.

Liu, F. and Sarkar, S. K. (2009). A note on estimating the false discovery rate under mixture model. *Journal of Statistical Planning and Inference* **140**, 1601–1609.

Lu, X, and Perkins, D. L. (2007). Re-sampling strategy to improve the estimation of number of null hypotheses in FDR control under strong correlation structures. *BMC Bioinformatics* **8**, 157.

Pounds, S. and Morris, S. W. (2003). Estimating the occurrence of false positives and false negatives in microarray studies by approximating and partitioning the empirical distribution of p-values. *Bioinformatics* **19**, 10, 1236–1242.

Pounds, S. and Cheng, C. (2004). Improving false discovery rate estimation. *Bioinformatics* **20**, 11, 1737–1745.

Romano, J. P., Shaikh, A. M.and Wolf, M. (2008). Control of the false discovery rate under dependence using the bootstrap and subsampling. *TEST* **17**, 417–442.

Sarkar, S. K. (1998). Some probability inequalities for ordered MTP_2 random variables: a proof of the simes conjecture. *Ann. Stat.* **26**, 2, 494–504.

Sarkar, S. K. (2002). Some results on false discovery rate in stepwise multiple testing procedures. *Ann. Stat.* **30**, 239–257.

Sarkar, S. K. (2004). FDR-controlling procedures and their false negatives rates. *Journal of Statistical Planning and Inference* **125**, 119–139.

Sarkar, S. K. (2006). False discovery and false non-discovery rates in single-step multiple testing procedures. *Ann. Stat.* **34**, 394–415.

Sarkar, S. K. (2007). Stepup procedures controlling generalized FWER and generalized FDR. *Ann. Stat.* **35**, 2405–2420.

Sarkar, S. K. (2008a). On the Simes inequality and its generalization. *IMS Collections Beyond Parametrics in Interdisciplinary Research: Festschrift in Honor of Professor Pranab K. Sen* **1**, 231–242.

Sarkar, S. K. (2008b). On methods controlling the False Discovery Rate (with discussions). *Sankhya* **70**, 135–168.

Sarkar, S. K. and Chang, C. K. (1997). The simes method for multiple hypothesis testing with positively dependent test statistics. *J. Am. Stat. Assoc.* **92**, 1601–1608.

Schweder, R. and Spjotvoll, E. (1982). Plots of p-values to evaluate many tests simultaneously. *Biometrika* **69**, 493–502.

Simes, R. J. (1986) An improved Bonferroni procedure for multiple tests of significance. *Biometrika* **73**, 751–754.

Storey, J. D. (2002). A direct approach to false discovery rates. *J. Roy. Stat. Soc. B* **64**, 479–498.

Storey, J. D. (2003). The positive false discovery rate: a bayesian interpretation and the q-value. *Ann. Stat.* **31**, 2013–2035.

Storey, J. D., Taylor, J. E. and Siegmund, D. (2004). Strong control, conservative point estimation, and simultaneous conservative consistency of false discovery rates: A unified approach. *J. Roy. Stat. Soc. B* **66**, 187–205.

Storey, J. D. and Tibshirani, R. (2003). Statistical Significance for genome-wide studies. *Proc. Nat. Acad. Sci. USA* **100**, 9440–9445.

Chapter 2

Adaptive Multiple Testing Procedures Under Positive Dependence

Wenge Guo*,§, Sanat K. Sarkar†,§ and Shyamal D. Peddada‡,§

**Department of Mathematical Sciences*
New Jersey Institute of Technology
Newark, New Jersey, 07102, USA
wenge.guo@njit.edu

†Department of Statistics, Temple University
Philadelphia, PA 19122, USA
sanat@temple.edu

‡Biostatistics Branch, National Institute of Environmental Health Sciences
Research Triangle Park, NC 27709, USA
peddada@niehs.nih.gov

In multiple testing, the unknown proportion of true null hypotheses among all null hypotheses that are tested often plays an important role. In adaptive procedures this proportion is estimated and then used to derive more powerful multiple testing procedures. Hochberg and Benjamini (1990) first presented adaptive Holm and Hochberg procedures for controlling the familywise error rate (FWER). However, until now, no mathematical proof has been provided to demonstrate that these procedures control the FWER in finite samples. In this paper, we present new adaptive Holm and Hochberg procedures and prove they can control the FWER in finite samples under some common types of positive dependence. Through a small simulation study, we illustrate that these adaptive procedures are more powerful than the corresponding non-adaptive procedures.

§The research of Wenge Guo is supported by NSF Grants DMS-1006021, the research of Sanat Sarkar is supported by NSF Grants DMS-0603868 and DMS-1006344, and the research of Shyamal Peddada is supported by the Intramural Research Program of the NIH, National Institute of Environmental Health Sciences (Z01 ES101744-04). The authors thank Gregg E. Dinse, Mengyuan Xu and the referee for carefully reading this manuscript and for their useful comments that improved the presentation.

2.1. Introduction

In this article, we consider the problem of simultaneously testing a finite number of null hypotheses H_i, $i = 1, \ldots, n$, based on their respective p-values P_i, $i = 1, \ldots, n$. A main concern in multiple testing is the multiplicity problem, namely, that the probability of committing at least one Type I error sharply increases with the number of hypotheses tested at a pre-specified level. There are two general approaches for dealing with this problem. The first one is to control the familywise error rate (FWER), which is the probability of one or more false rejections, and the second one is to control the false discovery rate (FDR), which is the expected proportion of Type I errors among the rejected hypotheses, proposed by Benjamini and Hochberg (1995). The first approach works well for traditional small-scale multiple testing , while the second one is more suitable for modern large-scale multiple testing problems.

Given the ordered p-values $P_{1:n} \leq \cdots \leq P_{n:n}$ with the associated null hypotheses $H_{1:n}, \cdots, H_{n:n}$, and a non-decreasing sequence of critical values $\alpha_1 \leq \cdots \leq \alpha_n$, there are two main avenues open for developing multiple testing procedures based on the marginal p-values – stepdown and stepup.

- A stepdown procedure based on these critical values operates as follows. If $P_{1:n} > \alpha_1$, do not reject any hypothesis. Otherwise, reject hypotheses $H_{1:n}, \cdots, H_{r:n}$, where $r \geq 1$ is the largest index satisfying $P_{1:n} \leq \alpha_1, \cdots, P_{r:n} \leq \alpha_r$. If, however, $P_{r:n} > \alpha_r$ for all $r \geq 1$, then do not reject any hypothesis. Thus, a stepdown procedure starts with the most significant hypothesis and continues rejecting hypotheses as long as their corresponding p-values are less than or equal to the corresponding critical values.
- A stepup procedure, on the other hand, operates as follows. If $P_{n:n} \leq \alpha_n$, then reject all null hypotheses; otherwise, reject hypotheses $H_{1:n}, \cdots, H_{r:n}$, where $r \geq 1$ is the smallest index satisfying $P_{n:n} > \alpha_n, \ldots, P_{r+1:n} > \alpha_{r+1}$. If, however, $P_{r:n} > \alpha_r$ for all $r \geq 1$, then do not reject any hypothesis. Thus, a stepup procedure begins with the least significant hypothesis and continues accepting hypotheses as long as their corresponding p-values are greater than the corresponding critical values until reaching the most significant hypothesis $H_{1:n}$.

If $\alpha_1 = \cdots = \alpha_n$, the stepup or stepdown procedure reduces to what is usually referred to as a single-step procedure.

For controlling the FWER, a number of widely used procedures are available, among which the Bonferroni, Holm (1979) and Hochberg (1988) procedures are relatively popular. The Bonferroni procedure is a single-step procedure with the critical values $\alpha_i = \alpha/n, i = 1, \ldots, n$. The Holm procedure is a stepdown procedure with the critical values $\alpha_i = \alpha/(n - i + 1), i = 1, \ldots, n$, and the Hochberg procedure is a stepup procedure based on the same set of critical values as Holm's. With the null p-values having the $U(0,1)$, or stochastically larger than the $U(0,1)$, distribution, the Bonferroni and Holm procedures both control the FWER at α without any further assumption on the dependence structure of the p-values. The Hochberg procedure controls the FWER at α when the null p-values are independent or positively dependent in the the following sense:

$$E\{\phi(P_1, \ldots, P_n) \mid P_i = u\} \uparrow u \in (0,1), \tag{2.1}$$

for each P_i and any increasing (coordinatewise) function ϕ. (Hochberg, 1988; Sarkar, 1998; Sarkar and Chang, 1997). The condition (2.1) is the positive dependence through stochastic ordering (PDS) condition defined by Block, Savits and Shaked (1985), although it is often referred to as the positive regression dependence on subset (of null p-values), the PRDS condition, considered in Benjamini and Yekutieli (2001) and Sarkar (2002) in the context of FDR. Also, it has been noted recently that this positive dependence condition can be replaced by the following weaker condition:

$$E\{\phi(P_1, \ldots, P_n) \mid P_i \le u\} \uparrow u \in (0,1). \tag{2.2}$$

The condition (2.1) or (2.2) is satisfied by a number of multivariate distributions arising in many multiple testing situations, for example, those of multivariate normal test statistics with positive correlations, absolute values of studentized independent normals, and multivariate t and F (Benjamini and Yekutieli, 2001; Sarkar, 2002).

Since these procedures are often conservative by a factor which is the unknown proportion of true null hypotheses, the conservativeness in these procedures could be reduced, and hence the power could potentially be increased, if an estimate of this proportion can be suitably incorporated into these procedures. With that idea in mind, Hochberg and Benjamini (1990) proposed adaptive Bonferroni, Holm and Hochberg procedures for controlling the FWER. However, it has not been proved yet that these adaptive FWER procedures actually can provide an ultimate control over the FWER. Recently, Guo (2009) introduced new adaptive Bonferroni and Holm procedures by simplifying those in Hochberg and Benjamini (1990).

He proved that, under a conditional independence model, while his adaptive Bonferroni procedure controls the FWER for finite samples, the adaptive Holm procedure approximately controls the FWER for large samples.

For controlling the FDR, the well-known procedure is that of Benjamini and Hochberg (1995). The same phenomenon in terms of conservativeness happens also with this procedure and a number of adaptive versions of it that control the FDR have been introduced in the literature; see Benjamini and Hochberg (2000), Storey *et al.* (2004), Benjamini *et al.* (2006), Ferreira and Zwinderman (2006), Sarkar (2006, 2009), Benjamini and Heller (2007), Farcomeni (2007), Blanchard and Roquain (2009), Wu (2008), Gavrilov *et al.* (2009), and Sarkar and Guo (2009). It is important to note that in the case of finite samples, the FDR control of these adaptive procedures has been proved only when the underlying test statistics are independent. Using a simulation study, Benjamini *et al.* (2006) demonstrated that some adaptive FDR procedures, such as Storey's, which control the FDR under independence, may fail to do so under dependence. Thus, developing an adaptive procedure controlling the FWER or FDR even under dependence in finite samples appears to be an important undertaking.

In this paper, we concentrate mainly on developing adaptive FWER procedures. We take a general approach to constructing such a procedure that controls the FWER under independence or positive dependence. This involves using a concept of adaptive global testing and the closure principle of Marcus *et al.* (1976). The closure principle is a useful tool to derive FWER controlling multiple testing procedures based on valid tests available for different possible intersections or global null hypotheses. In adaptive global testing, information about the number of true null hypotheses is extracted from the available p-values and incorporated into a procedure while testing an intersection or global null hypothesis and maintaining a control over the (global) type I error rate. We derive two such adaptive global tests, with one involving an estimate of the number of true null hypotheses considered in Hommel's (1988) FWER controlling procedure and the other based on an estimate of this number that can be obtained by applying the Benjamini and Hochberg's (1995) FDR controlling procedure. Both of these tests provide valid controls of the (global) type I error rate under independence or positive dependence, in the sense of (2.1) or (2.2), of the p-values. Based on these global tests and applying the closure principle, we derive alternative adaptive Holm and adaptive Hochberg procedures. We offer theoretical proofs of the FWER controls of these procedures in finite samples under independence or positive dependence in the sense of

(2.1) or (2.2) of the p-values. We provide numerical evidence through a small-scale simulation study that the present adaptive Holm and Hochberg procedures can be more powerful, as expected, than the corresponding non-adaptive procedures.

The paper is organized as follows. In Sec. 2.2, we introduce what we mean by an adaptive global test and present two such tests. In Sec. 2.3, we present the developments of our proposed new adaptive Holm and Hochberg procedures and prove that they control the FWER under independence or positive dependence in the sense of (2.1) or (2.2) of the p-values. A real-life application of our procedures and the results of a simulation study investigating the performances of our procedures relative to others are also presented in this section. Some concluding remarks are made in Sec. 2.4.

2.2. Adaptive Global Tests

In this section, we will present our idea of an adaptive global test. Given any family of null hypothesis $H_1, \ldots, H_m$, and the corresponding p-values P_i, $i = 1, \ldots, m$, consider testing the global null hypothesis $H_0 = \bigcap_{i=1}^m H_i$. We will focus on global tests where the rejection regions are of the form $\bigcup_{i=1}^m \{P_{i:m} \leq c_i\}$; that is, where each ordered p-value $P_{i:m}$ is compared with a cut-off point c_i, with $0 \leq c_1 \leq \ldots \leq c_m \leq 1$, and H_0 is rejected if $P_{i:m} \leq c_i$ holds for at least one i. Such a test has been referred to as a cut-off test (Bernhard *et al.*, 2004). It allows making decisions on the individual null hypotheses once the global null hypothesis is rejected, which is important since we need to develop in the next section multiple testing procedures based on it through the closure principle.

There are a number of such cut-off global tests available in the literature, such as the Bonferroni test, where $c_1 = \ldots = c_m = \alpha/m$, and the Simes test, where $c_i = i\alpha/m$, for $i = 1, \ldots, m$, (Simes, 1986). However, the idea of extracting information about the number, say m_0, of the true null hypotheses in the family of interest and incorporating that into the construction of a global cut-off test has not yet been seen in the literature. Why would such an adaptive global test make sense? Consider, for instance, the statistic $W_m(\lambda) = \sum_{i=1}^m I(P_i > \lambda)$ (with I being the indicator function), which is the number of insignificant p-values observed when the fixed rejection threshold $\lambda \in (0, 1)$ is chosen for each p-value. A high value of $W_m(\lambda)$ would indicate that m_0 is likely to be large, and hence would provide an evidence towards accepting the global null hypothesis. Similarly, a small value of $W_m(\lambda)$ would provide an evidence towards rejecting the null hypothesis. It is important to note that, although a test just based on $W_m(\lambda)$, for a fixed λ and controlling the type I error rate at a pre-specified level could be formulated for testing H_0, either exactly using the Binomial distribution when the p-values are independent, since in this case $W_m(\lambda) \sim \text{Bin}(m, 1-\lambda)$ under H_0, or approximately using, for example, a permutation test when the p-values have an unknown or more complicated dependence structure, this would not be helpful in terms of providing a cut-off test. Nevertheless, the value of $W_m(\lambda)$ can be factored into each P_i in a way that shrinks P_i towards a smaller value, making it more likely to be significant, if $W_m(\lambda)$ is small, and expands P_i to a larger value if $W_m(\lambda)$ is large. Of course, instead of $W_m(\lambda)$, we could use any other statistic, or a consistent estimate of m_0, a large value of which would indicate acceptance of H_0. This is how we will develop two global cut-off tests in the following.

First, we develop our adaptive Simes global test, borrowing the idea of estimating m_0 from Benjamini *et al.* (2006). Let $\mathbf{P}_m = (P_1, \ldots, P_m)$, and $W_1(\mathbf{P}_m)$ be the total number of accepted null hypotheses when the FDR controlling procedure of Benjamini and Hochberg (1995), the BH procedure, is applied to $\mathbf{P}_m$. Recall that the BH procedure is a stepup procedure based on the critical values of the original Simes global test. With

$$\widehat{m}_0^{(1)}(\mathbf{P}_m) = \max\left\{W_1(\mathbf{P}_m), 1\right\}, \tag{2.3}$$

we define the following:

Adaptive Simes Test. *Reject H_0 if $P_{i:m} \leq c_i$ for at least one $i = 1, \ldots, m$, where $c_i = i\alpha/\widehat{m}_0^{(1)}(\mathbf{P}_m)$.*

The fact that the adaptive Simes test controls the type I error rate at α under independence or positive dependence in the sense of (2.1) or (2.2) can be proved as follows. Let R_1 and R_2 be the total numbers of the null hypotheses rejected by the BH procedure and the stepup procedure with the same critical values as in the adaptive Simes test. Then, the type I error rate of the adaptive Simes test is give by

$$\text{pr}\left\{R_2 > 0\right\} = \text{pr}\left\{R_2 > 0, R_1 = 0\right\} + \text{pr}\left\{R_2 > 0, R_1 > 0\right\}, \tag{2.4}$$

with the probabilities being evaluated under H_0. Since $\widehat{m}_0^{(1)}(\mathbf{P}_m) = \max\{m - R_1, 1\}$, and $R_2 = 0$ with probability one if $R_1 = 0$,

$$\text{pr}\left\{R_2 > 0\right\} = \text{pr}\left\{R_2 > 0, R_1 > 0\right\} \leq \text{pr}\left\{R_1 > 0\right\},$$

which is less than or equal to α under independence or positive dependence in the sense of (2.1) or (2.2) of the p-values due to the well-known Simes' inequality (Simes, 1986; Sarkar, 1998; Sarkar and Chang, 1997).

We will now obtain an adaptive Bonferroni global test. We will do that through reinterpreting the Hommel procedure (Hommel, 1988) as an adaptive version of the Bonferroni procedure. The Hommel procedure is defined as follows. Let

$$W_2(\mathbf{P}_m) = \{j \in \{1, \ldots, m\} : P_{m-j+k:m} > k\alpha/j,\ k = 1, \ldots, j\}$$

and

$$\widehat{m}_0^{(2)}(\mathbf{P}_m) = \max\{W_2(\mathbf{P}_m), 1\}. \tag{2.5}$$

If $W_2(\mathbf{P}_m)$ is nonempty, reject H_i whenever $P_i \le \alpha/\widehat{m}_0^{(2)}(\mathbf{P}_m)$. If, however, $W_2(\mathbf{P}_m)$ is empty, reject all H_i, $i = 1, \ldots, m$. Notice that $\widehat{m}_0^{(2)}(\mathbf{P}_m)$ represents the maximum size of the subfamily of null hypotheses whose members are all declared to be true when applying the Simes test. In other words, $\widehat{m}_0^{(2)}(\mathbf{P}_m)$ provides an estimate of m_0, in terms of which the Hommel procedure can be interpreted as the following:

Adaptive Bonferroni Test. *Reject H_0 if $P_{i:m} \le c_i$ for at least one $i = 1, \ldots, m$, where $c_i = \alpha/\widehat{m}_0^{(2)}(\mathbf{P}_m)$.*

We can summarize the above discussions in the following proposition.

Proposition 2.1. *Given any family of null hypotheses H_i, $i = 1, \ldots, m$, consider testing the global null hypothesis $H_0 = \bigcap_{i=1}^m H_i$ using a cut-off test of the form $\bigcup_{i=1}^m \{P_{i:m} \le c_i\}$. When the p-values are independent or positively dependent in the sense of (2.1) or (2.2), the adaptive Simes and Bonferroni tests are valid level α tests.*

Remark 2.1. In the above adaptive tests, we use $\max\{1, i\}$ as the estimate of m_0 once the hypotheses $H_{m-i+k:m}$, $k = 1, \ldots, i$, are accepted, where this i, for the adaptive Simes test, is the index such that $P_{m-i+k:m} > (m-i+k)\alpha/m$ for all $k = 1, \ldots, i$, whereas, for the adaptive Bonferroni test, it is the largest index from 1 to m such that $P_{m-i+k:m} > k\alpha/i$ for all $k = 1, \ldots, i$. Since $(m-i+k)\alpha/m \ge k\alpha/i$ for $k = 1, \ldots, i$, the estimate of m_0 is more liberal in the adaptive Simes test than in the adaptive Hommel test, i.e., $\hat{m}_0^{(1)} \le \hat{m}_0^{(2)}$, implying that the adaptive Simes test is more powerful.

Remark 2.2. In the alternative adaptive Bonferroni procedure considered in Guo (2009), $c_i = \alpha/\widehat{m}_0$, $i = 1, \ldots, m$, where $\widehat{m}_0 = [W_m(\lambda) + 1]/(1-\lambda)$. It also provides a valid level α global test for testing H_0, but under a model that assumes independence of the p-values conditional on any (random) configurations of true and false null hypotheses.

2.3. Adaptive Multiple Testing Procedures

We now consider our main problem, which is, to simultaneously test the null hypotheses H_i, $i = 1, \ldots, n$, and develop newer adaptive versions of Holm's stepdown and Hochberg's stepup procedures that utilize information about the number of true null hypotheses suitably extracted from the data and ultimately maintain a control over the FWER at α. We first present these procedures. Then, we provide a real life application, and the results of a simulation study investigating the performances of our proposed adaptive procedures in relation to those of the corresponding conventional, non-adaptive procedures.

2.3.1. *The procedures*

We will develop our procedures using the following closure principle of Marcus *et al.* (1976) that is often used to construct FWER controlling procedures.

CLOSURE PRINCIPLE. *Suppose that for each $I \subseteq \{1, \ldots, n\}$ there is a valid level α global test for testing the intersection null hypothesis $\bigcap_{i \in I} H_i$. An individual null hypothesis H_i is rejected if for each $I \subseteq \{1, \ldots, n\}$ with $I \ni i$, $\bigcap_{j \in I} H_j$ is rejected by the corresponding global test.*

A multiple testing procedure satisfying the closure principle is termed a closed testing procedure. It controls the FWER at α. Many of the multiple testing procedures in the literature controlling the FWER are either closed testing procedures or can be presented as some versions of such a procedure. The level α adaptive global tests presented in the above section will be the key towards developing our proposed adaptive FWER controlling procedures based on the closure principle. Before we do that, we first need to introduce a few more additional notations.

Consider all possible sub-families of the null hypotheses, $\{H_i,\ i \in I_m\}$, $I_m \subseteq \{1, \ldots, n\}$, $m = 1, \ldots, n$, where I_m is of cardinality m. Define $\widehat{n}_0^{(1)}(\mathbf{P}_m)$ and $\widehat{n}_0^{(2)}(\mathbf{P}_m)$, the two estimates of the number of true null hypotheses in $\{H_i, i \in I_m\}$ based on the corresponding set of p-values $\mathbf{P}_m = \{P_i, i \in I_m\}$, as in (2.3) and (2.5), respectively. Since these estimates are symmetric and componentwise increasing in $\mathbf{P}_m$, and every ordered component of any m-dimensional subset of the p-values is smaller than the corresponding component of $\widetilde{\mathbf{P}}_m = (P_{n-m+1:n}, \ldots, P_{n:n})$, we have the following: $\widehat{n}_0^{(j)}(\widetilde{\mathbf{P}}_m) \geq \widehat{n}_0^{(j)}(\mathbf{P}_m)$, for any $\mathbf{P}_m$ and $j = 1, 2$. For convenience, we will denote $\widehat{n}_0^{(j)}(\widetilde{\mathbf{P}}_m)$ simply as $\widehat{n}_0^{(j)}(m)$, $j = 1, 2$.

It is important to note exactly how $\widehat{n}_0^{(j)}(m)$ is defined for $j = 1, 2$. Consider using the p-values $P_{n-m+1:n}, \ldots, P_{n:n}$ to test corresponding null hypotheses. Then, from (2.3), $\widehat{n}_0^{(1)}(m) = \max\{W_1(\widetilde{\mathbf{P}}_m), 1\}$, where $W_1(\widetilde{\mathbf{P}}_m)$ is the number of accepted null hypotheses in the stepup test involving these p-values and the critical values $j\alpha/m$, $j = 1, \ldots, m$. Similarly, from (2.5), $\widehat{n}_0^{(2)}(m) = \max\{W_2(\widetilde{\mathbf{P}}_m), 1\}$, where

$$W_2(\widetilde{\mathbf{P}}_m) = \{j \in \{1, \ldots, m\} : P_{n-j+k:n} > k\alpha/j,\ k = 1, \ldots, j\}.$$

It is easy to see that while $\widehat{n}_0^{(2)}(m)$ is increasing in m, $\widehat{n}_0^{(1)}(m)$ may not be so. If $\widehat{n}_0^{(1)}(m)$ is not increasing in m, we will make a minor modification of it as follows: Let $\widehat{n}_0^{(1)'}(1) = \widehat{n}_0^{(1)}(1)$ and $\widehat{n}_0^{(1)'}(m) = \max\left(\widehat{n}_0^{(1)'}(m-1), \widehat{n}_0^{(1)}(m)\right)$, for $2 \le m \le n$. Obviously, such modified $\widehat{n}_0^{(1)'}(m)$ is always increasing in m, and for each $m = 1, \ldots, n$, $\widehat{n}_0^{(1)'}(m) \ge \widehat{n}_0^{(1)}(m)$, with the equality holding when $\widehat{n}_0^{(1)}(m)$ is increasing in m.

Now, we present our adaptive Holm procedure in the following theorem.

Theorem 2.1. *Consider the the stepdown procedure with the critical values* $\alpha/\hat{n}_0^{(2)}(n-j+1)$, $j = 1, \ldots, n$. *It controls the FWER at* α *when the p-values are independent or positively dependent in the sense of (2.1) or (2.2).*

PROOF. Suppose that $P_{j:n}$ is the smallest among the p-values that correspond to the n_0 true null hypotheses. If $P_{j:n} \le \alpha/\hat{n}_0^{(2)}(n-j+1)$, then for any m-dimensional subset of the null hypotheses containing the true null hypothesis corresponding to $P_{j:n}$, the adaptive Bonferroni test with the critical constants $c_j = \alpha/\hat{n}_0^{(2)}(\mathbf{P}_m)$ rejects its intersection H_0^m, where $\mathbf{P}_m$ is the corresponding p-value vector of the m individual null hypotheses. Since under H_0^m, $m \le n_0 \le n-j+1$, then we have $P_{j:n} \le \alpha/\hat{n}_0^{(2)}(n-j+1) \le \alpha/\hat{n}_0^{(2)}(m) \le \alpha/\hat{n}_0^{(2)}(\mathbf{P}_m)$. Thus, if $P_{j:n} \le \alpha/\hat{n}_0^{(2)}(n-j+1)$, $H_{j:n}$ is rejected by the closed testing procedure based on the above Bonferroni test. Therefore, $\text{pr}\{P_{j:n} \le \alpha/\hat{n}_0^{(2)}(n-j+1)\}$ is less than or equal to the FWER of the closed testing procedure. By the closure principle and Proposition 2.1, $\text{pr}\{P_{j:n} \le \alpha/\hat{n}_0^{(2)}(n-j+1)\} \le \alpha$. Therefore, the FWER of the adaptive Holm procedure is less than or equal to α. □

Remark 2.3. In the alternative adaptive Holm's procedure of Guo (2009), $c_i = \alpha/\min\{n-i+1, \widehat{n}_0\}$, $i = 1, \ldots, n$, where $\widehat{n}_0 = [W_n(\lambda)+1]/(1-\lambda)$. It asymptotically (as $n \to \infty$) controls the FWER at α under a conditional independence model (Wu, 2008). The adaptive Holm procedure in Theorem

2.1, on the other hand, not only controls the FWER in finite samples but also under a more general type of dependence situation.

Next, we present our adaptive Hochberg procedure through the adaptive Simes test defined in the preceding section.

Theorem 2.2. *Consider the stepup procedure with the critical values* $\alpha/\hat{n}_0^{(1)'}(n-j+1), j = 1, \ldots, n$. *It controls the FWER at* α *when the p-values are independent or positively dependent in the sense of (2.1) or (2.2).*

PROOF. Let $i_0 = \max\{i : P_{i:n} \leq \hat{n}_0^{(1)'}(n-i+1)\}$. First, for any subset of m individual hypotheses such that the corresponding smallest p-value is $P_{i_0:n}$, the adaptive Simes test with the critical constants $c_j = j\alpha/\hat{n}_0^{(1)}(\mathbf{P}_m)$, $j = 1, \ldots, m$ rejects its intersection hypothesis, since $m \leq n - i_0 + 1$ and thus $P_{i_0:n} \leq \alpha/\hat{n}_0^{(1)'}(n - i_0 + 1) \leq \alpha/\hat{n}_0^{(1)'}(m) \leq \alpha/\hat{n}_0^{(1)}(m) \leq \alpha/\hat{n}_0^{(1)}(\mathbf{P}_m)$, where $\mathbf{P}_m$ is the corresponding p-value vector of the m hypotheses. Second, consider a different subset of m individual hypotheses with exactly k hypotheses whose p-value is less than $P_{i_0:n}$. It is easy to see that $\hat{n}_0^{(1)'}(j+1) \leq \hat{n}_0^{(1)'}(j)+1$ for any $1 \leq j \leq n-1$, thus $\hat{n}_0^{(1)'}(n-i_0+1+k) \leq \hat{n}_0^{(1)'}(n - i_0 + 1) + k$. Also, $m \leq n - i_0 + 1 + k$. Therefore,

$$
\begin{aligned}
P_{i_0:n} &\leq \frac{\alpha}{\hat{n}_0^{(1)'}(n-i_0+1)} \leq \frac{(k+1)\alpha}{\hat{n}_0^{(1)'}(n-i_0+1)+k} \\
&\leq \frac{(k+1)\alpha}{\hat{n}_0^{(1)'}(n-i_0+1+k)} \leq \frac{(k+1)\alpha}{\hat{n}_0^{(1)'}(m)} \qquad (2.6) \\
&\leq \frac{(k+1)\alpha}{\hat{n}_0^{(1)}(m)} \leq \frac{(k+1)\alpha}{\hat{n}_0^{(1)}(\mathbf{P}_m)}.
\end{aligned}
$$

In (2.6), the second inequality follows from the fact that $(k+1)\alpha/\hat{n}_0^{(1)'}(n - i_0 + 1) + k$ is increasing in k. Thus, in such situation, the adaptive Simes test also rejects its intersection hypothesis. Summarizing these two cases, the closed testing procedure based on the adaptive Simes tests will reject $H_{i_0:n}$.

For other null hypothesis $H_{i:n}$ with $i < i_0$, we only need to prove that for each subset of m individual hypotheses without containing $H_{i_0:n}$ for which $P_{i:n}$ is the $(k+1)$-smallest p-value less than $P_{i_0:n}$, its intersection hypothesis is rejected by the adaptive Simes test. Actually, by using the same arguments as (2.6), we can prove that $P_{i:n} \leq (k+1)\alpha/\hat{n}_0^{(1)}(\mathbf{P}_m)$. Thus, $H_{i:n}$ is also rejected by the closed testing procedure. By using the closure principle and proposition 2.1, the adaptive Hochberg procedure controls the

FWER at level α when the p-values are independent or positively dependent in the sense of (2.1) or (2.2). □

Remark 2.4. It is easy to see that for each $1 \leq i \leq n$, $\hat{n}_0^{(1)}(n-i+1) \leq n-i+1$, thus $\hat{n}_0^{(1)'}(n-i+1) \leq n-i+1$. Therefore, the adaptive Hochberg procedure is more powerful than the corresponding non-adaptive one.

2.3.2. *An application*

We revisit a dose-finding diabetes trial study analyzed in Dmitrienko *et al.* (2007). The trial compares three doses of an experimental drug versus placebo. The efficacy profile of the drug was studied using three endpoints: Haemoglobin A1c (primary), Fasting serum glucose (secondary), and HDL cholesterol (secondary). These endpoints were examined at each of the three doses, and the raw p-values are 0.005, 0.011, 0.018, 0.009, 0.026, 0.013, 0.010, 0.006, and 0.051. We pre-specify $\alpha = 0.05$. By using the conventional, non-adaptive Holm and Hochberg procedures, we see that two null hypotheses are rejected at level 0.05 for both these tests. In contrast, our proposed adaptive Holm and Hochberg procedures both reject seven null hypotheses at the same level.

2.3.3. *A simulation study*

We performed a small scale simulation study investigating the performances of our proposed adaptive Holm and Hochberg procedures in comparison with those of the corresponding conventional, non-adaptive Holm and Hochberg procedures. We made these comparisons in terms of the FWER control at the desired level and power, with the power being defined as the expected proportion of the false null hypotheses that are correctly rejected.

We generated $n = 50$ dependent normal random variables $N(\mu_i, 1)$, $i = 1, \ldots, n$, with a common correlation $\rho = 0.2$, and with n_0 of the 50 μ_i's being equal to 0 and the remaining equal to 3, and applied the four different procedures to test $H_i : \mu_i = 0$ against $K_i : \mu_i \neq 0$ simultaneously for $i = 1, \ldots, 50$ at level $\alpha = 0.05$. We repeated these steps $2,000$ times before calculating the proportion (estimated FWER) of times at least one true null hypothesis is falsely rejected and the average proportion (estimated power) of false null hypotheses that are rejected. Figures 2.1 and 2.2 present the estimated FWERs and powers, respectively, of the four procedures, each plotted against different values of n_0. As seen from Fig. 2.1, our suggested adaptive Holm and Hochberg procedures provide better

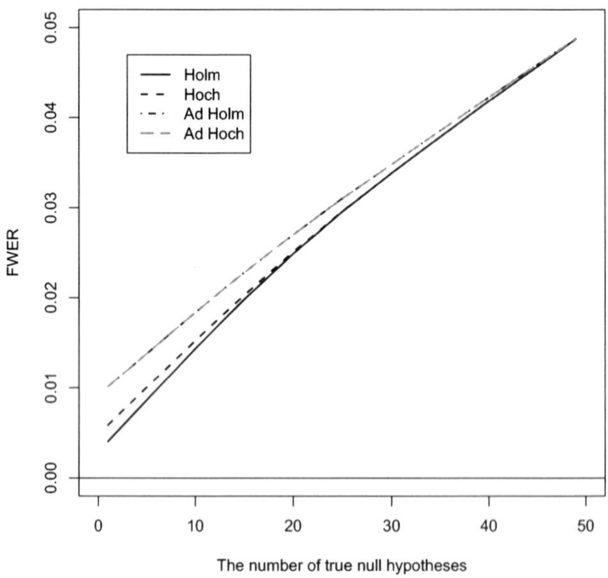

Fig. 2.1. Comparison of familywise error rates of four procedures: Holm (solid), Hochberg (small dashes), adaptive Holm (dot-dash), and adaptive Hochberg (dashes), with parameters $n = 50$, $\alpha = 0.05$.

control of the FWER than those of the conventional, non-adaptive Holm and Hochberg procedures, although with increasing number of true null hypotheses all procedures become less and less conservative. Figure 2.2 presents the comparisons in terms of power. As seen from this figure, our suggested adaptive Holm and Hochberg procedures have better power performances than the corresponding non-adaptive Holm and Hochberg procedures. Again, with increasing number of true null hypotheses, the difference in power gets smaller and closer to zero.

2.4. Concluding Remarks

A knowledge of the proportion of true null hypotheses among all the null hypotheses tested can be useful for developing improved versions of conventional FDR or FWER controlling procedures. A number of adaptive versions of an FDR or FWER controlling procedure exist in the literature, each attempts to improve the FDR or FWER procedure by extracting information about the number of true null hypotheses from the available

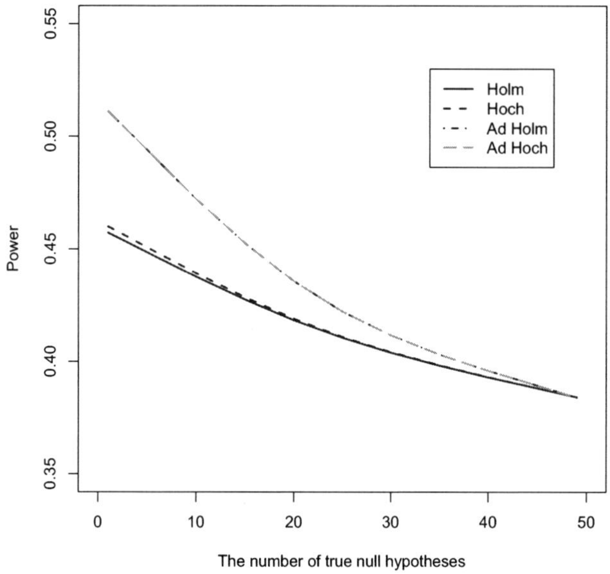

Fig. 2.2. Comparison of average power of four procedures: Holm (solid), Hochberg (small dashes), adaptive Holm (dot-dash), and adaptive Hochberg (dashes), with parameters $n = 50$, $\alpha = 0.05$.

data and incorporating that into the procedure. However, in finite sample settings, the ultimate control of the FDR or FWER for these adaptive procedures has only been proved under the assumption of independence or conditional independence of the p-values. In this article, we make an attempt for the first time, as far as we know, to develop adaptive FWER procedures that provide ultimate control over the FWER not only under independence but also under positive dependence of the p-values.

It is important to point out that there are some essential differences between adaptive and non-adaptive procedures. For example, for a non-adaptive single-step FWER controlling procedure, weak control implies strong control, but that conclusion does not hold for an adaptive single-step procedure. We explain that phenomenon through the following example.

Example 2.1. Consider an adaptive Bonferroni procedure for which $\hat{n}_0^{(1)}(n)$ is used as the estimate of the number of true null hypotheses. For convenience, we will denote $\hat{n}_0^{(1)}(n)$ simply as $\hat{n}_0^{(1)}$. By Proposition 2.1, the single-step procedure can weakly control the FWER. However, we can show that it cannot strongly control the FWER. Let $n = 6$ and $n_0 = 2$.

Suppose four false null p-values are zero and two true null p-values q_1 and q_2 are independent identically distributed $U(0,1)$. Let $q_{(1)} \leq q_{(2)}$ denote the ordered values of q_1 and q_2. Let R be the total number of rejections. Then,

$$\text{FWER} = \text{pr}\{q_{(1)} \leq \alpha/\hat{n}_0^{(1)}, 4 \leq R \leq 6\}.$$

Note that

$$\begin{aligned}
&\text{pr}\{q_{(1)} \leq \alpha/\hat{n}_0^{(1)}, R = 4\} = \text{pr}\{q_{(1)} \leq \alpha/2, q_{(1)} > 5\alpha/6, q_{(2)} > \alpha\} = 0, \\
&\text{pr}\{q_{(1)} \leq \alpha/\hat{n}_0^{(1)}, R = 5\} = \text{pr}\{q_{(1)} \leq \alpha, q_{(1)} \leq 5\alpha/6, q_{(2)} > \alpha\} \\
&\quad = \text{pr}\{q_{(1)} \leq 5\alpha/6, q_{(2)} > \alpha\},
\end{aligned}$$

and

$$\begin{aligned}
&\text{pr}\{q_{(1)} \leq \alpha/\hat{n}_0^{(1)}, R = 6\} = \text{pr}\{q_{(1)} \leq \alpha, q_{(2)} \leq \alpha\} \\
&\quad \geq \text{pr}\{q_{(1)} \leq 5\alpha/6, q_{(2)} \leq \alpha\}.
\end{aligned}$$

Thus

$$\text{FWER} = \sum_{r=4}^{6} \text{pr}\{q_{(1)} \leq \alpha/\hat{n}_0^{(1)}, R = r\} \geq \text{pr}\{q_{(1)} \leq 5\alpha/6\} > \alpha.$$

References

Benjamini, Y. and Heller, R. (2007). False discovery rate for spatial signals. *J. Am. Statist. Assoc.* **102**, 1272–1281.

Benjamini, Y. and Hochberg, Y. (1995). Controlling the false discovery rate: A practical and powerful approach to multiple testing. *J. R. Statist. Soc. B* **57**, 289–300.

Benjamini, Y. and Hochberg, Y. (2000). The adaptive control of the false discovery rate in multiple hypothesis testing with independent statiatics. *J. Educ. Behav. Statist.* **25**, 60–83.

Benjamini, Y., Krieger, K. and Yekutieli, D. (2006). Adaptive linear step-up procedures that control the false discovery rate. *Biometrika* **93**, 491–507.

Benjamini, Y. and Yekutieli, D. (2001). The control of the false discovery rate in multiple testing under dependency. *Ann. Statist.* **29**, 1165–1188.

Bernhard, G., Klein, M. and Hommel, G. (2004). Global and multiple test procedures using ordered p-values – A review. *Statist. Pap.* **45**, 1-14.

Blanchard, G. and Roquain, E. (2009). Adaptive FDR control under independence and dependence. *Journal of Machine Learning Research* **10**, 2837-2871.

Block, H. W., Savits, T. H. and Shaked, M. (1985). A concept of negative dependence using stochastic ordering. *Statist. Probab. Lett.* **3**, 81-86.

Dmitrienko, A., Wiens, B., Tamhane, A. and Wang, X. (2007). Global and Tree-structured gatekeeping tests in clinical trials with hierarchically ordered multiple objectives. *Statist. Med.* **26**, 2465–2478.

Farcomeni, A. (2007). Some results on the control of the false diiscovery rate under dependence. *Scandinavian Journal of Statistics* **34**, 275-297.

Ferreira, J. A. and Zwinderman A. H. (2006). On the Benjamini-Hochberg Method. *Annals of Statistics* **34**, 1827-1849.

Gavrilov, Y., Benjamini, Y. and Sarkar, S. K. (2009). An adaptive step-down procedures with proven FDR control under independence. *Ann. Statist.* **37**, 619–629.

Guo, W. (2009). A note on adaptive Bonferroni and Holm procedures under dependence. *Biometrika* **96**, 1012-1018.

Hochberg, Y. (1988). A sharper Bonferroni procedure for multiple testing of significance. *Biometrika* **75**, 800–802.

Hochberg, Y. and Benjamini, Y. (1990). More powerful procedures for multiple significance testing. *Statist. Med.* **9**, 811–818.

Holm, S. (1979). A simple sequentially rejective multiple testing procedure. *Scand. J. Statist.* **6**, 65–70.

Hommel, G. (1988). A stagewise rejective multiple test procedure based on a modified Bonferroni test. *Biometrika* **75**, 383-386.

Marcus, R., Peritz, E. and Gabriel, K. R. (1976). On closed testing procedures with special reference to ordered analysis of variance. *Biometrika* **63**, 655-660.

Sarkar, S. K. (1998). Some probability inequalities for ordered MTP_2 random variables(a proof of the Simes conjecture. *Ann. Statist.* **26**, 494-504.

Sarkar, S. K. (2002). Some results on false discovery rate in stepwise multiple testing procedures. *Ann. Statist.* **30**, 239–257.

Sarkar, S. K. (2006). False discovery and false non-discovery rates in single-step multiple testing procedures. *Ann. Statist.* **34**, 394–415.

Sarkar, S. K. (2008). On methods controlling the false discovery rate (with discussion). *Sankhya Ser A.* **70**, 135-168.

Sarkar, S. K. and Chang, C-K. (1997). The Simes method for multiple hypothesis testing with positively dependent test statistics. *J. Amer. Statist. Assoc.*, **92**, 1601-1608.

Sarkar, S. K. and Guo, W. (2009). On a generalized false discovery rate. *Ann. Statist.*, **37**, 337-363.

Simes, R. J. (1986). An improved Bonferroni procedure for multiple tests of significance. *Biometrika* **73**, 751-754.

Storey, J. D., Taylor, J. E. and Siegmund, D. (2004). Strong control, conservative point estimation and simultaneous conservative consistency of false discovery rates(a unified approach. *J. R. Statist. Soc. B* **66**, 187-205.

Wu, W. (2008). On false discovery control under dependence. *Ann. Statist.* **36**, 364–380.

Chapter 3

A False Discovery Rate Procedure for Categorical Data

Joseph F. Heyse

Biostatistics and Research Decision Sciences
Merck Research Laboratories, North Wales, PA 19454, USA
joseph_heyse@merck.com

Almost all multiple comparison and multiple endpoint procedures applied in experimental settings are designed to control the Family Wise Error Rate (FWER) at a prespecified level of α. Benjamini and Hochberg (1995) argued that in certain settings, requiring strict control of the FWER can be overly conservative. They suggested controlling the False Discovery Rate (FDR), defined as the expected proportion of true (null) hypotheses that are incorrectly rejected. When one or more of the hypotheses being tested uses a categorical data endpoint, it is possible to further increase the power of both FWER and FDR controlling procedures. Methods proposed by Tarone (1990) for FWER control and Gilbert (2005) for FDR control have increased power by using the discreteness in the distribution of the test statistic to essentially reduce the effective number of hypotheses considered for the multiplicity adjustment. A modified fully discrete FDR sequential procedure is introduced that uses the exact conditional distribution of potential study outcomes. The FDR control, and the potential gains in power were estimated using simulation. An application of the proposed FDR procedure in the setting of genetic data analysis is reviewed, and other potential uses of the method are discussed.

3.1. Introduction

When scientific investigations involve testing families of hypotheses there is a concern about inflated type I, or false positive errors. Multiple comparison procedures that control the Family Wise Error Rate (FWER) provide strong type I error control, but often allow a high rate of type II, or false negative errors. As a result, the study power sharply diminishes with increasing numbers of tests. The Bonferroni method is one popular FWER controlling procedure. Hochberg and Tamhane (1987), Hsu (1996), and Dmitrienko *et al.* (2005) provide comprehensive overviews of FWER controlling procedures.

For many practical applications it is preferable to apply a multiple comparison procedure that controls the risks of type I and type II errors more evenly. Examples may include genetic data analysis, or the evaluation of spontaneous adverse experience data in clinical trials. This objective can be addressed by methods that control the False Discovery Rate (FDR), defined as the expected proportion of rejected hypotheses that are false rejections of true null hypotheses (or false discoveries). Benjamini and Hochberg (1995) developed the first FDR controlling method, and showed that it provides large gains in power over FWER controlling methods. For categorical data, the test statistics are discrete, and the complete conditional null distribution can be enumerated, and used to improve the power of both FWER and FDR testing procedures. Mantel (1980) and Mantel *et al.* (1982) recognized that the power can be increased for multiple comparisons in rodent carcinogenicity studies which use categorical endpoints, defined as the presence or absence of tumors encountered in the study. In this application, treating the multiple tumor types/sites as independent, and conditional on the observed numbers of tumor bearing rodents, the null distribution of the test statistic is discrete. Levels of statistical significance for the multiple tumor types/sites are rarely equal to the prespecified level α, and for tumors with a low occurrence, it may not even be possible to reach the nominal unadjusted α level of statistical significance. Mantel (1980) introduced the idea, attributed to Tukey, of eliminating those tests for which rejection of the null hypothesis was not possible at the unadjusted α level, because only 1 or 2 tumor bearing rodents (out of 250) were observed for those tumor types/sites. Mantel *et al.* (1982) further improved upon this high level adjustment by using the complete null distribution, and Heyse and Rom (1988) and Westfall and Young (1989) considered the case for non-independent discrete endpoints using resampling procedures.

Tarone (1990) developed a modified Bonferroni method for discrete data using the ideas of Mantel (1980) and Tukey to reduce the essential dimensionality of the multiplicity problem. Gilbert (2005) used the same modification, and applied the Benjamini and Hochberg (1995) FDR controlling method to the reduced set of tests as a two-step procedure. First, the endpoints are identified that can potentially achieve a level of significance suitable for FWER, or FDR, control. Second, the Bonferroni, or Benjamini and Hochberg FDR, method is applied to the reduced set of endpoints. By construction the Tarone modification controls the FWER, and the Gilbert modification controls the FDR. Both have increased power relative to the original methods.

In this paper, a FDR controlling method is proposed that utilizes the full conditional null distribution for independent categorical endpoints. The fully discrete procedure controls the FDR at the prespecified level, and has power equal to or greater than both the Benjamini and Hochberg (1995) and Gilbert (2005) methods. A similar approach can be applied to the modified Bonferroni method of Tarone (1990), also with equal or greater power. The method can be applied in situations where some endpoints are categorical and some are continuous. When all endpoints are continuous the procedure is identical to the original Benjamini and Hochberg method.

A brief overview of the False Discovery Rate is given in Sec. 3.2, along with the Benjamini and Hochberg (1995) FDR controlling procedure. The proposed generalization for categorical data is described in Sec. 3.3. Section 3.4 summarizes the Tarone (1990) and Gilbert (2005) modified Bonferroni and FDR procedures for discrete data, along with a detailed illustration in Sec. 3.5. The methods are applied to an analysis of genetic variants using data from Gilbert (2005) (Sec. 3.6). The results of a simulation study are used to demonstrate the statistical error properties of the methods in Sec. 3.7. The paper ends with a few concluding remarks in Sec. 3.8.

3.2. False Discovery Rate

Consider a family of K hypotheses $F = \{H_1, H_2, \ldots, H_K\}$. Some of the K hypotheses are true null, and others are false. Associated with each hypothesis is a P-value determined from the tail probability for a suitably chosen test statistic. In these types of experimental situations, involving multiple statistical tests, there is the possibility of an inflated type I error unless an appropriate multiplicity adjustment is applied to the determination of statistical significance. This setting is depicted in Table 3.1.

Table 3.1. False Discovery Rate (Benjamini and Hochberg, 1995)

	Declared Insignificant	Declared Significant	Total
Number of True Hypotheses	U	V	K_0
Number of False Hypotheses	T	S	$K - K_0$
Total	$K - R$	R	K

Of the K hypotheses considered in the study, R are declared significant overall, of which V are truly null, and as such, falsely rejected. S hypotheses are correctly rejected. The Family Wise Error Rate (FWER) is defined as the probability that any true null hypothesis among the $H_i \in F$ is falsely rejected. Several multiple comparison methods are available to control the FWER at levels less than or equal to the nominal type I error α. One popular method uses the Bonferroni inequality to reject any H_i that has an associated P-value $P_i \leq \alpha/K$. Available stepwise procedures, such as Hochberg (1988), provide more powerful alternatives to the Bonferroni method. For nicely written overviews, the reader is referred to Hochberg and Tamhane (1987), Hsu (1996), and Dmitrienko *et al.* (2005).

Benjamini and Hochberg (1995) argued that in many experimental settings, controlling the FWER can be overly stringent. They proposed controlling the False Discovery Rate (FDR) as a more balanced alternative between Type I and Type II errors. Returning to Table 3.1, the FDR is defined as the expected proportion $E(V/R)$ of false discoveries, V, relative to the potential discoveries, R, declared significant. In studies where no hypotheses are declared significant, and therefore, there are no potential discoveries, the FDR is defined to be zero.

Benjamini and Hochberg (1995) developed a procedure for controlling the FDR at a level α, that is based on the ordering of the K observed P-values, $P_{(1)} \leq P_{(2)} \leq \cdots \leq P_{(K)}$. The hypothesis associated with the ordered $P_{(j)}$ is denoted by $H_{(j)}$ for $j = 1, 2, \ldots, K$. Define J as the largest value of j such that $P_{(j)} \leq (j/K)\alpha$,

$$J = \mathrm{Max}\{j : P_{(j)} \leq (j/K)\alpha\}. \tag{3.1}$$

The procedure rejects the J hypotheses $H_{(1)}, H_{(2)}, \ldots, H_{(J)}$ associated with the J smallest P-values. If $J = 0$ then no hypotheses are rejected. Benjamini and Hochberg (1995) proved that the procedure in (3.1) controls the FDR at levels less than or equal α for the K continuous tests.

A convenient form of the Benjamini and Hochberg procedure can be used that provides FDR adjusted P-values. Define

$$P_{[K]} = P_{(K)},$$

and

$$P_{[j]} = \mathrm{Min}\{P_{[j+1]}, (K/j)P_{(j)}\} \tag{3.2}$$

for values of $j \leq K-1$. Using this form of the Benjamini and Hochberg procedure, the hypotheses associated with values of $P_{[j]} \leq \alpha$ are rejected. Both procedures (3.1) and (3.2) will always reject the same set of J hypotheses.

Table 3.2. Exact null distribution of 6 responders with a balanced randomization

Number of Responders (X) in Group 2:1	Probability of Observing X Responders Assuming Null Hypothesis	Cumulative Probability
6:0	0.0156	0.0156
5:1	0.0938	0.1094
4:2	0.2344	0.3438
3:3	0.3125	0.6563
2:4	0.2344	0.8907
1:5	0.0938	0.9845
0:6	0.0156	1.0000

In the case that all tested hypotheses are true, that is, when $K_0 = K$, this procedure provides weak control of the FWER (Simes, 1986). However, when $K_0 < K$, and some hypotheses in F are false, the procedure does not control the FWER, but offers greater power. Hommel (1988) first showed that when $K_0 < K$ the procedure does not control the FWER. Hochberg (1988) constructed a FWER controlling procedure that uses a similar stepwise calculation as in Eq. (3.1), except that at each step, $P_{(j)}$ is compared to $[1/(K-j+1)]\alpha$, rather than $(j/K)\alpha$. The constants for the two methods are the same for $j = 1$ and $j = K$, but the constants are larger for values of j between 1 and $K(1 < j < K)$ for the FDR procedure. Therefore, the FDR procedure of Benjamini and Hochberg is potentially more powerful than FWER controlling procedures.

3.3. Modified FDR Procedure for Categorical Data

Construction of an FDR procedure for discrete data uses the term

$$(K/j)P_{(j)}$$

from the sequential calculations in Eq. (3.2). This term is appropriate for test statistics with continuous distributions, where all of the K tests can potentially yield a P-value equal to $P_{(j)}$. However, with categorical data the test statistics have discrete distributions, and the K tests cannot all yield P-values exactly at that level. As an example, consider the distribution in Table 3.2 which was computed from a hypothetical data set with a total of six positive responses across two groups with a balanced randomization. These calculations were based on a Fisher's exact test, and for simplicity, they are shown 1-sided.

The first point to note is that only the extreme single scenario of all responders being observed in Group 2 (i.e., a 6:0 split) yields an unadjusted significance level below $\alpha = 0.05$. When performing a multiplicity adjustment for endpoint j, values of $P_{(j)} > 0.0156$ would actually use $P_{(j)}$ in Eq. (3.2), which is an over adjustment. Also, P-values < 0.0156 are not achievable for this hypothetical endpoint distribution, and therefore, this dimension should not be included in the adjustment of $\mathrm{P}_{(j)}$ for values of $P_{(j)} < 0.0156$.

Define $Q_i(P)$ as the largest P-value achievable for hypothesis $i = 1, 2, \ldots, K$ that is less than or equal to P. $Q_i(P)$ is taken as zero when P-values $\leq P$ are not achievable for hypothesis i due to a low occurrence of responders or an extreme value of P. Using the distribution in Table 3.2 as an example, $Q_i(0.02) = 0.0156$ and $Q_i(0.01) = 0$.

FDR for categorical data can use a similar stepwise calculation

$$P^*_{[K]} = P_{(K)}$$
$$P^*_{[j]} = \mathrm{Min}\{P_{[j+1]}, [\sum_{i=1}^{K} Q_i(P_{(j)})]/j\} \tag{3.3}$$

for values of $j \leq K - 1$. Hypotheses with levels of $P^*_{[j]} \leq \alpha, j = 1, 2, \ldots, K$ are declared significant. Notice in Eq. (3.3) that the term $[\sum_{i=1}^{K} Q_i(P_{(j)})]/j$ replaced the term $(K/j)P_{(j)}$ in Eq. (3.2). The difference is due to the recognition that P-values of exactly $P_{(j)}$ are not possible for some endpoints because of the discrete nature of distributions. As a result, $0 \leq Q_i(P_{(j)}) \leq P_{(j)}$ for all $i, j = 1, \ldots, K$ due to the discreteness of the test statistics, and $Q_i(P_{(j)})$ will equal 0 if a P-value as extreme as $P_{(j)}$ is not possible for endpoint i. Comparing equations (3.2) and (3.3) we find that $Q_i(P_{(j)}) \leq P_{(j)}$ and therefore $P^*_{[j]} \leq P_{[j]}$.

The proof that the FDR procedure modified for discrete data controls the FDR at level α follows the same argument used by Gilbert (2005). Theorem 5.1 in Benjamini and Yekutieli (2001) proved that for independent test statistics the Benjamini and Hochberg procedure in equations (3.1) or (3.2) controls the FDR at levels less than or equal $(K_0/K)\alpha$ for both continuous and discrete test statistics. For continuous test statistics the FDR is controlled at exactly $(K_0/K)\alpha$. Because the tail probabilities are smaller for discrete test statistics the FDR is controlled at levels less than or equal to $(K_0/K)\alpha$ for categorical data. Equality holds for the continuous case because the K_0 P-values are uniformly distributed under the null hypotheses so that $\Pr\{P_{(i)} < (j/K)\alpha\} = (j/K)\alpha$. For categorical endpoints $\Pr\{P_{(i)} < (j/K)\alpha\}$ may be less than $(j/K)\alpha$ because of the discrete nature of the test statistic. The control of FDR and the gain in power are a result of this gap.

For the proposed fully discrete procedure, since $Q_i(P_{(j)}) \leq P_{(j)}$ the multiplicity adjusted P-values from Eq. (3.3) will always be less than or equal to those computed from Eq. (3.2), so that $P^*_{[j]} \leq P_{[j]}$. Since the procedure based on $P_{[j]}$ controls the FDR at levels less than or equal to $(K_0/K)\alpha$, the procedure based on $P^*_{[j]}$ will also control the FDR at levels less than or equal to $(K_0/K)\alpha$. The potential gain in power comes from the gap between $P^*_{[j]}$ and $P_{[j]}$.

Note that when all K hypotheses are based on continuous data, $Q_i(P_{(j)}) = P_{(j)}$ for every value of i and the adjustment procedure is identical to the original Benjamini and Hochberg method. However, when some endpoints are continuous, and some are categorical, we can simply use $Q_i(P_{(j)})$ for the categorical endpoints and $P_{(j)}$ for the continuous endpoints as appropriate in Eq. (3.3). As with the fully discrete procedure, the modified procedure will also control the FDR and potentially provide greater power than applying the original method.

3.4. Tarone and Gilbert Modified Procedures

The Tarone (1990) modified Bonferroni procedure, and the Gilbert (2005) modified Benjamini and Hochberg procedure are based on the ideas published by Mantel (1980), which he attributed to Tukey. Define α_i^* as the smallest P-value achievable for hypothesis test i. For the simple Fisher's exact test, α_i^* can be determined by computing a P-value for the most extreme possibility that all of the observed responses occurred in one of the two treatment groups. For the hypothetical example in Table 3.2, $\alpha^* = 0.0156$.

The Tarone (1990) method reduces the dimensionality of the multiplicity adjustment by eliminating from consideration those hypotheses for which rejection is not possible because of the low occurrence of responders. For integer m define $K(m)$ as the number of the K hypotheses for which $m\alpha_i^* < \alpha$, and M as the smallest value of m such that $K(m) \leq m$. R_M will be the index set of hypotheses that satisfy $M\alpha_i^* \leq \alpha$. The modified Bonferroni test rejects hypotheses contained in R_M and for which $P_i \leq \alpha/M$.

Gilbert (2005) extended Tarone's idea for controlling the FWER to controlling the FDR by essentially applying the Benjamini and Hochberg procedure in Eq. (3.1) to the reduced number of tests in the index set R_M. This is a two-step procedure. First, apply Tarone's method to identify the index set R_M as the tests appropriate for multiple significance testing and second, to apply the FDR procedure in Eq. (3.1) to the M tests defined by R_M. Gilbert showed that the Tarone modified procedure controls the FDR at levels less than or equal to α, and has power at least as great as the Benjamini and Hochberg (1995) procedure applied to all K P-values.

3.4.1. *Discrete modification to Bonferroni adjustment*

Using logic similar to the development of the discrete FDR in Eq. (3.3), a Bonferroni type adjustment can be constructed for fully discrete test statistics as

$$P_{[j]}^{+} = \sum_{i=1}^{K} Q_i(P_{(j)}). \tag{3.4}$$

Statistical significance would be declared for those tests with $P_{[j]}^{+} < \alpha$. This method is expected to have greater power than applying the Bonferroni method in the discrete setting. As with the Tarone modification, some endpoints are essentially eliminated when $Q_i(P_{(j)}) = 0$, and a further improvement is gained for endpoints when $Q_i(P_{(j)}) < P_{(j)}$ due to the discrete nature of the test statistic distribution.

3.5. Illustration

Tarone (1990) analyzed an experiment in which complementary DNA (cDNA) transcripts were produced from transcribed RNA obtained from cells grown under normal conditions and from cells grown under an unusual study condition. The cDNA transcripts from a gene of interest were sequenced and compared to the known nucleotide sequence to determine the number of individual nucleotide changes in the transcripts. The frequencies of the changes were compared from the control and study cells to evaluate differences in the transcribed RNA.

The data in Table 3.3 are from Tarone (1990, Table 1) which report the frequencies of nucleotide changes observed at nine sites. The DNA sequences examined in the experiment were 200 nucleotides in length. Most sites had none or only a few changes noted. The nine included for analysis were those with a sufficient number of changes to possibly detect statistical significance at the unadjusted one-sided $\alpha = 0.05$ level using the Fisher's

Table 3.3. Observed frequencies of nucleotide changes in cDNA transcripts from control and study cells (Tarone, 1990)

Ordered Nucleotide	Control (X_{0i}/N_{0i})	Study (X_{1i}/N_{1i})	1-Sided P-value	α_i^*	B-H FDR	T-G FDR	Discrete FDR
1	1/10	8/11	0.0058	0.00019	0.052	0.017	0.0097
2	1/11	3/9	0.217	0.026	0.655	ND	0.309
3	2/11	4/10	0.268	0.0039	0.655	0.402	0.484
4	1/10	3/10	0.291	0.043	0.655	ND	0.548
5	2/9	2/8	0.665	0.029	0.801	ND	0.716
6	2/11	2/10	0.669	0.035	0.801	ND	0.716
7	2/9	2/9	0.712	0.041	0.801	ND	0.716
8	2/9	2/9	0.712	0.041	0.801	ND	0.716
9	3/8	2/7	0.818	0.0070	0.818	0.818	0.818

Notes: P-values are 1-sided using Fisher's exact test. α_i^* is the most extreme possible significance level possible for ordered nucleotide i. B-H FDR is the Benjamini and Hochberg FDR procedure using Eq. (3.2). T-G FDR is the two step Tarone/Gilbert procedure based on the $M = 3$ element index set $R_3 = \{1, 3, 9\}$. ND is Not Defined. Discrete FDR used Eq. (3.3).

exact test, conditional on the fixed marginal totals, and assuming independence between sites. Tarone reported the data by nucleotide order number. The nucleotides in Table 3.3 are reported in order according to the 1-sided P-value.

As shown by the values of α_i^*, all of the $K = 9$ statistical tests had the potential to result in an unadjusted significance level of $\leq \alpha$, but only nucleotide site 1 had an unadjusted P-value $= P_{(1)} = 0.0058$ below the 0.05 level. Applying the Benjamini and Hochberg FDR procedure (B-H FDR) to the complete set of 9 sites gave an adjusted P-value of $P_{[1]} = 0.052$ which does not reach the critical α level. The two-step Tarone/Gilbert method identified $M = 3$ sites with index set $R_3 = \{1, 3, 9\}$ having the potential to reach a level of significance $\leq \alpha/3 = 0.017$. Applying the B-H FDR procedure to these three sites does establish an adjusted statistical significance of $P_{[1]} = 0.017$ for nucleotide site 1, which is below 0.05. Applying the fully discrete FDR procedure to $P_{(1)} = 0.0058$ gives $Q_1(P_{(1)}) = 0.0058$, $Q_3(P_{(1)}) = 0.0039$, and $Q_i(P_{(1)}) = 0$ for the remaining 7 sites, so that $P_{[1]}^* = 0.0097$. Note that $P_{[1]}^*$ is less than $P_{[1]}$ for the T-G FDR procedure since only two sites ($i = 1$, and $i = 3$) contributed, and because the contribution for site 3 was less than $P_{(1)}$. The T-G FDR procedure used 3×0.0058.

Tarone (1990) applied his modified Bonferroni procedure to the data in Table 3.3. This approach used $M = 3$ and the index set $R_3 = \{1, 3, 9\}$

to give an adjusted P-value of 0.017. Using Eq. (3.4) a fully discrete Bonferroni adjustment for $P_{(1)}$ is 0.0097.

3.6. Application: Genetic Variants of HIV

Gilbert (2005) was motivated by an application in genetics for his development of a modified False Discovery Rate procedure for categorical data. He compared the mutation rates at $K = 118$ positions on HIV amino acid sequences of 73 Southern African patients infected with subtype C virus to 73 North American patients infected with subtype B virus. The goal of the study was to identify the positions in gag 24 amino acid sequences at which the probability of a non-consensus amino acid differed between the sets of subtype C and B sequences. Letting p_{1i} and p_{2i} be the probabilities of a non-consensus amino acid at position i for subtype C and subtype B, the statistical testing problem is to consider the family of hypotheses $H_i : p_{1i} = p_{2i}$ for $i = 1, 2, \ldots, 118$ to identify positions of difference.

Fisher's exact test was used to compute 118 unadjusted two-sided P-values. Gilbert recognized the need to consider a multiplicity adjustment and developed a modified FDR because of the discrete nature of the data. For example, there were 5 or fewer non-consensus amino acids across both groups for which the most extreme possible P-value $\alpha_i^* > 0.05$ so that statistical significance could not be established even at the unadjusted $\alpha = 0.05$ level.

Applying the Benjamini and Hochberg (1995) FDR (B-H FDR) controlling procedure to the full analysis set of 118 positions identified 12 significant positions. Using the Tarone (1990) procedure reduced the dimensionality of the multiple testing to $M = 25$ positions from the original 118. Applying the B-H FDR to these 25 positions identified 15 significant positions. The fully discrete FDR procedure in Eq. (3.3) identified 20 significant positions.

3.7. Simulation Study

A simulation study was conducted to evaluate the statistical operating characteristics of the FDR controlling methods when applied to categorical data. The methods compared were the original Benjamini and Hochberg (1995) procedure (B-H FDR), the Gilbert (2005) modified FDR two step procedure that first uses Tarone (1990) to identify a candidate set of hypotheses for consideration (T-G FDR), and the fully discrete FDR introduced in this paper (Discrete FDR). The methods were compared on the basis of three statistical error properties:

(1) The rate of rejecting hypotheses when all hypotheses are true ($K_0 = K$).
(2) The rate of rejecting true hypotheses when some hypotheses are false ($K_0 < K$).
(3) The rate of rejecting false hypotheses when some hypotheses are true ($K - K_0 < K$).

The simulations considered $K = \{5, 10, 15, 20\}$ independent hypotheses with a specified number of true (K_0) and false ($K - K_0$).

For each simulated condition, data for $T = 10,000$ two-group experiments were generated using a binomial random variable, and compared using Fisher's exact test using a one-sided $\alpha = 0.05$. The control group binomial probability parameter was chosen randomly from a uniform distribution, U(0.01, 0.5), and the sample sizes per group used were $N = \{10, 25, 50, 100\}$. For false hypotheses the odds ratio was specified for an effect size using $OR = \{1.5, 2, 2.5, 3\}$.

The rate of rejected true hypotheses

$$r_t = (\#\text{of rejected true hypotheses})/K_0$$

and the rate of rejected false hypotheses

$$s_t = (\#\text{of rejected false hypotheses})/(K - K_0)$$

were computed for each simulated experiment $t = 1, 2, \ldots, T$. The average rejection rates $(\Sigma r_t)/T$ and $(\Sigma s_t)/T$ were reported as the basis for comparing methods.

Figure 3.1 shows the rate of rejecting true null hypotheses when all hypotheses are true ($K_0 = K$). A dashed line is included for reference at the prespecified rejection rate of $\alpha = 0.05$. The FDR was safely controlled for all three methods, and increased with increasing sample size. The fully discrete FDR was less conservative than the other two methods for each sample size and for every value of K. The increasing FDR with sample size is due to an increasing number of outcome events, and therefore the test statistics are less discrete. For this reason the differences between the methods is reduced at $N = 100$. Since $K_0 = K$ in this setting, the FDR control actually provides FWER control.

Figure 3.2 shows the rate of rejecting true null hypotheses when some of the hypotheses are false ($K_0 < K$) for $K = 10$ and $K = 20$ and varying

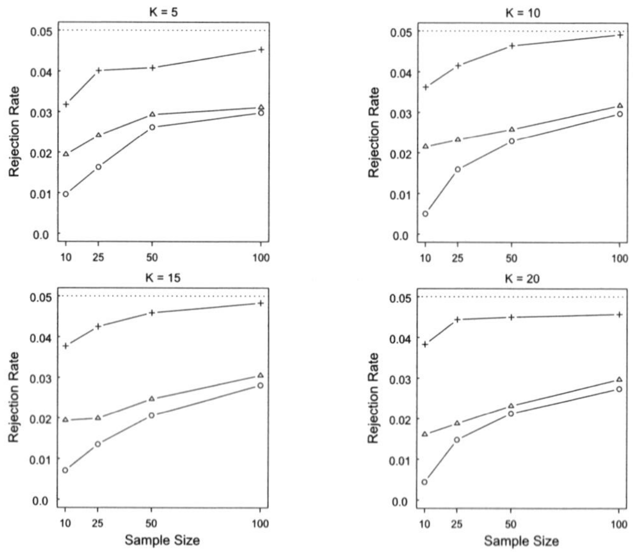

Fig. 3.1. Rate of rejecting true null hypotheses when all hypotheses are true ($K_0 = K$). The dashed line is for reference at $\alpha = 0.05$. Three methods are displayed: Benjamini and Hochberg FDR (○), Tarone/Gilbert modified FDR (Δ), and the fully discrete FDR (+).

numbers of true hypotheses. The dashed reference line is the theoretical upper bound $(K_0/K)\alpha$ for independent hypotheses. The fully discrete procedure is again less conservative than the both the B-H FDR and the T-G FDR procedures, and the differences are reduced at the increased sample sizes.

Figure 3.3 presents the rate of rejecting false hypotheses when some hypotheses are true for the two conditions ($K = 10, K - K_0 = 4$) and ($K = 20, K - K_0 = 8$). The rejection rate was uniformly higher for the discrete FDR procedure and increased for greater effect sizes as measured by the specified odds ratio, OR.

3.8. Concluding Remarks

In many experimental settings involving multiple hypotheses, the False Discovery Rate (FDR) can provide advantages to Family Wise Error Rate (FWER) control. When all of the independent hypotheses are true the FDR control is equal to FWER control, and when some hypotheses are false

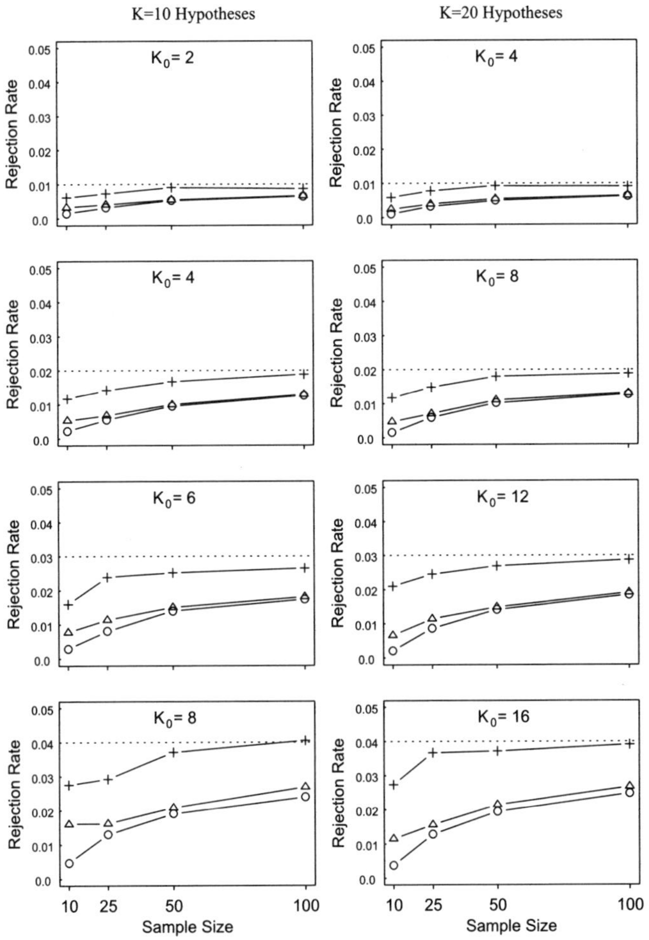

Fig. 3.2. Rate of rejecting true null hypotheses when some of the hypotheses are false ($K_0 < K$). Dashed reference line is the theoretical upper bound $(K_0/K)\alpha$ for $\alpha = 0.05$. Three methods are displayed: Benjamini and Hochberg FDR ($\circ$), Tarone/Gilbert modified FDR (Δ), and the fully discrete FDR ($+$).

the FDR provides greater power than FWER controlling procedures. This property of FDR is preferable in settings such as clinical adverse experience reporting (Mehrotra and Heyse, 2004), animal carcinogenicity studies, and when identifying pharmacogenetic associations.

It is well known that when analyzing categorical response variables, the test statistics have discrete distributions, and testing is conservative even

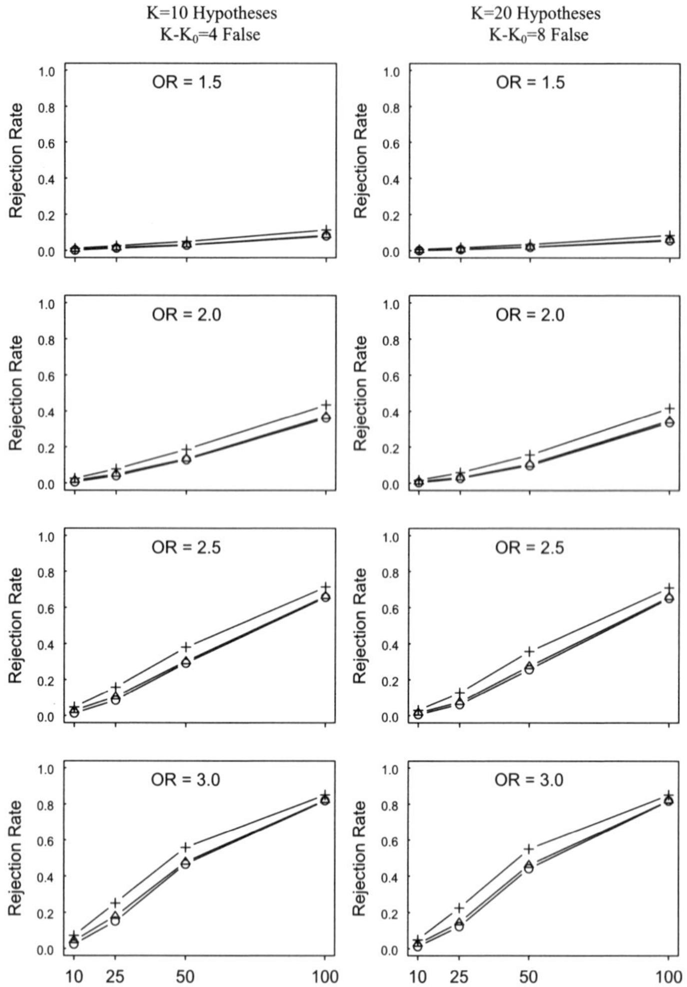

Fig. 3.3. Rate of rejecting false hypotheses when some hypotheses are true. Three methods are displayed: Benjamini and Hochberg FDR (○), Tarone/Gilbert modified FDR (Δ), and the fully discrete FDR (+). OR = Odds Ratio.

for single hypothesis experiments. This problem is compounded in multiple hypothesis testing situations, especially when some of the outcomes are infrequent and may not be able to produce statistically significant results even at the unadjusted α level. The proposed fully discrete FDR based on an exact conditional analysis of binomial data controls the FDR at α. The discrete nature of the testing distribution does result in a slightly conservative FDR control. The FDR control is less conservative for increasing sample sizes and increasing numbers of hypotheses due to the increased numbers of events observed.

The ICH-E9 guidelines for the statistical design and analysis of clinical trials recognized the concern over inflated type I errors when summarizing the results of the many clinical adverse experiences encountered in a study. However, they went on to regard a greater concern for type II errors. Controlling FDR is preferable in this application as it provides a more balanced alternative to Bonferroni type methods that address the FWER. Using a method that fully utilizes the exact distribution for the binary outcomes will further increase the power. The advantages of using the fully discrete FDR controlling procedure for genetic data was illustrated in the paper with the HIV genetic variants data from Gilbert (2005). Of the 118 positions considered on the amino acid sequences, most had too few mutations to even be considered for a multiplicity adjustment. For the others a gain in power was achieved using the complete discrete distribution.

The fully discrete procedure described in this paper assumes independence among the endpoints. This assumption may only approximately hold in practice. For example, Heyse and Rom (1988) showed that for the rodent carcinogenicity study, an analysis assuming independent tumor types gave very similar results to the analysis that properly handled the dependence, providing some rationale for using independence as a simplifying assumption. Benjamini and Yekutieli (2001) proved that the Benjamini and Hochberg procedure controls the FDR for statistics with positive regression dependence. This condition would be satisfied with non-negative correlations among the endpoints. Benjamini and Yekutieli (2001) also provided a simple modification of the Benjamini and Hochberg that controls the FDR for all forms of dependence. This approach can be readily applied in the discrete setting using the Gilbert (2005) modified procedure. Resampling procedures can also be considered. In general, the properties of the fully discrete FDR controlling procedure will follow closely those for the Benjamini and Hochberg procedure regarding dependence. This is certainly an area where simulations can help quantify the potential impact of dependence on the analysis.

Clearly, accounting for the discreteness in the distribution of the test statistic increases the power of the testing procedure relative to the Benjamini and Hochberg (1995) and the Gilbert (2005) two-step method. Similar approaches can also be applied to popular FWER controlling methods, such as the Bonferroni inequality, which would be expected to have greater power than Tarone's (1990) modified procedure.

References

Benjamini Y and Hochberg Y: Controlling the False Discovery Rate: a Practical and Powerful Approach to Multiple Testing. *Journal of the Royal Statistical Society, Series B*, **57**:289-300 (1995).

Benjamini Y and Yekutieli D: The Control of the False Discovery in Multiple Testing under Dependency. *Annals of Statistics*, **29**:1165-88 (2001).

Dmitrienko A, Molenberghs G, Chuang-Stein C, and Offen W: Chapter 2 in *Analysis of Clinical Trials Using SASR*: A Practical Guide. SAS Institute, placeCityCary, StateNC (2005).

Gilbert PB: A Modified False Discovery Rate Multiple-Comparisons Procedure for Discrete Data, Applied to Human Immunodeficienty Virus Genetics. *Appl. Statist*, **54**:143-158 (2005)

Heyse JF, Rom D: Adjusting for Multiplicity of Statistical Tests in the Analysis of Carcinogenicity Studies. *Biom J.* **30:**883-896 (1988).

Hochberg Y: A Sharper Bonferroni Procedure for Multiple Tests of Significance. *Biometrika*, **75:**800-802 (1988).

Hochberg Y and Tamhane AC: *Multiple Comparison Procedures.* Stateplace, New York: Wiley (1987).

Hommel G: A Stagewise Rejective Multiple Test Procedure Based on a Modified Bonferroni Test. *Biometrika*, **75**:383-386 (1988).

Hsu J: *Multiple Comparisons Procedures.* CityplaceLondon: Chapman and Hall (1996).

ICH Expert Working Group: ICH Harmonized Tripartite Guidelines in Statistical Principles for Clinical Trials. *Statistics in Medicine*, **18**:1905-1942 (1999).

Mantel N: Assessing Laboratory Evidence for Neoplastic Activity. *Biometrics*, **36:**381-399 (1980).

Mantel N: Tukey JW, Ciminera JL, and Heyse JF: Tumorigenicity Assays, Including Use of the Jackknife. *Biom J.* **24:**579-596 (1982).

Mehrotra DV and Heyse JF: Use of the False Discovery Rate for Evaluating Clinical Safety Data. *Statistical Methods in Medical Research*, **13**:227-238 (2004).

Simes RJ: An Improved Bonferroni Procedure for Multiple Tests of Significance. *Biometrika*, **73**:751-754 (1986).

Tarone RE: A Modified Bonferroni Method for Discrete Data. *Biometrics*, **46**:515-522 (1990).

Westfall PH and Young SS: P-Value Adjustments for Multiple Tests in Multivariate Binomial Models. *Journal of the American Statistical Association*, **84**:780-786 (1989).

Part II

Survival Analysis

Chapter 4

Conditional Nelson-Aalen and Kaplan-Meier Estimators with Müller-Wang Boundary Kernel

Xiaodong Luo*

Department of Psychiatry, Mount Sinai School of Medicine
New York, NY, 10029, USA
xiaodong.luo@mssm.edu

Wei-Yann Tsai

Department of Biostatistics, Columbia University
New York, NY, 10032, USA
wt5@columbia.edu

This paper studies the kernel assisted conditional Nelson-Aalen and Kaplan-Meier estimators. The presented results improve the existing ones in two aspects: (1) the asymptotic properties (uniform consistency, rate of convergence and almost sure iid representation) are extended to the entire support of the covariates by use of Müller-Wang boundary kernel; and (2) the order of the remainder terms in the iid representation is improved from $(\log n/(nh^d))^{3/4}$ to $\log n/(nh^d)$ thanks to the exponential inequality for U-statistics of order two. These results are useful for semiparametric estimation based on a first stage nonparametric estimation.

4.1. Introduction

Kernel assisted conditional cumulative hazard and survival function estimators were first proposed by Beran (1981). These estimators can be used as a base to check the model assumptions for the popular semiparametric models such as the proportional hazards model, the proportional odds ratio model and the accelerated failure time model (Gentleman and Crowley, 1991; Bowman and Wright, 2000). Also, these estimators can be applied in the context of censored quantile regression (Dabrowska, 1992; Bowman and Wright, 2000).

*Corresponding author.

The properties of the kernel assisted estimators have been intensively studied in the literature. For instance, Dabrowska derived the uniform rates of convergence (1989) and discussed estimation of the quantiles of the conditional survival function (1992), González-Manteiga and Cadarso-Suárez (1994) gave an almost sure representation of the estimators as a sum of iid random variables, and van Keilegom and Veraverbeke (1997) provided a bootstrap procedure to estimate the asymptotic bias and variance.

The above studies can be generalized in two aspects. First, the asymptotic properties of the kernel estimators should neither be restricted to the "central portion" of the support of the covariates as in Dabrowska (1989,1992), nor to the case of fixed design as in González-Manteiga and Cadarso-Suárez (1994) and van Keilegom and Veraverbeke (1997). Second, after an iid approximation, the remainder terms should be as small as $o(n^{-1/2})$ under milder conditions so that all of the theory on sum of iid random variables can be applied.

The need of this generalization arises in weighted estimating equations in the semiparametric estimation problems where the weight function is determined by the conditional survival function which needs to be estimated nonparametrically in advance. We refer this problem as the semiparametric estimation based on a first stage nonparametric estimation, in which the nonparametric estimators need to be estimated consistently over the entire support of the covariates and the remainder terms should be negligible after an iid approximation. In this paper, the first aspect is tackled with the help of the boundary kernel introduced by Müller and Wang (1994), and the order of the remainder terms is improved from $(\log n/(nh^d))^{3/4}$ to $(\log n/(nh^d))$ thanks to the exponential inequality for U-statistics derived by Giné, Latała and Zinn (2000) and Houdré and Reynaud-Bouret (2003).

Let T and C be the random variables representing the survival time and the censoring time, respectively. Let $\mathrm{Z} = (Z_1, \cdots, Z_d)$ be the d-dimensional vector of covariates of the joint distribution G and density g. Because of censoring, the observable random variables are $Y = \min(T, C)$ and $\delta = I(T \leq C)$. Let $S(t|\mathrm{z}) = P(T > t|\mathrm{Z} = \mathrm{z})$, $F_1(t|\mathrm{z}) = P(Y \leq t, \delta = 1|\mathrm{Z} = \mathrm{z})$, $F_2(t|\mathrm{z}) = P(Y \geq t|\mathrm{Z} = \mathrm{z})$, and

$$\Lambda(t|\mathrm{z}) = -\int_{(0,t]} \frac{S(ds|\mathrm{z})}{S(s-|\mathrm{z})}$$

be the conditional cumulative hazard function associated with $S(t|\mathrm{z})$. By the product-integral formula in Gill and Johansen (1990)

$$S(t|\mathrm{z}) = \prod_{s \le t} \{1 - \Lambda(ds|\mathrm{z})\}.$$

As in the literature, we assume that T and C are conditionally independent given Z to guarantee the identifiability of $\Lambda(t|\mathrm{z})$ and $S(t|\mathrm{z})$, under which the conditional cumulative hazard function can also be expressed as

$$\Lambda(t|\mathrm{z}) = \int_{(0,t]} \frac{F_1(ds|\mathrm{z})}{F_2(s|\mathrm{z})}$$

for any t satisfying $F_2(t|\mathrm{z}) > 0$.

Let $(Y_i, \delta_i, \mathrm{Z}_i)$, $j = 1, \cdots, n$ be a sample of iid random variables each having the same distribution as (Y, δ, Z). To estimate $\Lambda(t|\mathrm{z})$ and $S(t|\mathrm{z})$, Beran (1981) proposed the following kernel assisted conditional Nelson-Aalen and Kaplan-Meier estimators:

$$\hat{\Lambda}(t|\mathrm{z}) = \int_{(0,t]} \frac{\hat{F}_1(ds|\mathrm{z})}{\hat{F}_2(s|\mathrm{z})}$$

and

$$\hat{S}(t|\mathrm{z}) = \prod_{s \le t} \{1 - \hat{\Lambda}(ds|\mathrm{z})\},$$

where $\hat{F}_1(s|\mathrm{z})$ and $\hat{F}_2(s|\mathrm{z})$ are the Nadaraya-Watson kernel estimators of $F_1(s|\mathrm{z})$ and $F_2(s|\mathrm{z})$ given by

$$\hat{F}_1(s|\mathrm{z}) = \frac{(nh^d)^{-1} \sum_{j=1}^n I(Y_j \le s, \delta_j = 1) K(h^{-1}(\mathrm{z} - \mathrm{Z}_j))}{(nh^d)^{-1} \sum_{j=1}^n K(h^{-1}(\mathrm{z} - \mathrm{Z}_j))} \tag{4.1}$$

$$\hat{F}_2(s|\mathrm{z}) = \frac{(nh^d)^{-1} \sum_{j=1}^n I(Y_j \ge s) K(h^{-1}(\mathrm{z} - \mathrm{Z}_j))}{(nh^d)^{-1} \sum_{j=1}^n K(h^{-1}(\mathrm{z} - \mathrm{Z}_j))} \tag{4.2}$$

with the kernel function K and the bandwidth h.

Under some regularity conditions, Dabrowska (1989) gave the rates of uniform convergence for the estimate $\hat{S}(t|\mathrm{z})$ over the "central portion" of the distribution of Z. This result is very important but not sufficient for the applications in which the uniform convergence over the whole support of Z is needed. The situations include, for instance, the semiparametric estimation with the conditional survival function as the first stage nonparametric estimate. In this paper, we focus on the case of most practical importance that Z has a bounded support. We apply the boundary kernel introduced by Müller and Wang (1994) in the Nadaraya-Watson estimates of $\hat{F}_1(s|\mathrm{z})$ and $\hat{F}_2(s|\mathrm{z})$. This kernel enables us to handle the boundary effect of the kernel estimate and gives us the desired uniform convergence (or more precisely, the rates of uniform convergence) of $\hat{\Lambda}(\cdot|\mathrm{z})$ and $\hat{S}(\cdot|\mathrm{z})$ over the whole support of Z.

In applications, it is helpful to approximate the conditional Nelson-Aalen and Kaplan-Meier estimates with iid random processes. This topic has been studied in a vast literature, for example, Dabrowska (1992) for the Bahadur representation of the conditional quantile, González-Manteiga and Cadarso-Suárez (1994) with a generalized Kaplan-Meier estimator, and van Keilegom and Veraverbeke (1997) in fixed design nonparametric censored regression. Under the condition that $F_i(t|\mathrm{z})$, $i = 1, 2$ are differentiable in t, the remainder terms are shown to have order $O((\log n)^{3/4}/(nh^d)^{3/4})$ in all of the above papers. The differentiability condition seems auxiliary (unless in the study of the conditional quantiles) since the kernel smoothing is only applied to the covariates. In this paper, we will drop this condition and prove that the remainder terms are actually of order $O(\log n/(nh^d))$ with the help of the exponential inequality for U-statistics of order two given by Giné, Latała and Zinn (2000) and Houdré and Reynaud-Bouret (2003).

The rest of the paper is organized as follows. Section 4.2 reviews the boundary kernel and gives some extra properties not listed in Müller and Wang (1994). Section 4.3 derives the asymptotic properties: the rates of uniform convergence and the iid representaions with the remainder terms of the aforementioned order. Section 4.4 gives the proofs. And Sec. 4.5 brings up some discussions.

4.2. The Boundary Kernel

We only consider the univariate kernel since the multivariate kernel can be defined as the direct product of the univariate kernels. In viewing of this, we assume that the dimension of the covariate Z is 1, and, without loss of generality, we further assume that Z has support $[0, 1]$.

Let K_{z} be the kernel that depends on the point z where the estimate to be computed. A boundary kernel has an adjustment when z falls into the boundary regions–the area within one bandwidth from an endpoint. For any $0 < h < 1/2$, let $O_1 = [0, h]$, $O_2 = (h, 1 - h]$, and $O_3 = (1 - h, 1]$, following Müller and Wang (1994), this type of kernel may be defined as

$$K_{\rm z}(u) = \begin{cases} K_+({\rm z}/h, u) & {\rm z} \in O_1, \\ K_+(1, u) & {\rm z} \in O_2, \\ K_-((1-{\rm z})/h, u) & {\rm z} \in O_3, \end{cases} \tag{4.3}$$

where $K_+, K_- : [0,1] \times [-1,1] \to I\!R$ are k-th order boundary kernels satisfying

$$K_+(q, \cdot) \in \mathcal{M}_k([-1, q]),$$
$$K_-(q, u) = K_+(q, -u),$$

where $\mathcal{M}_k([a_1, a_2]) = \Big\{ f : f$ is of bounded variation on its support

$$[a_1, a_2], \int f^2(u)du < \infty, \int f(u)du = 1, \int f(u)u^j du = 0, 1 \leq j < k,$$

and $\left| \int f(u)u^k du \right| < \infty \Big\}.$ (4.4)

$$K_+(q, \cdot) \text{ is } \alpha \text{ times continuously differentiable on } [-1, q], \tag{4.5}$$

$$K_+^{(j)}(q, -1) = 0, \qquad 0 \leq j < \alpha,$$
$$K_+^{(j)}(q, q) = 0, \qquad 0 \leq j < \beta, \tag{4.6}$$

where the condition (4.6) do not apply if there is no j in the indicated range.

According to Müller and Wang (1994), a class of boundary kernels $K_+(q, \cdot)$ satisfying (4.4)-(4.6) for any $\alpha, \beta \geq 0$, $\alpha + \beta > 0$, for all $q \in [0, 1]$, is given by the following polynomials of degree $\alpha + \beta + k - 1$:

$$K_{+\alpha\beta}(q, u) = \begin{cases} \left(\dfrac{2}{1+q} \right)^{\alpha+\beta+1} (1+u)^\alpha (q-u)^\beta \\ \qquad \times \displaystyle\sum_{j=0}^{k-1} P_{j\alpha\beta}\left(2\frac{1+u}{1+q} - 1 \right) P_{j\alpha\beta}\left(\frac{1-q}{1+q} \right), \text{if} -1 \leq u \leq q, \\ 0, \text{elsewhere.} \end{cases}$$

where $(P_{j\alpha\beta})_{j \geq 0}$ are the normalized Jacobi polynomials on $[-1, 1]$ generated from the weight function $W_{\alpha\beta}(u) = (1+u)^\alpha (1-u)^\beta$, i.e.

$$P_{j\alpha\beta}(u) = \binom{j+\beta}{j}^{-1} \frac{(-1)^j}{2^j j!} (1+u)^{-\alpha}(1-u)^{-\beta} \frac{d^j}{du^j}[(1+u)^{\alpha+j}(1-u)^{\beta+j}]$$

see Szegö (1975, Chap. 4.1).

Besides the properties stated in Müller and Wang (1994), we derive some properties of the boundary kernel $K_+(q,\cdot)$ which are useful in our paper.

PROPERTY 1. $\sup_{q\in[0,1],u\in\mathbb{R}}|K_{+\alpha\beta}(q,u)| \le C_0$, where $C_0 = C_0(k,\alpha,\beta)$ is a positive constant.

PROPERTY 2. For any $\alpha \ge 1$, $x \ge 0$ and $0 \le q, q_1 \le 1$, $|K_{+\alpha\beta}(q,q-x) - K_{+\alpha\beta}(q_1,q_1-x)| \le C_1|q-q_1|$, where $C_1 = C_1(k,\alpha,\beta)$ is a positive constant not depending on x, q and q_1.

PROPERTY 3. For any $\alpha \ge 1$, $\beta \ge 1$, and $u, u_1 \in \mathbb{R}$, $|K_{+\alpha\beta}(1,u) - K_{+\alpha\beta}(1,u_1)| \le C_2|u-u_1|$, where $C_2 = C_2(k,\alpha,\beta)$ is a positive constant not depending on u and u_1.

PROPERTY 4. For any $\alpha \ge 1$ and $\beta \ge 1$, $K_{+\alpha\beta}(\cdot,\cdot)$ is a continuous function on $\mathbb{R}^2$.

PROPERTY 5. For any $\alpha \ge 2$ and $\beta \ge 2$, $K_{+\alpha\beta}(\cdot,\cdot)$ is a continuously differentiable function on $\mathbb{R}^2$.

4.3. The Estimators

From the definition of $\hat{\Lambda}(t|\mathrm{z})$, it is easy to see that the denominators in (4.1) and (4.2) will be cancelled out as long as they are nonzero. Therefore, it is convenient to work on

$$\hat{H}_1(t,\mathrm{z}) = \frac{1}{nh^d}\sum_{j=1}^{n} I(Y_j \le t, \delta_j = 1)K_{\mathrm{z}}\left(\frac{\mathrm{z}-\mathrm{Z}_j}{h}\right)$$

and

$$\hat{H}_2(t,\mathrm{z}) = \frac{1}{nh^d}\sum_{j=1}^{n} I(Y_j \ge t)K_{\mathrm{z}}\left(\frac{\mathrm{z}-\mathrm{Z}_j}{h}\right).$$

For $i = 1,2$, let $H_i = E\hat{H}_i$ and $H_i^0(t,\mathrm{z}) = g(\mathrm{z})F_i(t|\mathrm{z})$. To save notations, we still use $\hat{\Lambda}(t|\mathrm{z})$ and $\hat{S}(t|\mathrm{z})$ to denote respectively the conditional Aalen-Nelson and Kaplan-Meier estimators when the boundary kernel function is used, i.e.

$$\hat{\Lambda}(t|\mathrm{z}) = \int_{(0,t]} \frac{\hat{H}_1(ds,\mathrm{z})}{\hat{H}_2(s,\mathrm{z})} \tag{4.7}$$

and

$$\hat{S}(t|\mathrm{z}) = \prod_{s\le t}\{1 - \hat{\Lambda}(ds|\mathrm{z})\}. \tag{4.8}$$

Note that the functions $\hat{H}_i$, $i = 1, 2$ may be negative as a result of using the boundary kernel and therefore $\hat{\Lambda}(t|\mathrm{z})$ and $\hat{S}(t|\mathrm{z})$ may not be the proper cumulative hazard function and survival function in the sense that they may not be nonnegative nondecreasing functions. Since the negativity and non-monotonicity of $\hat{\Lambda}(t|\mathrm{z})$ and $\hat{S}(t|\mathrm{z})$ are not the major concerns of this paper and may be tolerable in the first step of semiparametric estimation, we will not discuss this issue further. As it will become clear later, the introduction of the k-th order boundary kernel corrects the bias of the estimates in the boundary region hence ensures the uniform convergence over the entire support of Z.

Assume that Z has bounded support. Let $[a_i, b_i]$, $i = 1, \cdots, d$ be the supports of each of the component of Z. Without loss of generality, we assume $a_1 = a_2 = \cdots = a_d = 0$ and $b_1 = b_2 = \cdots = b_d = 1$. Let $J = [a_1, b_1] \times \cdots \times [a_d, b_d]$ and $I = [0, u_F] \times J$, where u_F is some positive constant. Let $l_H = \inf_{(t,\mathrm{z}) \in I} H_2^0(t, \mathrm{z})$, $l_g = \inf_{\mathrm{z} \in J} g(\mathrm{z})$, and $u_g = \sup_{\mathrm{z} \in J} g(\mathrm{z})$. Clearly, $\sup_{(t,\mathrm{z}) \in I} H_i^0(t, \mathrm{z}) \le u_g$, $i = 1, 2$. We need the following assumptions:

(A1) $l_H > 0$ and $0 < l_g \le u_g < \infty$.
(A2) The functions $H_i^0(t, \mathrm{z})$, $i = 1, 2$, have bounded continuous k times partial derivatives with respect to z on I.

Remark 4.1. Assumption (A1) will be used to bound the tails of $\hat{H}_i(t, \mathrm{z}) - H_i(t, \mathrm{z})$, $i = 1, 2$. Assumption (A2) will be needed to guarantee asymptotic unbiasedness of the estimators $H_i(t, \mathrm{z})$ to $H_i^0(t, \mathrm{z})$, $i = 1, 2$. Note that, given z, the functions $H_i^0(t, \mathrm{z})$, $i = 1, 2$, may have discontinuity points in t.

In the sequel, we will choose the parameters $\alpha \ge 1$ and $\beta \ge 1$ for the k-th order boundary kernel in (4.7). Let the bandwidth $h = o(1)$ and satisfy

$$a_n \triangleq \left(\frac{\log n}{nh^d} \right)^{1/2} = o(1), \tag{4.9}$$

we have the following theorem.

Theorem 4.1. *Under assumptions (A1)-(A2) and with h satisfying (4.9),*

$$\sup_{(t,z) \in I} |\hat{\Lambda}(t|z) - \Lambda(t|z)| = O(a_n + h^k) \qquad a.s. \tag{4.10}$$

$$\sup_{(t,z) \in I} |\hat{S}(t|z) - S(t|z)| = O(a_n + h^k) \qquad a.s. \tag{4.11}$$

Remark 4.2.

We obtain the same rate of uniform convergence after adjusting the boundary effect using Müller-Wang boundary kernel. For the uniform convergence, the optimal bandwidth h_{uc} can be chosen, up to a positive constant, as $h_{uc} = (\log n/n)^{1/(2k+d)}$, resulting in the best rate of uniform convergence $(\log n/n)^{k/(2k+d)}$.

To study the iid approximations of the proposed estimators, define

$$\hat{L}_\Lambda(t, \mathrm{z}) = \int_{[0,t]} \frac{\hat{H}_1(ds, \mathrm{z}) - H_1^0(ds, \mathrm{z})}{H_2^0(s, \mathrm{z})} - \int_{[0,t]} \frac{\hat{H}_2(s, \mathrm{z}) - H_2^0(s, \mathrm{z})}{[H_2^0(s, \mathrm{z})]^2} H_1^0(ds, \mathrm{z}),$$

$$\hat{L}_S(t, \mathrm{z}) = -\int_{(0,t]} S(u - |\mathrm{z})\hat{L}_\Lambda(du, \mathrm{z})\frac{S(t|\mathrm{z})}{S(u|\mathrm{z})},$$

$$\tilde{L}_S(t, \mathrm{z}) = -\int_{(0,t]} S(u - |\mathrm{z})[\hat{\Lambda}(du|\mathrm{z}) - \Lambda(du|\mathrm{z})]\frac{S(t|\mathrm{z})}{S(u|\mathrm{z})}.$$

Apparently, $\hat{L}_\Lambda(t, \mathrm{z})$ and $\hat{L}_S(t, \mathrm{z})$ are averages of iid random processes and we shall show later that $\sup_I |\hat{L}_S(t, \mathrm{z}) - \tilde{L}_S(t, \mathrm{z})| = O(b_n)$ a.s., with $b_n = a_n^2 + h^k$.

Theorem 4.2. *Under assumptions (A1)-(A2) and with h satisfying (4.9), the conditional Nelson-Aalen estimate* $\hat{\Lambda}(t|z)$ *can be written as*

$$\hat{\Lambda}(t|z) - \Lambda(t|z) = \hat{L}_\Lambda(t, z) + R_\Lambda(t, z)$$

such that

$$\sup_{(t,z)\in I} |R_\Lambda(t, z)| = O(b_n) \qquad a.s.$$

Theorem 4.3. *Under assumptions (A1)-(A2) and with h satisfying (4.9), the conditional Kaplan-Meier estimate* $\hat{S}(t|z)$ *can be written as*

$$\hat{S}(t|z) - S(t|z) = \hat{L}_S(t, z) + R_S(t, z) \tag{4.12}$$

such that

$$\sup_{(t,z)\in I} |R_S(t, z)| = O(b_n) \qquad a.s.$$

Furthermore, $\sup_I |\hat{L}_S(t, z) - \tilde{L}_S(t, z)| = O(b_n)$ *a.s., thus in (4.12)* $\hat{L}_S$ *can be replaced by* $\tilde{L}_S$.

Remark 4.3. The remainder terms here have order $b_n = a_n^2 + h^k$ which is smaller than the order $b_n' = a_n^{3/2} + h^k$ in Dabrowska (1992), González-Manteiga and Cadarso-Suárez (1994), and van Keiledom and Veraverbeke

(1997). This improvement is significant since we only need $k > d$ instead of $k > 3d/2$ to guarantee a $o(n^{-1/2})$ rate. The $o(n^{-1/2})$ rate is very important in the semiparametric estimation when we need to plug in the nonparametric estimates of $\Lambda(t|\mathrm{z})$ and(or) $S(t|\mathrm{z})$. The $o(n^{-1/2})$ rate can be achieved, for example, as follows. When $k > d$, choose a real number r such that $d < r < k$ and set $h = l_0 n^{-\frac{1}{2r}}$ for any fixed $l_0 > 0$. The so-chosen h gives us $a_n^2 = \frac{\log n}{nh^d} = o(n^{-\frac{1}{2}})$ and $h^k = o(n^{-\frac{1}{2}})$. In particular, the optimal rate for the iid representation is given by $(\log n/n)^{k/(k+d)}$, which is achieved by the optimal bandwidth $h_{iid} = (\log n/n)^{1/(k+d)}$.

4.4. Proofs

4.4.1. *Preliminaries and notation*

For any $0 \le s \le t \le u_F$, let $\mathcal{E} : s = t_0 < t_1 < \cdots < t_p = t$ be a partition of $(s,t]$. For any function f, define its variation norm on $(s,t]$ and its supremum norm on $[s,t]$ as

$$||f||^v_{(s,t]} = \sup_{\mathcal{E}} \sum_{i=1}^{p} |f(t_i) - f(t_{i-1})| \quad \text{and} \quad ||f||^\infty_{[s,t]} = \sup_{u \in [s,t]} |f(u)|.$$

Also, let $||f||_{[s,t]} = ||f||^\infty_{[s,t]} + ||f||^v_{(s,t]}$. A function f is *caldag* if and only if it is right continuous and of left limit. We first state the following lemma which is derived from Lemma 5 of Gill and Johansen (1990).

Lemma 4.1. *Let f_1, f_2 and f be cadlag functions such that f_1 and f_2 are of bounded variation on $[0, u_F]$, where f may have unbounded variation. Then for $0 \le s \le t \le u_F$,*

$$\left| \int_{(s,t]} f_1(u) df(u) \right| \le 2||f||^\infty_{[s,t]} ||f_1||_{[s,t]},$$

$$\left| \int_{(s,t]} f_1(u-) f_2(u) df(u) \right| \le 4||f||^\infty_{[s,t]} ||f_1||_{[s,t]} ||f_2||_{[s,t]}.$$

The following exponential inequality for U-statistics of order two, from Giné, Latała and Zinn (2000) and Houdré and Reynaud-Bouret (2003), is a generalization of Bernstein's Inequality to U-statistics and useful in establishing Theorems 4.2 and 4.3.

Lemma 4.2. *Let $W_1, \cdots, W_n$ be independent random p-vectors defined on a probability space $(\Omega, \mathcal{F}, P)$ and $\xi_{ij} : I\!R^p \times I\!R^p \to I\!R$, $i,j = 1, \cdots, n$ are*

Borel measurable functions such that $\xi_{ij}(W_i, W_j) = \xi_{ji}(W_j, W_i)$ *and*

$$E(\xi_{ij}(W_i, W_j)|W_i) = E(\xi_{ij}(W_i, W_j)|W_j) = 0.$$

Define

$$\Xi_n = \sum_{1\le i<j\le n} \xi_{ij}(W_i, W_j).$$

Then, there exists some positive constant G *such that, for all* $x > 0$,

$$P\Big(\|\Xi_n\| > x\Big) \le G\exp\left\{-\frac{1}{G}\min\Big(\frac{x^2}{G_3^2}, \frac{x}{G_4}, \frac{x^{2/3}}{G_2^{2/3}}, \frac{x^{1/2}}{G_1^{1/2}}\Big)\right\}$$

where

$$G_1 = \max_{1\le i<j\le n} \|\xi_{ij}\|_\infty, \quad G_3^2 = \sum_{1\le i<j\le n} E\xi_{ij}^2,$$

$$G_2^2 = \max\left\{\sup_{t,i}\Big[\sum_{j=i+1}^{n} E(\xi_{ij}^2(W_i, W_j)|W_i = t)\Big], \sup_{t,j}\Big[\sum_{i=1}^{j-1} E(\xi_{ij}^2(W_i, W_j)|W_j = t)\Big]\right\},$$

and

$$G_4 = \sup\left\{E\Big[\sum_{1\le i<j\le n} \xi_{ij}(W_i, W_j)a_i(W_i)b_j(W_j)\Big] : E\Big[\sum_{i=1}^{n-1} a_i^2(W_i)\Big] \le 1, E\Big[\sum_{j=2}^{n} b_j^2(W_j)\Big] \le 1\right\}.$$

Let $F(t) = \Pr(Y \le t)$ be the marginal distribution of Y. For any $\epsilon_1 > 0$, define $t_0 = -\infty$ and $t_m = \sup\{t > t_{m-1}, F(t) - F(t_{m-1}) \le \epsilon_1\}$, we have $F(t_m) - F(t_{m-1}) \ge \epsilon_1$ and $F(t_m-) - F(t_{m-1}) \le \epsilon_1$. Insert 0 and u_F into the sequence if they are not the cut points. The so-chosen sequence $0 = t_0 < t_1 < \cdots < t_M = \infty$ forms a partition of the real line with $M = O(1/\epsilon_1)$. Let $\mathcal{T}_m = [t_{m-1}, t_m) \cap [0, u_F]$, $m = 1, \cdots, M$.

For any $\epsilon_2 > 0$, we can partition the region $[0, 1]$ as $0 = x_0 < x_1 < \cdots < x_L = 1$ such that both h and $1 - h$ are cut-points and $x_l - x_{l-1} \le \epsilon_2$ for $l = 1, \cdots, L$ with $L = O(1/\epsilon_2)$. Let $\mathcal{X}_l = (x_{l-1}, x_l]$, $l = 1, \cdots, L$. By Properties 2 and 3 of the kernel function, we have, for any $\xi \in [0, 1]$,

$$\sup_{z\in\mathcal{X}_l} |K_z(\frac{z-\xi}{h}) - K_{x_l}(\frac{x_l-\xi}{h})| \le (C_1 \vee C_2)(\frac{\epsilon_2}{h}),$$

where $x \vee y = \max\{x, y\}$ and $x \wedge y = \min\{x, y\}$ for any $x, y \in \mathbb{R}$. For the multivariate kernel defined for $\mathrm{z} = (z_1, \cdots, z_d)$ and $\mathrm{u} = (u_1, \cdots, u_d)$ as $K_{\mathrm{z}}(\mathrm{u}) = \prod_{c=1}^{d} K_{z_c}(u_c)$, we may repeat the above partition d times to form a partition of $[0, 1]^d$ with the cubes of size $O(\epsilon_2)$. To save notation, we still use $\mathcal{X}_l$, $l = 1, \cdots, L$ to denote each of the cubes and use x_l, $l = 1, \cdots, L$ to denote the uppermost point of the cube $\mathcal{X}_l$. Notice that the number of the cubes, L, is of order $O(1/\epsilon_2^d)$ instead. And if z falls into a cube $\mathcal{X}_l$, then we have the similar relationship as in the univariate case, i.e. for any $\boldsymbol{\xi} \in [0, 1]^d$

$$\sup_{\mathrm{z} \in \mathcal{X}_l} \left| K_{\mathrm{z}}(\frac{\mathrm{z} - \boldsymbol{\xi}}{h}) - K_{\mathrm{x}_l}(\frac{\mathrm{x}_l - \boldsymbol{\xi}}{h}) \right| \le d(C_1 \vee C_2) C_0^{d-1} (\frac{\epsilon_2}{h}).$$

Let $\hat{A}_p(\mathrm{z}) = (nh^d)^{-1} \sum_{i=1}^{n} |K_{\mathrm{z}}(\frac{\mathrm{z} - \mathrm{z}_i}{h})|^p$, $A_p(\mathrm{z}) = EA_p(\mathrm{z})$ and $\mu_{K_p} = u_g(2C_0^p)^d$, $p = 1, \cdots, 4$. Clearly, $\sup_J A_p(\mathrm{z}) \le \mu_{K_p}$, $p = 1, \cdots, 4$. For any fixed $\lambda > \frac{16}{3}$, we choose $\epsilon_1 = \lambda n^{-1} \log n = \lambda a_n^2$ and $\epsilon_2 = \epsilon_1 h$ and create the partition of $[0, u_F]$ and J with the so-chosen ϵ_1 and ϵ_2. Let $\hat{A}(m) = n^{-1} \sum_{i=1}^{n} I(t_{m-1} < Y_i < t_m)$ and $A(m) = E\hat{A}(m)$, $m = 1, \cdots, M$. Clearly, $\max_{1 \le m \le M} A(m) \le \epsilon_1$. Define

$$A_n = \{\max_{1 \le m \le M} \hat{A}(m) > 2\epsilon_1\}, B_n = \{\sup_J \hat{A}_1(\mathrm{z}) > 4E_1\}$$

$$C_n = \{\sup_J \hat{A}_2(\mathrm{z}) > 4E_2\}, D_n = \{\inf_I \hat{H}_2(t, \mathrm{z}) < \frac{l_H}{2}\}$$

where E_1 and E_2 are two constants satisfying

$$E_1 > \max\left\{1, \frac{\mu_{K_1}}{2}, 2\lambda d(C_1 \vee C_2) C_0^{d-1}, 2(2+d)\left[\mu_{K_2} + \frac{\mu_{K_1} + C_0^d}{3}\right]\right\}$$

$$E_2 > \max\left\{1, \frac{\mu_{K_2}}{2}, 4\lambda d(C_1 \vee C_2) C_0^{2d-1}, 2(2+d)\left[\mu_{K_4} + \frac{\mu_{K_2} + C_0^{2d}}{3}\right]\right\}.$$

For any positive integers d and k, let

$$C_H(k) = \sup_I \left| \frac{\partial^{(k)} H_1^0(t, \mathrm{z})}{\partial \mathrm{z}^k} \right| \vee \sup_I \left| \frac{\partial^{(k)} H_2^0(t, \mathrm{z})}{\partial \mathrm{z}^k} \right|$$

$$\lambda_0(d, k) = \frac{(2C_0)^d d^k C_H(k)}{k!}.$$

4.4.2. *Proof of Theorem 4.1*

Theorem 4.1 can be established through Lemmas 4.3–4.5. First, we discuss the properties of $\hat{H}_i$, $i = 1, 2$. Note that we have the variance-bias decomposition

$$\hat{H}_i(t,\mathrm{z}) - H_i^0(t,\mathrm{z}) = [\hat{H}_i(t,\mathrm{z}) - H_i(t,\mathrm{z})] + [H_i(t,\mathrm{z}) - H_i^0(t,\mathrm{z})]$$

for $i = 1, 2$. The bias terms $H_i(t,\mathrm{z}) - H_i^0(t,\mathrm{z})$ are deterministic functions in (t,z) which rely on the smoothness of the target functions H_i^0 and the kernel function K_{Z}. And the variance terms $\hat{H}_i(t,\mathrm{z}) - H_i(t,\mathrm{z})$ are averages of iid random processes which can be bounded using Bernstein's inequality.

Lemma 4.3. *Under condition (A1)–(A2) and with h satisfying (4.9), we have for $i = 1, 2$,*

$$\sup_{(t,z)\in I} |H_i(t,z) - H_i^0(t,z)| = O(h^k)$$

$$\sup_{(t,z)\in I} |\hat{H}_i(t,z) - H_i(t,z)| = O(a_n) \qquad a.s.$$

Proof. We will illustrate the proof for $i = 1$. Without confusion, we abbreviate $K_+ = K_{+\alpha\beta}$. First, we calculate the bias term $H_1(t,\mathrm{z}) - H_1^0(t,\mathrm{z})$. Without loss of generality, we assume in this calculation that the dimension of the covariate Z is one and has support $[0,1]$. We write $H_1(t,\mathrm{z}) = \sum_{j=1}^3 H_{1j}(t,\mathrm{z})$, where $H_{1j}(t,\mathrm{z}) = H_1(t,\mathrm{z})I(\mathrm{z} \in O_j)$, $j = 1, 2, 3$. We calculate

$$\begin{aligned} H_{11}(t,\mathrm{z}) &= h^{-1}E\{I(Y \le t, \delta = 1)K_+(1, \frac{\mathrm{z}-\mathrm{Z}}{h})\}I(0 \le \mathrm{z} \le h) \\ &= h^{-1}\int_0^1 H_1^0(t,\xi)K_+(\frac{\mathrm{z}}{h}, \frac{\mathrm{z}-\xi}{h})d\xi I(0 \le \mathrm{z} \le h) \\ &= \int_{-1}^{\frac{\mathrm{z}}{h}} H_1^0(t,\mathrm{z}-hu)K_+(\frac{\mathrm{z}}{h}, u)du I(0 \le \mathrm{z} \le h) \\ &= \{H_1^0(t,\mathrm{z}) + R_{11}(t,\mathrm{z})h^k\}I(0 \le \mathrm{z} \le h) \end{aligned} \tag{4.13}$$

where

$$\sup_I |R_{11}(t,\mathrm{z})| \le \lambda_0(1,k)$$

The last equality in (4.13) is by the differentiability assumption (A2) on H_1^0 and the property of the k-th order kernel K_+. And for the same reason, we have $H_{1j}(t,\mathrm{z}) = \{H_1^0(t,\mathrm{z}) + R_{1j}(t,\mathrm{z})h^k\}I_{O_j}$ with $\sup_I |R_{1j}(t,\mathrm{z})| \le \lambda_0(1,k)$, $j = 2, 3$. Therefore, we have

$$\sup_I |H_1(t,\mathrm{z}) - H_1^0(t,\mathrm{z})| = O(h^k).$$

Next we will bound the variance term $\hat{H}_1(t,\mathrm{z}) - H_1(t,\mathrm{z})$. We will be discussing the general case when $d \ge 1$. First, note that there exists some

$\theta > 0$ such that for $n \geq 3$,

$$
\begin{aligned}
\Pr(A_n) &\leq \sum_{m=1}^{M} \Pr\left\{\hat{A}(m) > 2\epsilon_1\right\} \\
&\leq \sum_{m=1}^{M} \Pr\left\{\hat{A}(m) - A(m) > \epsilon_1\right\} \\
&\leq \sum_{m=1}^{M} \exp\left\{-\frac{1}{2}\frac{n\epsilon_1^2}{\epsilon_1 + \frac{\epsilon_1}{3}}\right\} \qquad \text{(by Bernstein's inequality)} \\
&= \sum_{m=1}^{M} \exp\left\{-\frac{3}{8}n\epsilon_1\right\} \\
&= O(\frac{1}{\epsilon_1}) \exp\left\{-\frac{3}{8}n\epsilon_1\right\} \\
&= O(n^{-(1+\theta)}) \qquad \text{(by the selection of } \epsilon_1 \text{ and } \lambda) \qquad (4.14)
\end{aligned}
$$

i.e. $\Pr(A_n, \text{i.o.}) = 0$ by Borel-Cantelli's Lemma.

Second, note that

$$
\sup_{(t,\mathrm{z})\in I} |\hat{H}_1(t,\mathrm{z}) - H_1(t,\mathrm{z})| = \max_{1\leq m\leq M, 1\leq l\leq L} \sup_{(t,\mathrm{z})\in\mathcal{T}_m\times\mathcal{X}_l} |\hat{H}_1(t,\mathrm{z}) - H_1(t,\mathrm{z})|
$$

where the supremum over an empty set is defined as $-\infty$. With the selection of ϵ_1 and ϵ_2, we have

$$
\begin{aligned}
&\sup_{(t,\mathrm{z})\in\mathcal{T}_m\times\mathcal{X}_l} |\hat{H}_1(t,\mathrm{z}) - H_1(t,\mathrm{z})| \\
&\leq |\hat{H}_1(t_{m-1}, \mathrm{x}_l) - H_1(t_{m-1}, \mathrm{x}_l)| + \frac{3C_0^d\epsilon_1}{h^d} + \frac{2d(C_1 \vee C_2)C_0^{d-1}\epsilon_2}{h^{d+1}} \quad \text{on } A_n^C.
\end{aligned}
$$

Furthermore, for any positive number λ_1 satisfying

$$
\lambda_1 > \max\left\{1, 3\lambda C_0^d, 2d(C_1 \vee C_2)C_0^{d-1}, 2(3+d)\left[\mu_{K_2} + \frac{\mu_{K_1} + C_0^d}{3}\right]\right\},
$$

we have

$$
\begin{aligned}
&\Pr(\sup_{(t,\mathrm{z})\in I} |\hat{H}_1(t,\mathrm{z}) - H_1(t,\mathrm{z})| > 3\lambda_1 a_n) \\
&\leq \Pr(\sup_{(t,\mathrm{z})\in I} |\hat{H}_1(t,\mathrm{z}) - H_1(t,\mathrm{z})| I_{D_n^C} > 3\lambda_1 a_n) + \Pr(D_n), \qquad (4.15)
\end{aligned}
$$

and by Bernstein's inequality,

$$\begin{aligned}
&\Pr(\sup_{(t,\mathrm{z})\in I} |\hat{H}_1(t,\mathrm{z}) - H_1(t,\mathrm{z})| I_{D_n^C} > 3\lambda_1 a_n) \\
&\leq \sum_{m=1}^{M}\sum_{l=1}^{L} \Pr(\sup_{(t,\mathrm{z})\in \mathcal{T}_m\times\mathcal{X}_l} |\hat{H}_1(t,\mathrm{z}) - H_1(t,\mathrm{z})| I_{D_n^C} > 3\lambda_1 a_n) \\
&\leq \sum_{m=1}^{M}\sum_{l=1}^{L} \Pr(|\hat{H}_1(t_{m-1},\mathrm{x}_l) - H_1(t_{m-1},\mathrm{x}_l)| > \lambda_1 a_n) \\
&\leq O(n^{2+d}) n^{-\frac{\lambda_1}{2[\mu_{K_2}+(\mu_{K_1}+C_0^d)/3]}} \\
&= O(n^{-(1+\theta)})
\end{aligned} \tag{4.16}$$

for some $\theta > 0$. Combining (4.14)–(4.16) together with Borel-Cantelli's Lemma, we obtain $\sup_{(t,\mathrm{z})\in I} |\hat{H}_1(t,\mathrm{z}) - H_1(t,\mathrm{z})| = O(a_n)$, a.s..

For the proof of $\hat{H}_2$, we only need to change from $\mathcal{T}_m$ to $\mathcal{T}_{2m} = [t_{m-1}, t_m) \cap [0, u_F)$, $m = 1, \cdots, M$ since H_2^0 is a left continuous function in t. The detail is omitted here. □

We next study the properties of $\hat{A}_p$, $p = 1, 2$. Note that in the region $\mathcal{X}_l$,

$$|\hat{A}_1(\mathrm{z}) - \hat{A}_1(\mathrm{x}_l)| \leq d(C_1 \vee C_2) C_0^{d-1} \frac{\epsilon_2}{h^{d+1}}$$
$$|A_1(\mathrm{z}) - A_1(\mathrm{x}_l)| \leq d(C_1 \vee C_2) C_0^{d-1} \frac{\epsilon_2}{h^{d+1}}$$

therefore, for $n \geq 3$

$$\begin{aligned}
&\Pr(\sup_J \hat{A}_1(\mathrm{z}) > 4E_1) \\
&\leq \Pr(\sup_J |\hat{A}_1(\mathrm{z}) - A_1(\mathrm{z})| > 2E_1) \qquad \text{(by the choice of } E_1) \\
&\leq \sum_{l=1}^{L} \Pr(\sup_{\mathcal{X}_l} |\hat{A}_1(\mathrm{z}) - A_1(\mathrm{z})| > 2E_1) \\
&\leq \sum_{l=1}^{L} \Pr(|\hat{A}_1 \mathrm{x}_l - A_1 \mathrm{x}_l| > E_1) \qquad \text{(by the choice of } E_1)
\end{aligned}$$

$$\le 2\sum_{l=1}^{L} \exp\left\{-\frac{1}{2}\frac{nE_1^2}{\frac{\mu_{K_2}}{h^d}+\left(\mu_{K_1}+\frac{C_0^d}{h^d}\right)\frac{E_1}{3}}\right\} \quad \text{(by Bernstein's inequality)}$$

$$\le O(n^{1+d})\exp\left\{-\frac{1}{2}\frac{nh^dE_1}{\mu_{K_2}+(\mu_{K_1}+C_0^d)/3}\right\} \quad \text{(by the choice of } E_1\text{)}$$

$$\le O(n^{1+d})\exp\left\{-\frac{1}{2}\frac{E_1\log n}{\mu_{K_2}+(\mu_{K_1}+C_0^d)/3}\right\} \quad \text{(by (4.9))}$$

$$= O(n^{-(1+\theta)}) \quad \text{(by the choice of } E_1\text{)} \tag{4.17}$$

for some $\theta > 0$. Similarly, we can show

$$\Pr(\sup_J \hat{A}_2(\mathrm{z}) > 4E_2) = O(n^{-(1+\theta)}). \tag{4.18}$$

Furthermore, let

$$n_0 = \inf\left\{n : n \ge 3, 3\lambda_1 a_n \le \frac{l_H}{4} \text{ and } \lambda_0(d,k)h^k \le \frac{l_H}{4}\right\}$$

we have for any $n \ge n_0$,

$$\begin{aligned}\Pr(D_n) &= \Pr(\inf_I \hat{H}_2(t,\mathrm{z}) < \frac{l_H}{2})\\ &\le \Pr(\sup_I |\hat{H}_2(t,\mathrm{z}) - H_2^0(t,\mathrm{z})| > \frac{l_H}{2})\\ &\le \Pr(\sup_I |\hat{H}_2(t,\mathrm{z}) - H_2(t,\mathrm{z})| > \frac{l_H}{4})\\ &\le \Pr(\sup_I |\hat{H}_2(t,\mathrm{z}) - H_2(t,\mathrm{z})| > 3\lambda_1 a_n)\\ &= O(n^{-(1+\theta)}).\end{aligned} \tag{4.19}$$

We have the following lemma.

Lemma 4.4. *On* $B_n^C \cap D_n^C$

$$|\hat{\Lambda}(t|z) - \Lambda(t|z)| \le \lambda_2\Big[\sup_I |\hat{H}_2(t,z) - H_2^0(t,z)| + \sup_I |\hat{H}_1(t,z) - H_1^0(t,z)|\Big]$$

where $\lambda_2 = \max\{2(\frac{2}{l_H} + \frac{16E_1}{l_H^2}), \frac{2u_g}{l_H^2}\}$.

Proof. We write

$$\begin{aligned}\hat{\Lambda}(t|\mathrm{z}) - \Lambda(t|\mathrm{z}) &= \int_{0,t]} \frac{\hat{H}_1(ds,\mathrm{z}) - H_1^0(ds,\mathrm{z})}{\hat{H}_2(s,\mathrm{z})}\\ &\quad + \int_{0,t]} \left[\frac{1}{\hat{H}_2(s,\mathrm{z})} - \frac{1}{H_2^0(s,\mathrm{z})}\right] H_1^0(ds,\mathrm{z}).\end{aligned}$$

On $B_n^C \cap D_n^C$, we calculate

$$\Big\|\frac{1}{\hat{H}_2(\cdot,\mathrm{z})}\Big\|_{[0,t]}^{\infty} \le \frac{2}{l_H}, \qquad \Big\|\frac{1}{\hat{H}_2(\cdot,\mathrm{z})}\Big\|_{(0,t]}^{v} \le \frac{16E_1}{l_H^2}$$

and

$$\Big|\frac{1}{\hat{H}_2(s,\mathrm{z})} - \frac{1}{H_2^0(s,\mathrm{z})}\Big| \le \frac{2}{l_H^2}\Big|\hat{H}_2(s,\mathrm{z}) - H_2^0(s,\mathrm{z})\Big|$$

Therefore, we have

$$\Big|\int_{0,t]} \frac{\hat{H}_1(ds,\mathrm{z}) - H_1^0(ds,\mathrm{z})}{\hat{H}_2(s,\mathrm{z})}\Big| \le 2\Big(\frac{2}{l_H} + \frac{16E_1}{l_H^2}\Big)||\hat{H}_1(\cdot,\mathrm{z}) - H_1(\cdot,\mathrm{z})||_{[0,t]}^{\infty}$$

by applying Lemma 4.1, and

$$\Big|\int_{0,t]}\Big[\frac{1}{\hat{H}_2(s,\mathrm{z})} - \frac{1}{H_2^0(s,\mathrm{z})}\Big]H_1^0(ds,\mathrm{z})\Big| \le \frac{2u_g}{l_H^2}||\hat{H}_2(\cdot,\mathrm{z}) - H_2(\cdot,\mathrm{z})||_{[0,t]}^{\infty}.$$

□

Lemma 4.4 and Lemma 4.3, combined with (4.17) and (4.19), give us the desired rate of uniform convergence of $|\hat{\Lambda}(t|\mathrm{z}) - \Lambda(t|\mathrm{z})|$. To establish the same rate for $|\hat{S}(t|\mathrm{z}) - S(t|\mathrm{z})|$, it suffices to prove the following lemma.

Lemma 4.5. *On* $B_n^C \cap D_n^C$

$$|\hat{S}(t|z) - S(t|z)| \le \lambda_3 \sup_I |\hat{\Lambda}(t|z) - \Lambda(t|z)|$$

where $\lambda_3 = 4\{\exp(\frac{8E_1}{l_H}) + \frac{8E_1}{l_H}\exp(\frac{16E_1}{l_H})\}\{\exp(\frac{u_g}{l_H}) + \frac{u_g}{l_H}\exp(\frac{2u_g}{l_H})\}$.

Proof. Define

$$\hat{\Lambda}_0(t|\mathrm{z}) = \frac{2}{l_H}\frac{1}{nh^d}\sum_{i=1}^{n} I(Y_i \le t)|K_{\mathrm{z}}(\frac{\mathrm{z}-\mathrm{z}_i}{h})| \quad \text{and} \quad \Lambda_0(t|\mathrm{z}) = \frac{1}{l_H}H_1^0(t,\mathrm{z})$$

On D_n^C, $\hat{\Lambda}(\cdot|\mathrm{z})$ is dominated by $\hat{\Lambda}_0(\cdot|\mathrm{z})$ in the sense that for any $0 \le s \le t \le u_F$

$$|\hat{\Lambda}(t|\mathrm{z}) - \hat{\Lambda}(s|\mathrm{z})| \le \hat{\Lambda}_0(t|\mathrm{z}) - \hat{\Lambda}_0(s|\mathrm{z}),$$

see Gill and Johansen (1990). And $\hat{\Lambda}_0(\cdot|\mathrm{z})$ can be viewed as a nonnegative additive interval function. An additive interval function is a function $\alpha(s,t)$, $0 \le s \le t \le u_F$, having the properties (see Page 1507 in Gill and Johansen 1990):

$$\alpha(s,t) = \alpha(s,u) + \alpha(u,t) \qquad \text{for all } s \le u \le t,$$
$$\alpha(s,s) = 0 \qquad \text{for all } s,$$
$$\alpha(s,t) = 0 \qquad \text{as } t \downarrow s \text{ for all } s$$

Also, on B_n^C, $\hat{\Lambda}_0(\cdot|\mathrm{z})$ is of bounded variation with the total variation

$$||\hat{\Lambda}_0(\cdot|\mathrm{z})||_{(0,u_F]}^v \leq \frac{8E_1}{l_H}$$

Clearly, this bound is not dependent on z, i.e. uniform in z. Similarly, one may find that $\Lambda(\cdot|\mathrm{z})$ is dominated by $\Lambda_0(\cdot|\mathrm{z})$, with $\Lambda_0(\cdot|\mathrm{z})$ being a nonnegative additive interval function and having the total variation

$$||\Lambda_0(\cdot|\mathrm{z})||_{(0,u_F]}^v \leq \frac{\mu_g}{l_H}$$

uniformly in z.

With these and repeatly use (20) and (21) in Gill and Johansen (1990), we have on $B_n^C \cap D_n^C$,

$$||\hat{S}(\cdot|\mathrm{z})||_{[0,t]}^\infty \leq \exp\{\frac{8E_1}{l_H}\} \tag{4.20}$$

$$||\hat{S}(\cdot|\mathrm{z})||_{(0,t]}^v \leq \hat{\Lambda}_0(t|\mathrm{z}) \exp\{\frac{16E_1}{l_H}\} \leq \frac{8E_1}{l_H} \exp\{\frac{16E_1}{l_H}\} \tag{4.21}$$

and

$$\left\|\frac{S(t|\mathrm{z})}{S(\cdot|\mathrm{z})}\right\|_{[0,t]}^\infty \leq \exp\{\frac{u_g}{l_H}\} \tag{4.22}$$

$$\left\|\frac{S(t|\mathrm{z})}{S(\cdot|\mathrm{z})}\right\|_{(0,t]}^v \leq \Lambda_0(t|\mathrm{z}) \exp\{\frac{2u_g}{l_H}\} \leq \frac{u_g}{l_H} \exp\{\frac{2u_g}{l_H}\} \tag{4.23}$$

Use Duhamel's equation for the difference of two product integrals (Theorem 6 in Gill and Johansen, 1990), Lemma 4.1, and the inequalities (4.20)–(4.23), we have

$$\begin{aligned}
|\hat{S}(t|\mathrm{z}) - S(t|\mathrm{z})| &= \Big|\prod_{(0,t]}\{1 - \hat{\Lambda}(ds|\mathrm{z})\} - \prod_{(0,t]}\{1 - \Lambda(ds|\mathrm{z})\}\Big| \\
&= \Big|\int_{(0,t]} \hat{S}(u - |\mathrm{z})[\hat{\Lambda}(du|\mathrm{z}) - \Lambda(du|\mathrm{z})]\frac{S(t|\mathrm{z})}{S(u|\mathrm{z})}\Big| \\
&\leq 4 \sup_{u\in[0,t]} |\hat{\Lambda}(u|\mathrm{z}) - \Lambda(u|\mathrm{z})| \times ||\hat{S}(\cdot|\mathrm{z})||_{[0,t]} \times \left\|\frac{S(t|\mathrm{z})}{S(\cdot|\mathrm{z})}\right\|_{[0,t]} \\
&\leq \lambda_3 \sup_I |\hat{\Lambda}(t|\mathrm{z}) - \Lambda(t|\mathrm{z})| \qquad \text{on } B_n^C \cap D_n^C
\end{aligned}$$

□

In the sequel, for any $x > 0$, we use the notation $O^*(x)$ to denote the quantity that is bounded by the product of a universal constant and x, i.e.

$$|O^*(x)| \leq Cx$$

where C is a positive constant not depending on t,z and n.

4.4.3. *Proof of Theorem 4.3*

We express explicitly the remainder term $R_\Lambda(t,\mathrm{z}) = \varsigma(t,\mathrm{z}) - \tau(t,\mathrm{z})$, where

$$\varsigma(t,\mathrm{z}) = \int_{(0,t]} \frac{[\hat{H}_2(s,\mathrm{z}) - H_2^0(s,\mathrm{z})]^2}{[H_2^0(s,\mathrm{z})]^2\hat{H}_2(s,\mathrm{z})}\hat{H}_1(ds,\mathrm{z})$$

$$\tau(t,\mathrm{z}) = \int_{(0,t]} \frac{\hat{H}_2(s,\mathrm{z}) - H_2^0(s,\mathrm{z})}{[H_2^0(s,\mathrm{z})]^2}(\hat{H}_1 - H_1^0)(ds,\mathrm{z})$$

It is easy to see that on $B_n^C \cap D_n^C$,

$$|\varsigma(t,\mathrm{z})| \le \frac{8E_1}{l_H^3}\sup_I |\hat{H}_2(t,\mathrm{z}) - H_2^0(t,\mathrm{z})|^2$$

and from Lemma 4.3

$$\sup_{(t,\mathrm{z})\in I} |\varsigma(t,\mathrm{z})| = O((a_n + h^k)^2) = O(a_n^2 + h^k) \qquad \text{a.s.}$$

So, it remains to show the following lemma.

Lemma 4.6. *Under assumptions (A1)-(A2) and with h satisfying (4.9),*

$$\sup_{(t,z)\in I} |\tau(t,z)| = O(a_n^2 + h^k) \qquad a.s.$$

Proof. We will use the exponential inequality for U-statistics in Lemma 4.2 to establish this. To this end, we write

$$\tau(t,\mathrm{z}) = \tau_1(t,\mathrm{z}) - \tau_2(t,\mathrm{z}) - \tau_3(t,\mathrm{z}) + \tau_4(t,\mathrm{z})$$

where

$$\tau_1(t,\mathrm{z}) = \frac{1}{(nh^d)^2}\sum_{i=1}^n\sum_{j=1}^n \frac{\delta_i I(Y_i \le t \wedge Y_j)}{[H_2^0(Y_i,\mathrm{z})]^2} K_{\mathrm{z}}(\frac{\mathrm{z}-\mathrm{z}_i}{h})K_{\mathrm{z}}(\frac{\mathrm{z}-\mathrm{z}_j}{h})$$

$$\tau_2(t,\mathrm{z}) = \frac{1}{nh^d}\sum_{i=1}^n \frac{\delta_i I(Y_i \le t)}{H_2^0(Y_i,\mathrm{z})} K_{\mathrm{z}}(\frac{\mathrm{z}-\mathrm{z}_i}{h})$$

$$\tau_3(t,\mathrm{z}) = \frac{1}{nh^d}\sum_{i=1}^n K_{\mathrm{z}}(\frac{\mathrm{z}-\mathrm{z}_i}{h}) \int_{(0,t\wedge Y_i]} \frac{H_1^0(ds,\mathrm{z})}{H_2^0(s,\mathrm{z})}$$

$$\tau_4(t,\mathrm{z}) = \int_{(0,t]} \frac{H_1^0(ds,\mathrm{z})}{H_2^0(s,\mathrm{z})}$$

It is easy to see that, for $\mathrm{z} \in \mathcal{X}_l$,

$$
\begin{aligned}
|\tau_1(t,\mathrm{z}) - \tau_1(t,\mathrm{x}_l)| &= O^*(\frac{\epsilon_2}{h^{d+1}}) = O^*(a_n^2) \qquad \text{on } B_n^C \\
|\tau_2(t,\mathrm{z}) - \tau_2(t,\mathrm{x}_l)| &= O^*(\frac{\epsilon_2}{h^{d+1}}) = O^*(a_n^2) \\
|\tau_3(t,\mathrm{z}) - \tau_3(t,\mathrm{x}_l)| &= O^*(\frac{\epsilon_2}{h^{d+1}}) = O^*(a_n^2) \\
|\tau_4(t,\mathrm{z}) - \tau_4(t,\mathrm{x}_l)| &= O^*(\epsilon_2) = O^*(a_n^2)
\end{aligned}
$$

Thus,

$$
\sup_{t\in[0,u_F],\mathrm{z}\in\mathcal{X}_l} |\tau(t,\mathrm{z}) - \tau(t,\mathrm{x}_l)| = O^*(a_n^2)
$$

Also we have, for $t \in \mathcal{T}_m$,

$$
\begin{aligned}
|\tau_1(t,\mathrm{x}_l) - \tau_1(t_{m-1},\mathrm{x}_l)| &= O^*(a_n^2) \qquad \text{on } A_n^C \cap B_n^C, \\
|\tau_2(t,\mathrm{x}_l) - \tau_2(t_{m-1},\mathrm{x}_l)| &= O^*(a_n^2) \qquad \text{on } A_n^C.
\end{aligned}
$$

To bound τ_3 and τ_4, note that,

$$
\begin{aligned}
&\sup_{\mathrm{z}\in J} |H_1^0(t_m-,\mathrm{z}) - H_1^0(t_{m-1},\mathrm{z})| \\
&\le \sup_{\mathrm{z}\in J} |H_1^0(t_m-,\mathrm{z}) - H_1(t_m-,\mathrm{z})| + \sup_{\mathrm{z}\in J} |H_1^0(t_{m-1},\mathrm{z}) - H_1(t_{m-1},\mathrm{z})| \\
&\qquad + \sup_{\mathrm{z}\in J} |H_1(t_m-,\mathrm{z}) - H_1(t_{m-1},\mathrm{z})| \\
&= O^*(h^k) + O^*(\frac{\epsilon_1}{h^d}) \qquad \text{(by Lemma 4.3)}, \\
&= O^*(a_n^2 + h^k)
\end{aligned}
$$

and thus

$$
\int_{(t_{m-1},t_m)} \frac{H_1^0(ds,\mathrm{z})}{H_2^0(s,\mathrm{z})} = O^*(a_n^2 + h^k).
$$

Also, it is easy to see that if $t \in [t_{m-1}, t_m) \cap [0, u_F]$, then either $t_{m-1} = u_F$ or $t_{m-1} < t_m \le u_F$. The case when $t_{m-1} = u_F$ is trivial so we only consider when $t_{m-1} < t_m \le u_F$,

$$
\begin{aligned}
&\sup_{t\in\mathcal{T}_m} |\tau_3(t,\mathrm{x}_l) - \tau_3(t_{m-1},\mathrm{x}_l)| \\
&= \sup_{t\in\mathcal{T}_m} \Big| \frac{1}{nh^d} \sum_{i=1}^n K_{\mathrm{x}_l}(\frac{\mathrm{x}_l - \mathrm{Z}_i}{h}) [\int_{(0,t\wedge Y_i]} \frac{H_1^0(ds,\mathrm{x}_l)}{H_2^0(s,\mathrm{x}_l)} \\
&\qquad - \int_{(0,t_{m-1}\wedge Y_i]} \frac{H_1^0(ds,\mathrm{x}_l)}{H_2^0(s,\mathrm{x}_l)}] \Big|
\end{aligned}
$$

$$
\begin{aligned}
&\le \sup_{t\in\mathcal{T}_m} \frac{1}{nh^d}\sum_{i=1}^n \Big|K_{\mathrm{x}_l}(\frac{\mathrm{x}_l - \mathrm{Z}_i}{h})\Big| I(t_{m-1} < Y_i < t)\int_{(t_{m-1},Y_i]} \frac{H_1^0(ds,\mathrm{x}_l)}{H_2^0(s,\mathrm{x}_l)} \\
&\quad + \sup_{t\in\mathcal{T}_m} \frac{1}{nh^d}\sum_{i=1}^n \Big|K_{\mathrm{x}_l}(\frac{\mathrm{x}_l - \mathrm{Z}_i}{h})\Big| I(Y_i \ge t)\int_{(t_{m-1},t]} \frac{H_1^0(ds,\mathrm{x}_l)}{H_2^0(s,\mathrm{x}_l)} \\
&\le \int_{(t_{m-1},t_m)} \frac{H_1^0(ds,\mathrm{x}_l)}{H_2^0(s,\mathrm{x}_l)} \times \hat{A}_1(\mathrm{x}_l) \\
&= O^*(a_n^2 + h^k) \qquad \text{on } B_n^C,
\end{aligned}
$$

and

$$
\sup_{t\in\mathcal{T}_m} |\tau_4(t,\mathrm{x}_l) - \tau_4(t_{m-1},\mathrm{x}_l)| = O^*(a_n^2 + h^k).
$$

Therefore

$$
\sup_{(t,\mathrm{z})\in\mathcal{T}_m\times\mathcal{X}_l} |\tau(t,\mathrm{z})| \le |\tau(t_{m-1},\mathrm{x}_l)| + O^*(a_n^2 + h^k). \tag{4.24}
$$

Now we examine $\tau(t_{m-1},\mathrm{x}_l)$. We have

$$
\tau_1(t_{m-1},\mathrm{x}_l) = \tau_{11}(t_{m-1},\mathrm{x}_l) + O^*(\frac{1}{nh^d}) \qquad \text{on } C_n^C,
$$

where

$$
\tau_{11}(t_{m-1},\mathrm{x}_l) = \frac{1}{(nh^d)^2}\sum_{i=1}^n\sum_{j\ne i} \frac{\delta_i I(Y_i \le t_{m-1}\wedge Y_j)}{[H_2^0(Y_i,\mathrm{x}_l)]^2} K_{\mathrm{x}_l}(\frac{\mathrm{x}_l - \mathrm{Z}_i}{h}) K_{\mathrm{x}_l}(\frac{\mathrm{x}_l - \mathrm{Z}_j}{h})
$$

We will construct a U-statistic from $\tau_{11}(t_{m-1},\mathrm{x}_l)$ and use Lemma 4.2 to bound it. Let $W_i = (Y_i, \delta_i, \mathrm{z}_i)$, $i = 1, \cdots, n$. Set for $i, j = 1, \cdots, n$,

$$
\begin{aligned}
\nu_{ij} &= \frac{\delta_i I(Y_i \le t_{m-1}\wedge Y_j)}{(nh^d)^2[H_2^0(Y_i,\mathrm{x}_l)]^2} K_{\mathrm{x}_l}(\frac{\mathrm{x}_l - \mathrm{Z}_i}{h}) K_{\mathrm{x}_l}(\frac{\mathrm{x}_l - \mathrm{Z}_j}{h}) \\
f_{ij} &= \nu_{ij} - E(\nu_{ij}|W_j) - E(\nu_{ij}|W_i) + E(\nu_{ij}) \\
g_{ij} &= f_{ij} + f_{ji}
\end{aligned}
$$

Then

$$
\tau_{11}(t_{m-1},\mathrm{x}_l) = \sum_{i=1}^n\sum_{j\ne i} \nu_{ij}
$$

and set

$$
U = \sum_{i=1}^n\sum_{j\ne i} f_{ij} = \sum_{1\le i<j\le n} g_{ij}
$$

We calculate that

$$E(\nu_{ij}|W_i) = \frac{\delta_i I(Y_i \le t_{m-1})}{n^2 h^d [H_2^0(Y_i, \mathrm{x}_l)]^2} K_{\mathrm{x}_l}(\frac{\mathrm{x}_l - \mathrm{Z}_i}{h})\{H_2^0(Y_i, \mathrm{x}_l) + O^*(h^k)\}$$

$$E(\nu_{ij}|W_j) = \frac{K_{\mathrm{x}_l}(\frac{\mathrm{x}_l - \mathrm{Z}_j}{h})}{n^2 h^d}\{\int_{(0,t_{m-1}\wedge Y_j]} \frac{H_1^0(ds, \mathrm{x}_l)}{H_2^0(s, \mathrm{x}_l)} + O^*(h^k)\}$$

$$E(\nu_{ij}) = \frac{1}{n^2}\int_{(0,t_{m-1}]} \frac{H_1^0(ds, \mathrm{x}_l)}{H_2^0(s, \mathrm{x}_l)} + \frac{O^*(h^k)}{n^2}$$

With these, we have

$$\sum_{i=1}^n \sum_{j\neq i} E(\nu_{ij}|W_i) = \tau_2(t_{m-1}, \mathrm{x}_l) + O^*(\frac{1}{n} + h^k) \qquad \text{on } B_n^C$$

$$\sum_{i=1}^n \sum_{j\neq i} E(\nu_{ij}|W_j) = \tau_3(t_{m-1}, \mathrm{x}_l) + O^*(\frac{1}{n} + h^k) \qquad \text{on } B_n^C$$

$$\sum_{i=1}^n \sum_{j\neq i} E(\nu_{ij}) = \tau_4(t_{m-1}, \mathrm{x}_l) + O^*(\frac{1}{n} + h^k)$$

In summary, we have, on $A_n^C \cap B_n^C \cap C_n^C$,

$$\tau(t_{m-1}, \mathrm{x}_l) = U + O^*(a_n^2 + \frac{1}{nh^d} + \frac{1}{n} + h^k) = U + O^*(a_n^2 + h^k) \quad (4.25)$$

Clearly, U is a U-statistic and can be bounded by the inequality in Lemma 4.2. To this end, we further calculate that

$$G_1 = O(\frac{1}{(nh^d)^2}), \qquad G_2^2 = O(\frac{1}{(nh^d)^3})$$

$$G_3^2 = O(\frac{1}{(nh^d)^2}) \quad \text{and} \quad G_4 = O(\frac{1}{nh^d})$$

Then, set $b_n = \lambda_4 a_n^2$ for a large enough λ_4, we have

$$\min\left(\frac{b_n^2}{G_3^2}, \frac{b_n}{G_4}, \frac{b_n^{2/3}}{G_2^{2/3}}, \frac{b_n^{1/2}}{G_1^{1/2}}\right) \ge G_0 \log n$$

for some $G_0 > 0$ which can be chosen large enough according to λ_4. Using Lemma 4.2, we have for some positive constant G,

$$\Pr(|U| > b_n) \le G \exp\left\{-\frac{G_0}{G} n^{\frac{2(r-d)-\eta}{4r}}\right\} = Gn^{-\frac{G_0}{G}}.$$

Thus, we have

$$
\begin{aligned}
&\Pr\{\sup_I |\tau(t,\mathrm{z})| > \lambda_4(a_n^2 + h^k)\} \\
&= \Pr\{\sup_I |\tau(t,\mathrm{z})| I_{A_n^C \cap B_n^C \cap C_n^C} > \lambda_4(a_n^2 + h^k)\} + \Pr(A_n \cup B_n \cup C_n) \\
&\leq \sum_{m=1}^{M} \sum_{l=1}^{L} \Pr\{\sup_{\mathcal{T}_m \times \mathcal{X}_l} |\tau(t,\mathrm{z})| I_{A_n^C \cap B_n^C \cap C_n^C} > \lambda_4(a_n^2 + h^k)\} \\
&+ \Pr(A_n \cup B_n \cup C_n),
\end{aligned}
\tag{4.26}
$$

and combine with (4.24) and (4.25),

$$
\begin{aligned}
&\sum_{m=1}^{M} \sum_{l=1}^{L} \Pr\{\sup_{\mathcal{T}_m \times \mathcal{X}_l} |\tau(t,\mathrm{z})| I_{A_n^C \cap B_n^C \cap C_n^C} > \lambda_4(a_n^2 + h^k)\} \\
&\leq \sum_{m=1}^{M} \sum_{l=1}^{L} \Pr(|U| > b_n) \\
&\leq O(n^{2+d}) n^{-\frac{G_0}{G}}
\end{aligned}
\tag{4.27}
$$

where λ_4 can be chosen large enough such that G_0 makes (4.27) of $O(n^{-(1+\theta)})$ for some $\theta > 0$. Therefore, combine with (4.26),(4.27), (4.14), (4.17) and (4.18) and use Borel-Cantelli's Lemma, we have

$$
\sup_I |\tau(t,\mathrm{z})| = O(a_n^2 + h^k) \qquad \text{a.s.}
$$

as desired. □

4.4.4. *Proof of Theorem 4.3*

By Duhamel's equation (Gill and Johansen, 1990), we have

$$
\begin{aligned}
L_1(t,\mathrm{z}) &= \hat{S}(t|\mathrm{z}) - S(t|\mathrm{z}) - \hat{L}(t|\mathrm{z}) \\
&= \int_{(0,t]} [\hat{S}(u-|\mathrm{z}) - S(u-|\mathrm{z})][\hat{\Lambda}(du|\mathrm{z}) - \Lambda(du|\mathrm{z})] \frac{S(t|\mathrm{z})}{S(u|\mathrm{z})} \\
&= L_{11}(t,\mathrm{z}) + O((a_n + h^k)^2) \qquad \text{a.s.}
\end{aligned}
$$

where

$$
L_{11}(t,\mathrm{z}) = \int_{(0,t)} [\hat{S}(u-|\mathrm{z}) - \hat{S}(u-|\mathrm{z})][\hat{\Lambda}(du|\mathrm{z}) - \Lambda(du|\mathrm{z})] \frac{S(t|\mathrm{z})}{S(u|\mathrm{z})}
$$

The $O((a_n+h^k)^2)$ remainder term is due to the rates of uniform convergence in Theorem 4.1. Using the integration-by-parts formula, we have

$$
L_{11}(t,\mathrm{z}) = -L_2(t,\mathrm{z}) + L_3(t,\mathrm{z})
$$

where

$$L_2(t,\mathrm{z}) = [\hat{S}(t-|\mathrm{z}) - S(t-|\mathrm{z})] \int_{(0,t)} [\hat{\Lambda}(du|\mathrm{z}) - \Lambda(du|\mathrm{z})] \frac{S(t|\mathrm{z})}{S(u|\mathrm{z})}$$

$$L_3(t,\mathrm{z}) = \int_{(0,t)} \int_{(0,u]} [\hat{\Lambda}(ds|\mathrm{z}) - \Lambda(ds|\mathrm{z})] \frac{S(t|\mathrm{z})}{S(s|\mathrm{z})} [\hat{S}(du|\mathrm{z}) - S(du|\mathrm{z})]$$

Clearly, $\sup_I |L_2(t,\mathrm{z})| = O((a_n + h^k)^2)$, a.s.. Using Duhamel's equation for $\hat{S}(t|\mathrm{z}) - S(t|\mathrm{z})$, we have

$$L_3(t,\mathrm{z}) = L_{31}(t,\mathrm{z}) - L_{32}(t,\mathrm{z})$$

where

$$L_{31}(t,\mathrm{z}) = \int_{(0,t)} \Big\{ \int_{(0,u]} [\hat{\Lambda}(ds|\mathrm{z}) - \Lambda(ds|\mathrm{z})] \frac{S(t|\mathrm{z})}{S(s|\mathrm{z})} \Big\} \times \Big\{ \int_{(0,u)} \hat{S}(s-|\mathrm{z})[\hat{\Lambda}(ds|\mathrm{z}) - \Lambda(ds|\mathrm{z})] \frac{S(u-|\mathrm{z})}{S(s|\mathrm{z})} \Big\} \Lambda(du|\mathrm{z})$$

and

$$\begin{aligned} L_{32}(t,\mathrm{z}) &= \int_{(0,t)} \Big\{ \int_{(0,u]} [\hat{\Lambda}(ds|\mathrm{z}) - \Lambda(ds|\mathrm{z})] \frac{S(t|\mathrm{z})}{S(s|\mathrm{z})} \Big\} \hat{S}(u-|\mathrm{z})[\hat{\Lambda}(du|\mathrm{z}) \\ &\quad -\Lambda(du|\mathrm{z})] \\ &= \int_{(0,t)} \hat{S}(u-|\mathrm{z}) \Big\{ \int_{(0,u]} [\hat{\Lambda}(ds|\mathrm{z}) - \Lambda(ds|\mathrm{z})] \frac{S(u|\mathrm{z})}{S(s|\mathrm{z})} \Big\} [\hat{\Lambda}(du|\mathrm{z}) \\ &\quad -\Lambda(du|\mathrm{z})] \frac{S(t|\mathrm{z})}{S(u|\mathrm{z})} \\ &= \int_{(0,t)} \hat{S}(u-|\mathrm{z}) L_4(du,\mathrm{z}) \frac{S(t|\mathrm{z})}{S(u|\mathrm{z})} \end{aligned}$$

with

$$L_4(u,\mathrm{z}) = \int_{(0,u]} \Big\{ \int_{(0,e]} [\hat{\Lambda}(ds|\mathrm{z}) - \Lambda(ds|\mathrm{z})] \frac{S(e|\mathrm{z})}{S(s|\mathrm{z})} \Big\} [\hat{\Lambda}(de|\mathrm{z}) - \Lambda(de|\mathrm{z})]$$

Clearly, $\sup_I |L_{31}(t,\mathrm{z})| = O((a_n + h^k)^2)$, a.s.. And use Lemma 4.1, we have for some positive constant λ_5,

$$\sup_I |L_{32}(t,\mathrm{z})| \le \lambda_5 ||L_4(\cdot,\mathrm{z})||^{\infty}_{[0,u_F]}$$

Using the integration-by-parts formula again, we find that

$$L_4(t,\mathrm{z}) = L_{41}(t,\mathrm{z}) + L_{42}(t,\mathrm{z}) - L_{43}(t,\mathrm{z})$$

where

$$L_{41}(t,\mathrm{z}) = \int_{(0,t]} [\hat{\Lambda}(ds|\mathrm{z}) - \Lambda(ds|\mathrm{z})] \frac{S(t|\mathrm{z})}{S(s|\mathrm{z})} \times [\hat{\Lambda}(t|\mathrm{z}) - \Lambda(t|\mathrm{z})]$$

$$L_{42}(t,\mathrm{z}) = \int_{(0,t]} [\hat{\Lambda}(u-|\mathrm{z}) - \Lambda(u-|\mathrm{z})] \times \int_{(0,u)} [\hat{\Lambda}(ds|\mathrm{z}) - \Lambda(ds|\mathrm{z})] \frac{S(u-|\mathrm{z})}{S(s|\mathrm{z})} \Lambda(du|\mathrm{z})$$

$$L_{43}(t,\mathrm{z}) = \int_{(0,t]} [\hat{\Lambda}(u-|\mathrm{z}) - \Lambda(u-|\mathrm{z})][\hat{\Lambda}(du|\mathrm{z}) - \Lambda(du|\mathrm{z})]$$

Apparently, $\sup_I |L_{41}(t,\mathrm{z})| = O((a_n + h^k)^2)$, a.s. and $\sup_I |L_{42}(t,\mathrm{z})| = O((a_n + h^k)^2)$, a.s.. And by Theorem 4.2,

$$\sup_I |L_{43}(t,\mathrm{z}) - L_5(t,\mathrm{z})| = O(a_n^2 + h^k) \qquad \text{a.s.}$$

with

$$L_5(t,\mathrm{z}) = \int_{(0,t]} \hat{L}_\Lambda(u,\mathrm{z})\hat{L}_\Lambda(du,\mathrm{z})$$

and the conclusion follows from the following Lemma 4.7.

Further, use Lemma 4.1, we have for some positive constant λ_6,

$$\begin{aligned}
&|\hat{L}_S(t,\mathrm{z}) - \tilde{L}_S(t,\mathrm{z})| \\
&= \left| \int_{(0,t]} S(u-|\mathrm{z})[\hat{\Lambda}(du|\mathrm{z}) - \Lambda(du|\mathrm{z}) - \hat{L}_\Lambda(du,\mathrm{z})] \frac{S(t|\mathrm{z})}{S(u|\mathrm{z})} \right| \\
&\le \lambda_6 \sup_I |\hat{\Lambda}(t|\mathrm{z}) - \Lambda(t|\mathrm{z}) - \hat{L}_\Lambda(t,\mathrm{z})| \\
&= O(a_n^2 + h^k) \qquad \text{a.s.,}
\end{aligned}$$

which completes the proof.

Lemma 4.7. *Under assumptions (A1)-(A2) and with h satisfying (4.9),*

$$\sup_I |L_5(t,z)| = O(a_n^2 + h^k) \qquad a.s.$$

Proof. The proof is essentially the same as that of Lemma 4.6. It consists of partition of I, approximation of $L_5(t,\mathrm{z})$ with $L_5(t_{m-1},\mathrm{x}_l)$ in $\mathcal{T}_m \times \mathcal{X}_l$, construction of a U-statistic through $L_5(t_{m-1},\mathrm{x}_l)$ and bounding the tail probability of the U-statistic using Lemma 4.2. We would like to leave the detailed proof to interested readers. □

4.5. Discussions

Since we do not require that the continuity in t, the estimators here can be applied to the case when the follow-up time is discreet or mixed.

In this paper, we only discuss the fixed bandwidth. The case of a varying bandwidth can be discussed, in which the order of the bandwidth is given by the results in this paper and the constant can be determined locally.

Caution should be taken when the sample size is small or the dimension of z is large, as in the regular smoothing problem.

Acknowledgments

We thank an anonymous referee for the helpful suggestions. Wei Yann Tsai with Department of Statistics, National Cheng Kung University, Tainan, Taiwan, and was partially supported by National Center for Theoretical Sciences (South), Taiwan.

References

Aalen, O. O. (1978). Nonparametric Inference for a Family of Counting Processes, *The Annals of Statistics* **6**, 701–726.

Beran, R. (1981). Nonparametric Regression with Randomly Censored Survival Data, *Technical Report*, Univ. California, Berkeley.

Bowman, A. W. and Wright, E. M. (2000). Graphical Exploration of Covariate Effects on Survival Data Through Nonparametric Quantile, *Biometrics* **56**, 563–570.

Dabrowska, D. M. (1989). Uniform Convergence of The Kernel Conditional Kaplan-Meier Estimate, *The Annals of Statistics* **17**, 1157–1167.

Dabrowska, D. M. (1992). Nonparametric Quantile Regression with Censored Data, *Sankhyā Ser. A (The Indian Journal of Statistics)* **54**, 252–259.

Gentleman, R. and Crowley, J. (1991). Graphical Methods for Censored Data, *Journal of the American Statistical Association* **86**, 678–683.

Gill, R. D. and Johansen, S. (1990). A Survey of Product-integration with a View Toward Application in Survival Analysis, *The Annals of Statistics* **18**, 1501–1555.

Giné, E., Latała, R. and Zinn, J. (2000). Exponential and moment inequalities for U-statistics, *High Dimensional Probability II. Progress in Probability* **47**, 13–38. Birkhauser, Boston, Boston, MA.

Gonzalez-Manteiga, W. and Cadarso-Suarez, C. (1994). Asymptotic Properties of A Generalized Kaplan-Meier Estimator with Some Applications, *Journal of Nonparametric Statistics* **4**, 65–78.

Houdré, C. and Reynaud-Bouret P. (2003). Exponential Inequalities, with Constants, for U-statistics of Order Two, *Stochastic inequalities and applications. Progress in Probability* **56**, 55–69. Birkhauser, Basel.

Iglesian-Pérez, M. C. (2003). Strong Representation of A Conditional Quantile Function Estimator with Truncated and Censored Data, *Statistics & Probability Letters* **65**, 79–91.

Kaplan, E. L. and Meier, P. (1958). Nonparametric Estimation from Incomplete Observations, *Journal of American Statistical Association* **53**, 457–481.

Major, P. and Rejtő, L. (1988). Strong Embedding of The Estimator of The Distribution Function under Random Censorship, *The Annals of Statistics* **16**, 1113–1132.

Müller, H-G. and Wang, J-L. (1994). Hazard Rate Estimation Under Random Censoring With Varying Kernels and Bandwidths, *Biometrics* **50**, 61–76.

Nadaraya, E. E. (1964). On Estimating Regression, *Theory of Probability and Its Applications* **9**, 141–142.

Nelson, W. (1972). Theory and Applications of Hazard Plotting for Censored Failure Data, *Technometrics* **14**, 945–966.

Stute, W. (1994). Strong and Weak Representations of Cumulative Hazard Kaplan-Meier Estimators on Increasing Sets, *Journal of Statistical Planning and Inference* **42**, 315–329.

Szegö, G. (1975). Orthogonal Polynomials, Providence, Rhode Island: American Mathematical Society.

van Keilegom, I. and Veraverbeka, N. (1997). Estimation and Bootstrap with Censored Data in Fixed Design Nonparametric Regression, *Annals of the Institute of Statistical Mathematics* **49**, 467–491.

Watson, G. S. (1964). Smooth Regression Analysis, *Sankhyā Ser. A* **26**, 359–372.

Chapter 5

Regression Analysis in Failure Time Mixture Models with Change Points According to Thresholds of a Covariate

Jimin Lee
University of North Carolina Asheville, Asheville, NC 28804 USA
jlee@unca.edu

Thomas H. Scheike
University of Copenhagen, Copenhagen DK-1014, Denmark
ts@biostat.ku.dk

Yanqing Sun
University of North Carolina at Charlotte, Charlotte, NC 28223 USA
yasun@uncc.edu

We use a mixture model or cure model to simultaneously model the probability of a patient who will never experience the failure and the risk of death or onset of the disease for those at risk of eventual failure. The former is modeled through a nonstandard logistic regression model and the latter is modeled using a nonstandard Cox model. We allow the regression coefficients in the both models to change according to the unknown thresholds of covariates. We develop the semiparametric maximum likelihood estimation procedures through the EM algorithm. We also formulate the test statistics to test the existence of a change point in the covariate effects and its location for the latency survival model as well for the cure probability. A simulation study is conducted to check the performance of the proposed estimation and testing procedures. The procedure is demonstrated through an application to the melanoma survival study.

5.1. Introduction

In survival analysis it is usually assumed that if complete follow-up were possible for all individuals, each would eventually experience the event of interest. In recent years, there has been considerable interest in modelling right-censored failure time data with potentially cured patients. This kind of data often arises from clinical follow-up studies where there exists a positive portion of subjects in the population who would never experience the event of interest. These subjects are usually referred to as 'cured,' while the remaining subjects who are susceptible to the event are referred to as 'uncured.' In some situations, some of these survivors are suspected actually to be cured in the sense that, no further events are observed even after an extended follow-up. A cancer patient is considered cured if the patient is not at risk of dying from the cancer. In some localized tumor cases, all the tumor cells have been killed by the radiation and it is extremely unlikely there will be any recurrences after a number of years, say 5 years, of the treatment. In other situations, a patient may have lived cancer free beyond the longest observed cancer free duration. In this case, the cure may as well be considered as long-term survival. The chance of being cured and the number of years of survival from diagnosis are of great interest to cancer patients and the medical community alike. Patients with colorectal cancer are often cured; the cure fraction for the localized colorectal cancer is in the range of 74.2%–79.3%, 40.4%–50.4% for the regional and 4.6%–6.7% for the distant, Yu and Tiwari (2005). The cure fraction for the breast cancer patients is estimated to be at least around 30%, Yu *et al.* (2003). As medical science progresses, the chance of being cured and the years of survival will likely improve.

In a cure model, the population is a mixture of susceptible and nonsusceptible individuals. All susceptible subjects would eventually experience the event if there were no censoring, while the nonsusceptible ones are cured from the event of interest. Let T_0 be the survival time of interest and let V be a binary variable where $V = 1$ indicates that the patient will never experience the event and $V = 0$ indicates that the patient will experience the event eventually. Let $T_0 = \infty$ if $V = 1$. Let Z be a p-dimensional covariate that may be associated with the chance of cure and the risk of eventual failure. We assume a positive cure fraction $c(Z) = P(V = 1|Z) > 0$. The survival probability of event-free at time t is then given by

$$S(t|Z) = P(T_0 > t|Z) = c(Z) + (1 - c(Z))\, G(t|Z), \tag{5.1}$$

where $G(t|Z) = P(T_0 > t|V = 0, Z)$ is the conditional survival function for those who eventually experience the event, often called latency distribution.

Many authors have studied model (5.1) by considering various models for the cure fraction $c(Z)$ and the latency distribution $G(t|Z)$. In Berkson and Gage (1952), $1-c(Z) = p$ is a nonnegative constant and an exponential distribution is used for survival function of T_0 with $S(t) = P(T_0 > t) = \exp(-\lambda t)$. Such a parametric approach was extended by Farewell (1982) to a logistic regression model for the cure fraction and a Weibull distribution model without covariates for the latency distribution. A logistic regression model for the cure fraction entails that

$$\text{logit}(c(Z)) = a^T Z, \tag{5.2}$$

where $\text{logit}(c) = \log\{c/(1-c)\}$ and a is a vector of coefficients. Under the Weibull regression model for the uncured, the conditional hazard function associated with the latency distribution $G(t|Z)$ can be described by

$$h(t|Z) = h_0(t) \exp\left(\alpha^T Z\right), \tag{5.3}$$

where $h_0(t) = \lambda\nu(\lambda t)^{\nu-1}$ is the hazard function for a Weibull distribution with λ, $\nu > 0$. Theoretical and empirical properties of the Weibull extension were fully studied in Farewell (1982).

Larson and Dinse (1985) used the proportional hazards model for the latency with a step function for the baseline hazard. Lo *et al.* (1993) used a similar model, but with the baseline hazard determined by piecewise linear splines. Yamaguchi (1992) used a general class of accelerated failure time models for the latency distribution. Other parametric mixture models have been considered in the literature for survival data with potentially cured patients. A parametric mixture model typically uses the logistic regression model for the cure rate and the parametric models such as lognormal, loglogistic and Weibull distributions are widely used to model the survival time of the uncured subjects. Parametric methods are parsimonious and easy to interpret. However, they can be sensitive to model misspecifications. Furthermore, there is often little physical evidence in a clinical study to suggest and justify a specific parametric model.

More attention has been paid on semiparametric mixture modelling approaches. Kuk and Chen (1992) developed a semiparametric model where the long term survival depends on the covariate through a logistic link, and the latency period depends on covariates in a proportional hazards structure with unspecified baseline hazard function. The proportional hazards model for the uncured is an extension of model (5.3) by letting $h_0(t)$ to be an unknown and unspecified hazard function. A main difficulty in fitting models (5.2) and (5.3) is that data is usually not completely observed in clinical follow-up studies. If an individual is observed to have experienced the event before the end of the follow-up, then an event time is recorded as finite and it is known that the individual is uncured with

$V = 0$ and $T_0 \leq C$. However, if an individual has not experienced the event by the end of the study, then V and T_0 are not observed. Instead, a censoring time C is observed and it is only known that either the individual is cured or the individual is uncured and will experience the event in the future. In this case, one observes (T, δ, Z) instead of (V, T, δ, Z), where $T = \min\{T_0, C\}$ and $\delta = I(T_0 \leq C)$. When the cure fraction is present, the observed event time no longer exhibits the proportional hazards property. Consequently the likelihood function for the ordinary Cox regression model becomes invalid. To deal with this difficulty, Kuk and Chen (1992) developed a Monte Carlo algorithm to approximate a rank-based likelihood function, thereby enabling them to perform maximum likelihood estimation. The proportional hazards cure model was further studied by Peng and Dear (2000) and Sy and Taylor (2000) using an EM algorithm approach. The asymptotic properties of the maximum likelihood estimator for the semiparametric logistic and proportional hazards mixture models were studied by Fang *et al.* (2005). Alternative semiparametric methods have been developed recently. Lu and Ying (2004) considered a general class of semiparametric transformation cure models. The model combined a logistic regression for the probability of event occurrence with a class of transformation models for the time of event occurrence. Included as special cases were the proportional hazards cure model and the proportional odds cure model. Estimating equations were proposed, which can be solved using an iterative algorithm simultaneously.

The latency model (5.3) assumes that the conditional hazard functions are proportional for different covariate values. The model (5.2) implies that the logarithm of odds ratio is constant at two different levels of a covariate of equal difference. In practice, the assumption of proportional hazards is not always adequate in the whole range of a covariate and the covariate may be dichotomized according to a threshold that may be fixed or estimated from the data.

An important generalization of the proportional hazards model is to allow the baseline function to depend on the strata defined by the covariates whose effect on the hazard is not proportional. The proportional hazards model holds for the subjects within each stratum. The hazard for an individual who belongs to stratum k is therefore

$$h_k(t|Z) = h_{k0}(t) \exp\left(\alpha^T Z\right), \quad k = 1, \cdots, K.$$

Luo and Boyett (1997) studied a model where a constant is added to the regression on a covariate Z_1 after a change point in another variable Z_2.

Assume that conditionally on Z the hazard rate of a survival time T_0 has the form $h_\theta(t|Z) = h_0(t)\exp(r_\theta(Z))$, where $r_\theta(Z) = \alpha^T Z_1 + \beta I(Z_2 \leq \zeta)$ and $I(\cdot)$ is an indicator function.

Jespersen (1986) studied a test of no change-point. Pons (2003) studied a nonregular Cox model with a change point according to the unknown threshold of a covariate. Pons (2003) extended the model studied by Luo and Boyett (1997) by taking into account the situation where the effects of some covariates may change according to the threshold of another covariate. The colorectal cancer, prostate cancer and breast cancer are among the most common cancers in US. Because of the progress in cancer detecting methods, e.g., new imaging technologies, tumor markers and biopsy procedures, the incidence for these three cancer sites experienced dramatic change during the last three decades. Tiwari *et al.* (2005) used the Bayesian method for a change point Poisson model to analyze the age-adjusted cancer rates in the US for the three types of cancers for the period from 1973 and 1999 and showed how these rates have changed over years with a focus on identifying change-points. For example, one may consider a change-point variable as the year of diagnosis. In the melanoma study in Andersen *et al.* (1993), the tumor thickness may be considered as a change-point variable with thresholds for tumor size to be estimated.

In this paper, we study a semiparametric mixture model to simultaneously model the probability of a patient who will never experience the failure and the risk of death or onset of the disease for those at risk of eventual failure. The former is modeled through an nonstandard logistic regression model and latter is modeled using a nonstandard Cox model, where the regression coefficients in both the models are allowed to change according to the unknown thresholds of covariates. The rationale for this generalization is that the chance of cure from a disease and the time length of survival with the disease maybe greatly affected by certain biomarker such as tumor thickness and by advances in medical sciences. These factors may also change the effects of other covariates on the cure probability and the length of survival. This model is described in Sec. 5.2.

Our approach is based on the semiparametric maximum likelihood using the EM algorithm. Both parametric and nonparametric components of the models are estimated in Sec. 5.3. The hypothesis testing procedures for testing existence of change point and the value of the threshold are proposed in Sec. 5.4. Section 5.5 presents some Monte Carlo simulations conducted to evaluate the proposed estimation and hypothesis testing procedures. In Sec. 5.6 we illustrate the proposed methods with an application to the melanoma survival data.

5.2. Model Descriptions

Let $Z = (Z_1^T, Z_2^T, Z_3)^T$ be a vector of covariates, where Z_1 and Z_2 are respectively p and q dimensional and Z_3 is a one-dimensional random variable. We consider the following change point structure of Pons (2003) for the latency hazard function $h(t|Z)$:

$$h(t|Z) = h_0(t)\exp\left(r_\vartheta(Z)\right) \tag{5.4}$$

where $h_0(t)$ is an unknown baseline hazard function and $r_\vartheta(Z) = \alpha^T Z_1 + \beta^T Z_2 + \gamma^T Z_2 I(Z_3 \le \zeta)$. Let $\xi = (\alpha^T, \beta^T, \gamma^T)^T$, $\vartheta = (\zeta, \xi^T)^T$ and $\bar{Z}(\zeta) = \left(Z_1^T, Z_2^T, Z_2^T I(Z_3 \le \zeta)\right)^T$. Then $r_\vartheta(Z) = \xi^T \bar{Z}(\zeta)$. We assume that the regression parameter α belongs to a bounded compact set of $\mathbb{R}^p$ and β and γ belong to bounded compact sets of $\mathbb{R}^q$. The threshold ζ is a parameter lying in a bounded interval $[\zeta_1, \zeta_2]$ strictly included in the support of Z_3.

The model (5.4) is useful in practice. In the study of the risk factors on survival with melanoma, tumor thickness is dichotomized according to the predetermined values Andersen *et al.* (1993). However, in practice, it may not be clear how tumor thickness alters the effects of other covariates on survival. The thresholds for tumor size need to be estimated. Motivated by similar applications, we consider the following logistic regression model for the cure fraction:

$$\text{logit}(c(X, \theta)) = \rho_\theta(X), \tag{5.5}$$

where $\rho_\theta(X) = a_0 + a^T X_1 + b^T X_2 + g^T X_2 I(X_3 \le \phi)$, $\theta = (\phi, \psi^T)^T$ and $\psi = (a_0, a^T, b^T, g^T)^T$. The covariate vector $X^T = \left(1, X_1^T, X_2^T, X_3^T\right)^T$ needs not to be the same as Z although one can let $Z = X$ in many situations. Denoting $\bar{X}(\phi) = \left(1, X_1^T, X_2^T, X_2^T I(X_3 \le \phi)\right)^T$, we can express $\rho_\theta(X) = \psi^T \bar{X}(\phi)$.

5.3. The EM Algorithm

Assume that the censoring random variable C is independent of T_0 given the covariates Z and X. Let $(T_{0i}, C_i, Z_i, X_i, V_i)$, $i = 1, \ldots, n$, be iid copies of (T_0, C, Z, X, V). The observed data consists of $\mathcal{D} = \{(T_i, \delta_i, Z_i, X_i), i = 1, \ldots, n\}$, where $T_i = \min(T_{0i}, C_i)$, and $\delta_i = 1$ if the individual i experienced the event at time T_i, i.e, $T_{0i} \le C_i$ while $\delta_i = 0$ if the individual i had not experienced the event by time T_i, i.e, $T_{0i} > C_i$. When $\delta_i = 0$, individual i may experience the event at some future time or will never experience the event. We assume that V_i is independent of C_i given (Z_i, X_i). The full data consists of $(T_i, \delta_i, Z_i, X_i, V_i)$, $i = 1, ..., n$, where, hypothetically, we can assure the status of V_i of each individual, knowing whether or not it will experience the event of interest even though it is censored.

The likelihood function for the observed data $(T_i, \delta_i, Z_i, X_i)$, $i = 1, \cdots, n$, under the models (5.4) and (5.5) can be written as

$$L(\theta, \vartheta, h_0(\cdot)) \\ = \prod_{i=1}^{n} \left[(1-c(x_i,\theta))h(t_i,\vartheta)S(t_i,\vartheta)\right]^{\delta_i} \left[c(x_i,\theta)+(1-c(x_i,\theta))S(t_i,\vartheta)\right]^{1-\delta_i},$$

where $h(t_i, \vartheta) = h(t_i|z_i)$ is given by (5.4), $S(t_i, \vartheta) = \{S_0(t_i)\}^{\exp(r_\vartheta(z_i))}$ and $S_0(t)$ is the survival function corresponding to the baseline hazard function $h_0(t)$. Here and throughout the paper, the notation (t_i, z_i, x_i, v_i) may also be used for (T_i, Z_i, X_i, V_i) for ease of the presentation. If all individuals will eventually experience the event of interest, that is, the cure fraction $c(x_i, \theta) = 0$, then the likelihood function L becomes the one used by Pons (2003) in the Cox regression model with a change-point according to a threshold in a covariate. The maximum partial log-likelihood estimators for $\vartheta = (\zeta, \xi^T)^T$ and $\xi = (\alpha^T, \beta^T, \gamma^T)$ can then be obtained, Pons (2003). However, when the cure fraction $c(x_i, \theta)$ is greater than zero, L no longer resembles the likelihood function for the proportional hazard model based on ordinary right censored survival data. The baseline hazard function can not be eliminated from the likelihood. This is caused by the additive term in second part of the function L. Various methods were proposed to maximize the joint semiparametric likelihood function. Kuk and Chen (1992) proposed a Monte Carlo simulation approach to estimate parameters in the model. Peng and Dear (2000) pointed that Kuk and Chen's method is inconvenient since their method depends on a Monte Carlo approximation of the marginal likelihood function. Peng and Dear (2000) and Sy and Taylor (2000) proposed a full data likelihood approach and used the EM algorithm to compute the maximum likelihood estimators.

The full data likelihood function for $(T_i, \delta_i, Z_i, X_i, V_i)$, $i = 1, \cdots, n$, equals

$$L_c(\theta, \vartheta, h_0(\cdot)) \\ = \prod_{i=1}^{n} \left[c(x_i,\theta)\right]^{v_i} \left[(1 - c(x_i,\theta))(h(t_i,\vartheta)S(t_i,\vartheta))^{\delta_i}(S(t_i,\vartheta))^{1-\delta_i}\right]^{1-v_i},$$

The full data log-likelihood is

$$l_c(\theta, \vartheta, h_0(\cdot)) \\ = \sum_{i=1}^{n} v_i \log[c(x_i,\theta)] \\ +(1 - v_i) \log\left[(1 - c(x_i,\theta))(h(t_i,\vartheta)S(t_i,\vartheta))^{\delta_i}(S(t_i,\vartheta))^{1-\delta_i}\right].$$

While it is certain that $V_i = 0$ when $\delta_i = 1$, it is often not clear the cure status V_i when $\delta_i = 0$ in practice. So V_i is only partially observable. Let $\mathcal{O}_i = (T_i, \delta_i, Z_i, X_i)$ and $v_i^* = E(V_i|\mathcal{O}_i)$. Note that $v_i^* = 0$ if $\delta_i = 1$. For $\delta_i = 0$,

$$v_i^* = \frac{c(x_i, \theta)}{c(x_i, \theta) + (1 - c(x_i, \theta))S(t_i, \vartheta)}, \tag{5.6}$$

where $S(t_i, \vartheta) = \exp\left[-H_0(t_i)\exp\{r_\vartheta(z(t_i))\}\right]$ and $H_0(t) = -\log\{S_0(t)\}$ is the cumulative baseline hazard function. The conditional expectation of $l_c(\theta, \vartheta, h_0(\cdot))$ given the observed data is

$$\begin{aligned} l^*(\theta, \vartheta, h_o(\cdot)) &= E(l_c(\theta, \vartheta, h_0(\cdot))|\mathcal{D}) \\ &= \sum_{i=1}^{n} v_i^* \log\{c(x_i, \theta)\} + (1 - v_i^*)\log\{1 - c(x_i, \theta)\} \\ &\quad + (1 - v_i^*)[\delta_i \log\{h(t_i, \vartheta)\} + \log\{S(t_i, \vartheta)\}]. \end{aligned}$$

We note that $l^*(\theta, \vartheta, h_0(\cdot))$ can be written as $l^*(\theta, \vartheta, h_0(\cdot)) = l_1^*(\theta) + l_2^*(\vartheta, h_0(\cdot))$, where

$$l_1^*(\theta) = \sum_{i=1}^{n} v_i^* \log\{c(x_i, \theta)\} + (1 - v_i^*)\log\{1 - c(x_i, \theta)\}, \tag{5.7}$$

$$l_2^*(\vartheta, h_0(\cdot)) = \sum_{i=1}^{n} (1 - v_i^*)[\delta_i \log\{h(t_i, \vartheta)\} + \log\{S(t_i, \vartheta)\}]. \tag{5.8}$$

Thus the maximization of $l^*(\theta, \vartheta, h_0(\cdot))$ can be carried out by maximizing $l_1^*(\theta)$ and $l_2^*(\vartheta, h_0(\cdot))$, separately. The maximization of $l_1^*(\theta)$ is similar to that used for the ordinary logistic regression. It involves maximizing a function of parameters, although it is discontinuous in ϕ. The $l_2^*(\vartheta, h_0(\cdot))$ is a function of parameter ϑ and the baseline function $h_0(\cdot)$.

Next, we discuss the procedure for maximizing (5.8) with respect to the parameters ϑ and the nonparametric baseline function $h_0(\cdot)$. Let $Y_i(t) = I(T_i \geq t)$ be the at-risk process. Replace $h_0(t)dt$ by $dH_0(t)$ and consider the estimator for $H_0(t)$ to be piecewise constant with jump type discontinuity at observed failure times. The $l_2^*(\vartheta, h_0(\cdot))$ is proportional to

$$l_2^*(\vartheta, h_0(\cdot)) \propto \sum_{i=1}^{n} (1 - v_i^*)[\delta_i \log\{dH_0(t_i)\} + \log\{S(t_i, \vartheta)\}].$$

Taking the derivative of $l_2^*(\vartheta, h_0(\cdot))$ with respect to the jump size $dH_0(t)$ at time t for each fixed value of ϑ and setting it as zero, we have

$$\frac{\partial l_2^*(\vartheta, h_0(\cdot))}{\partial(dH_0(t))}$$
$$= \sum_{i=1}^n I(t_i = t)(1 - v_i^*)\delta_i / dH_0(t) - (1 - v_i^*)Y_i(t)\exp(r_\vartheta(z_i)) = 0.$$

Solving the equation for $dH_0(t)$ yields

$$d\widetilde{H}_0(t, \vartheta) = \frac{\sum_{i=1}^n I(t_i = t)\delta_i}{\sum_{i=1}^n (1 - v_i^*)Y_i(t)\exp(r_\vartheta(z_i))}. \tag{5.9}$$

Thus $\widetilde{H}_0(t, \vartheta)$ can be expressed as

$$\widetilde{H}_0(t, \vartheta) = \int_0^t \frac{\sum_{i=1}^n dN_i(s)}{S^{(0)}(s, \vartheta)}, \tag{5.10}$$

where $N_i(s) = I(T_i \le s, \delta_i = 1)$ and $S^{(0)}(s, \vartheta) = \sum_{i=1}^n (1 - v_i^*)Y_i(s)\exp(r_\vartheta(z_i))$.

Plugging the expression (5.9) for $dH_0(t)$ in (5.8), we obtain the profile likelihood for ϑ:

$$\tilde{l}_2(\vartheta) = \sum_{i=1}^n (1 - v_i^*)[\delta_i \log\{d\widetilde{H}_0(t_i, \vartheta)\} + \delta_i r_\vartheta(z_i) - \widetilde{H}_0(t_i, \vartheta)\exp\{r_\vartheta(z_i)\}].$$

Recall that $\vartheta = (\zeta, \xi^T)^T$ and $\theta = (\phi, \psi^T)^T$, and ζ and ϕ are the change point parameters in the survival model (5.4) and the logistic model (5.5) respectively. Both $l_1^*(\theta)$ and $\tilde{l}_2(\vartheta)$ are not continuous in ζ and ϕ. To maximize $\tilde{l}_2(\vartheta)$ over ϑ, one approach is to perform a partial grids search, Pons (2003). Let Ξ is the bounded compact set for the range of ξ. For fixed ζ, let $\hat{\xi}(\zeta) = \text{argmax}_{\xi \in \Xi} \tilde{l}_2(\zeta, \xi)$, which can be found using Newton–Raphson method. The profile likelihood for ζ is then $l_2(\zeta) = \tilde{l}_2(\zeta, \hat{\xi}(\zeta))$ and its maximizer can be obtained through grid search based on

$$\hat{\zeta} = \inf\{\zeta \in [\zeta_1, \zeta_2] : \max\{l_2(\zeta^-), l_2(\zeta)\} = \sup_{\zeta \in [\zeta_1, \zeta_2]} l_2(\zeta)\},$$

where $l_2(\zeta^-)$ denotes the left-hand limit of $l_2(\zeta)$ at ζ. Let $\hat{\xi} = \hat{\xi}(\hat{\zeta})$ and $\hat{\vartheta} = (\hat{\zeta}, \hat{\xi}^T)^T$. Then $\tilde{l}_2(\vartheta)$ is maximized at $\hat{\vartheta}$. The estimator $\hat{\theta} = (\hat{\phi}, \hat{\psi}^T)^T$ can be derived similarly through the maximization of $l_1^*(\theta)$. The estimator of the cumulative baseline function is given by $\widehat{H}_0(t) = \widetilde{H}_0(t, \hat{\vartheta})$. The baseline survival function is estimated by $\widehat{S}_0(t) = e^{-\widehat{H}_0(t)}$.

The procedure of the EM algorithm can be summarized as follows. The EM algorithm starts with the initial values $\theta^{(0)}$, $\vartheta^{(0)}$ and $H_0^{(0)}(\cdot)$. Let $\theta^{(r)}$, $\vartheta^{(r)}$ and $H_0^{(r)}(\cdot)$ be the estimators at the rth iteration. The E-step in the

$(r+1)$th iteration calculates the conditional expectation v_i^* given by (5.6) at the current values $\theta^{(r)}$, $\vartheta^{(r)}$ and $H_0^{(r)}(\cdot)$. The M-step in the $(r+1)$th iteration maximizes (5.7) and (5.8) separately to obtain $\theta^{(r+1)}$, $\vartheta^{(r+1)}$ and $H_0^{(r+1)}(\cdot)$. The algorithm is iterated until it converges.

In order to obtain good estimates for θ and ϑ, it is important for $\hat{S}_0(t_{(k)})$ to approach zero, where $t_{(k)}$ is the last observed event time. Taylor (1995) suggested imposing the constraint $S_0(t_{(k)}) = 0$ in the special case of the proportional hazard mixture model with $\beta = 0$. The constraint occurs automatically when the weights V^* for censored observations after $t_{(k)}$ are set to one in the E step, essentially classifying them as nonsusceptible. The estimator with this constraint converges faster than the unconstrained MLE's. The unconstrained MLE's can be quite unstable. Heuristically, this constraint implicates existence of a nonsusceptible group and that there is sufficient follow-up beyond the time when most of the events occur.

The standard errors of the estimated parameters are not directly available under the EM algorithm. We use the bootstrap technique, Davison and Hinkley (1997), to calculate the variances of the estimators. The bootstrap method is conceptually simple and easy to implement. For each simulation sample $\{(T_i, \delta_i, Z_i, X_i), i = 1, \ldots, n\}$, a bootstrap sample is obtained by randomly select n quadruples with replacement from this simulation sample. Bootstrap estimates are the estimates of the parameters based on the bootstrap sample using the EM algorithm. The bootstrap standard errors of the estimated parameters are the sample standard errors of a number of, say 500, bootstrap estimates.

Finally, we list a few steps for estimating the smooth parameters, $\hat{\xi}(\zeta)$ and $\hat{\psi}(\phi)$, using Newton–Raphson method. For each ζ, $\hat{\xi}(\zeta)$ can be found using Newton–Raphson method. Let

$$S_k(t, \vartheta) = \sum_{i=1}^{n} (1 - v_i^*) Y_i(t) \exp\left(r_\vartheta(z_i)\right) \left(\frac{\partial r_\vartheta(z_i)}{\partial \xi}\right)^{\otimes k},$$

for $k = 0, 1, 2$, where $a^{\otimes 0} = 1$, $a^{\otimes 1} = a$ and $a^{\otimes 2} = aa^T$ for a vector a. Given V^* at the current values of the parameters and for fixed ζ, it can be shown that

$$\frac{\partial \tilde{l}_2(\zeta, \xi)}{\partial \xi} = \sum_{i=1}^{n} \int_0^{\tau} (1 - v_i^*) \left[\bar{z}_i(\zeta) - \frac{S_1(s, \vartheta)}{S_0(s, \vartheta)}\right] dN_i(s).$$

and

$$\frac{\partial^2 \tilde{l}_2(\zeta, \xi)}{\partial \xi^2} = -\sum_{i=1}^{n} \int_0^{\tau} (1 - v_i^*) \left[\frac{S_2(s, \vartheta)}{S_0(s, \vartheta)} - \left(\frac{S_1(s, \vartheta)}{S_0(s, \vartheta)}\right)^{\otimes 2}\right] dN_i(s).$$

The maximization of $\tilde{l}_2(\zeta, \xi)$ for fixed ζ can be carried out by repeating the following iterations until convergence:

$$\xi^{(r+1)} = \xi^{(r)} - \left(\frac{\partial^2 \tilde{l}_2(\zeta, \xi^{(r)})}{\partial \xi^2}\right)^{-1} \frac{\partial \tilde{l}_2(\zeta, \xi^{(r)})}{\partial \xi}.$$

Similarly, given V^* at the current values of the parameters and for fixed ϕ, it can be shown that

$$\frac{\partial l_1^*(\phi, \psi)}{\partial \psi} = \sum_{i=1}^{n} \bar{x}_i(\phi) \left(v_i^* - \frac{e^{\rho_\theta(x_i)}}{1 + e^{\rho_\theta(x_i)}}\right)$$

and

$$\frac{\partial^2 l_1^*(\phi, \psi)}{\partial \psi^2} = -\sum_{i=1}^{n} \frac{(\bar{x}_i(\phi))^{\otimes 2} e^{\rho_\theta(x_i)}}{(1 + e^{\rho_\theta(x_i)})^2}.$$

The maximization of $l_1^*(\phi, \psi)$ for fixed ϕ can be carried out by repeating the following iterations until convergence:

$$\psi^{(r+1)} = \psi^{(r)} - \left(\frac{\partial^2 l_1^*(\phi, \psi^{(r)})}{\partial \psi^2}\right)^{-1} \frac{\partial l_1^*(\zeta, \psi^{(r)})}{\partial \psi}.$$

5.4. Hypothesis Tests of Change-points

In this section, we present some simple tests of hypotheses to test the existence of a change point in the covariate effects and its location for the latency survival model as well for the cure fraction. For the latency hazard function $h(t|Z)$ given in (5.4), the test of no change-point can be formulated through testing the hypotheses

$$H_{A0}: \ \gamma = 0 \quad \text{versus} \quad H_{A1}: \ \gamma \neq 0.$$

A simple test for H_{A0} considers the test statistic $W_1 = \widehat{\gamma}/\widehat{s.d.}(\hat{\gamma})$, where $\widehat{s.d.}(\hat{\gamma})$ is the estimated standard error of $\hat{\gamma}$ through the bootstrapping. The null hypothesis H_{A0} is rejected at the significant level α if $W_1 < C^1_{\alpha/2}$ or $W_1 > C^1_{1-\alpha/2}$ where $C^1_{\alpha/2}$ and $C^1_{1-\alpha/2}$ are the lower and upper $\alpha/2$ percentiles of the bootstrap copies of W_1 respectively.

Similarly, the null hypothesis of no change-point for the cure fraction model (5.5) is formulated by

$$H_{B0}: \ g = 0 \quad \text{versus} \quad H_{B1}: \ g \neq 0.$$

And a simple test statistic is given by $W_2 = \widehat{g}/\widehat{s.d.}(\hat{g})$, where $\widehat{s.d.}(\hat{g})$ is the estimated standard error of $\hat{g}$ using the bootstrap method. The null hypothesis H_{B0} is rejected at the significant level α if $W_2 < C^2_{\alpha/2}$ or $W_2 > C^2_{1-\alpha/2}$ where $C^2_{\alpha/2}$ and $C^2_{1-\alpha/2}$ are the lower and upper $\alpha/2$ percentiles of the bootstrap copies of W_2 respectively.

Furthermore, in case of $\gamma \neq 0$, the location of a change-point for the latency survival model can be formulated by testing the hypotheses

$$H_{C0} : \zeta = \zeta_0 \quad \text{versus} \quad H_{C1} : \zeta \neq \zeta_0.$$

A simple test for H_{C0} considers the test statistic, $W_3 = (\widehat{\zeta} - \zeta_0)/\widehat{s.d.}(\hat{\zeta})$, where $\widehat{s.d.}(\hat{\zeta})$ is the estimated standard error of $\hat{\zeta}$ using the bootstrap method. The null hypothesis H_{C0} is rejected at the significant level α if $W_3 < C^3_{\alpha/2}$ or $W_3 > C^3_{1-\alpha/2}$ where $C^3_{\alpha/2}$ and $C^3_{1-\alpha/2}$ are the lower and upper $\alpha/2$ percentiles of the bootstrap copies of W_3 respectively.

The locations of change-point for the cure fraction in case of $g \neq 0$ can be formulated by

$$H_{D0} : \phi = \phi_0 \quad \text{versus} \quad H_{D1} : \phi \neq \phi_0.$$

A simple test for H_{D0} considers the test statistic, $W_4 = (\widehat{\phi} - \phi_0)/\widehat{s.d.}(\hat{\phi})$, where $\widehat{s.d.}(\hat{\phi})$ is the estimated standard error of $\hat{\phi}$ using the bootstrap method. The null hypothesis H_{D0} is rejected at the significant level α if $W_4 < C^4_{\alpha/2}$ or $W_4 > C^4_{1-\alpha/2}$ where $C^4_{\alpha/2}$ and $C^4_{1-\alpha/2}$ are the lower and upper $\alpha/2$ percentiles of the bootstrap copies of W_4 respectively.

5.5. Simulation Studies

In this section, we conduct a simulation study to check the performance of the proposed estimation and testing procedures. Let Z_1 be a binary random variable representing the indicator of the treatment group that allocates half of sample size to each group. Let Z_2 be a random variable generated from the uniform distribution on [0,1]. Consider the cure fraction model

$$\text{logit}(c(Z, \theta)) = a_0 + bZ_1 + gZ_1 I(Z_2 \leq \phi),$$

where $a_0 = -1, b = 1, g = 0.5, \phi = 0.5$. This corresponds to a cure rate of 26.89% in the control group and 56.2% for the treatment group. For the treatment group, the cure rate is 62.25% if $Z_2 \leq \phi$ and 50.0% otherwise.

The latency hazard function $h(t|Z)$ is taken to be

$$h(t|Z) = h_0(t) \exp\left(\beta Z_1 + \gamma Z_1 I(Z_2 \leq \zeta)\right), \quad 0 \leq t \leq \tau,$$

where the short-term effect parameters are set at $\beta = -0.1733$, $\gamma = 1$, $\zeta = 0.5$ and $h_0(t) = 1$. We take $\tau = 5$. The censoring times are generated

Table 5.1. The simulation summaries of the proposed estimators based on 500 replications, where SSE is the sample standard error of the estimates, ESE is the average of the estimated standard errors, and CP indicates the coverage probabilities.

sample size	parameter	true value	estimates	bias	ESE	SSE	CP
$n = 200$	β	-0.173	-0.138	0.036	0.294	0.304	95.4
	γ	1.000	1.014	0.014	0.425	0.380	97.8
	ζ	0.500	0.444	-0.059	0.157	0.144	95.8
	a_0	-0.200	-0.237	-0.037	0.241	0.207	96.2
	b	-0.300	-0.281	0.019	0.286	0.222	97.0
	g	-0.400	-0.386	0.014	0.410	0.241	99.6
	ϕ	0.500	0.541	0.041	0.382	0.368	97.4
$n = 300$	β	-0.173	-0.129	0.045	0.246	0.245	94.4
	γ	1.000	0.966	-0.034	0.324	0.290	97.4
	ζ	0.500	0.462	-0.038	0.134	0.111	96.2
	a_0	-0.200	-0.242	-0.042	0.190	0.160	98.2
	b	-0.300	-0.274	0.026	0.202	0.171	95.0
	g	-0.400	-0.361	0.039	0.255	0.138	98.4
	ϕ	0.500	0.528	0.028	0.283	0.245	98.0
$n = 400$	β	-0.173	-0.157	0.016	0.212	0.197	97.6
	γ	1.000	0.994	-0.006	0.267	0.244	98.8
	ζ	0.500	0.476	-0.024	0.107	0.087	94.4
	a_0	-0.200	-0.248	-0.048	0.161	0.131	95.4
	b	-0.300	-0.261	0.039	0.167	0.113	92.8
	g	-0.400	-0.353	0.047	0.185	0.105	96.8
	ϕ	0.500	0.517	0.017	0.244	0.183	97.8

according to an exponential distribution with mean of 3.57. With this choice of the parameter, the expected censoring proportion including those cured is 0.4289 for the control group, 0.6859 for the treatment group.

The performance of the proposed estimation procedure is examined at sample size $n = 200, 300$ and 400. The biases and estimated variances based on 500 replications are shown in Table 5.1, where the sample standard error (SSE) of the estimates and the average of the estimated standard errors (ESE) based on 500 bootstrap samples are listed side by side for comparison. The coverage probabilities (CP) of the corresponding bootstrap confidence intervals at 95% nominal level are also listed. Results from Table 5.1 indicate that the proposed estimators have small biases and their coverage probabilities are mostly close to the nominal level. A few elevated coverage probabilities may be caused by the additional variation due to the constrains imposed for model identifiability; see Sec. 5.7 for further discussions.

Figure 5.1 displays 50 estimated baseline survival functions of model (5.5) for sample size n = 400. The estimated curves (the gray lines) are close to the true baseline survival function (the solid line).

We also conduct further simulation studies to investigate the size and power of proposed tests. Table 5.2 shows the size and power of the test for no change point in the latency hazard function with $\gamma = 0, 0.8$ and 1.0. The results of the size and power of the test for no change point in the cure fraction of are given in Table 5.3 with $g = 0, -0.6$ and -1.2. We also show the size and power of the test for the location of change-point in the latency distribution with the null hypothesis of $\zeta = 0.5$ in Table 5.4. Power studies are conducted under the alternative model with $\zeta = 0.7$ and $\zeta = 0.9$. Similarly, Table 5.5 shows the size and power of the test for the location of change point in the cure model with the null hypothesis of $\phi = 0.5$ and the alternative hypothesis with $\phi = 0.7$ and $\phi = 0.9$. The significant level of $\alpha = 0.05$ is used for all the tests.

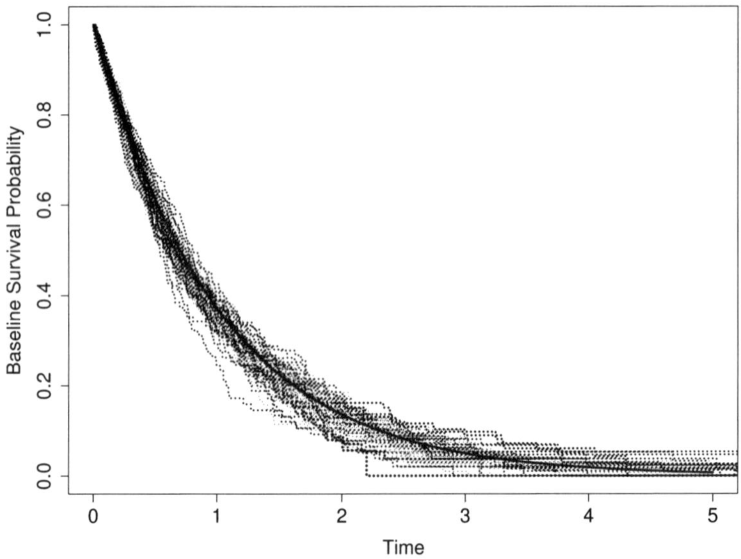

Fig. 5.1. Graphical displays of 50 estimated baseline survival functions for sample size $n = 400$. The solid line is the true baseline survival function. The gray lines are 50 estimated baseline survival function.

Table 5.2. Empirical sizes and powers of the test for $H_{A0} : \gamma = 0$ versus $H_{A1} : \gamma \neq 0$ for the latency survival model at nominal level $\alpha = 0.05$.

	parameter	$n = 200$	$n = 300$	$n = 400$	$n = 500$
size	$\gamma = 0.0$	0.046	0.062	0.051	0.052
power	$\gamma = 0.8$	0.852	0.780	0.856	0.932
power	$\gamma = 1.0$	0.778	0.894	0.962	0.996

Table 5.3. Empirical sizes and powers of the test for $H_{B0} : g = 0$ versus $H_{B1} : g \neq 0$ for the cure probability model at nominal level $\alpha = 0.05$.

	parameter	$n = 200$	$n = 300$	$n = 400$	$n = 500$
size	$g = 0.0$	0.012	0.026	0.037	0.050
power	$g = -0.6$	0.568	0.824	0.838	0.848
power	$g = -1.2$	0.574	0.900	0.964	0.980

Table 5.4. Empirical sizes and powers of the test for $H_{C0} : \zeta = 0.5$ versus $H_{C1} : \zeta \neq 0.5$ for the latency survival model at nominal level $\alpha = 0.05$.

	parameter	$n = 200$	$n = 300$	$n = 400$	$n = 500$
size	$\zeta = 0.5$	0.038	0.038	0.052	0.057
power	$\zeta = 0.7$	0.807	0.920	0.959	0.974
power	$\zeta = 0.9$	0.967	0.970	0.978	0.986

Table 5.5. Empirical sizes and powers of the test for $H_{D0} : \phi = 0.5$ versus $H_{D1} : \phi \neq 0.5$ for the cure probability model at nominal level $\alpha = 0.05$.

	parameter	$n = 200$	$n = 300$	$n = 400$	$n = 500$
size	$\phi = 0.5$	0.010	0.032	0.024	0.046
power	$\phi = 0.7$	0.570	0.766	0.798	0.862
power	$\phi = 0.9$	0.807	0.920	0.960	0.974

Tables 5.2–5.5 show that the observed sizes of the tests are reasonably close to the nominal level 0.05. The power in Table 5.2 is increased as γ increases. As $|g|$ increases, the power in Table 5.3 is also increased. Overall, the powers are consistent and satisfactory.

5.6. Application to the Melanoma Survival Study

We now illustrate the proposed method with an analysis of the failure time data from malignant melanoma survival study Andersen *et al.* (1993). Among these 205 patients with malignant melanoma operated on at Odense University Hospital in the period 1962–1977, 57 patients died from malignant melanoma, 14 patients died from other causes, and the remaining 134 patients were alive on January 1, 1978. The patients who did not die from malignant melanoma are considered as censored. The data set include the time to death from malignant melanoma since operation, the censoring status and censoring time, and the covariates including tumor thickness (mean: 2.92 mm; standard deviation: 0.96 mm), ulceration status (90 present and 115 not present), age (mean: 52 years; standard deviation: 17 years) and sex (79 male and 126 female).

Approximately 65.4% of patients were alive at the end of follow-up. There were about 16.6% (34 out of 205) patients survived beyond the longest observed failure time of 9.15 years. There is a good chance that some of them will not die from malignant melanoma related death. Here we analyze the data with a mixture model using the proposed method. For identifiability of the mixture model, we consider the patients who have survived beyond the longest observed failure time as cured. This is attained by setting the baseline survival function to zero after the last observed failure time.

First we fit the data using the mixture model with two covariates, tumor thickness in mm and sex (1 for males and 0 for females), and with no change points. Consider the latency hazard model

$$h(t|Z) = h_0(t) \exp(\alpha \text{ thickness } + \beta \text{ sex })$$

and the logistic regression model for the cure probability

$$\text{logit}(c(Z,\theta)) = a_0 + a \text{ thickness } + b \text{ sex } .$$

The estimates are given in Table 5.6.

The log relative risk for sex is 0.605 under the latency hazard model. The death was more likely to occur in the male group than in the female group. The estimated odds of cure for male equal $\exp(-0.381) = 0.68$ times the estimated odds for female. The estimated odds of cure were 32% lower for the male group.

Andersen *et al.* (1993) have stratified the patients into three groups using the cut points 2 mm and 5 mm for thickness and shown that the effects of tumor thickness on survival are different for different genders. We

Table 5.6. The summary of the estimates of covariate effects under the mixture model with no change points.

covariate	parameter	estimates	$\widehat{s.d.}$	Wald test-statistics	p-value
			Latency Survival Model		
thickness	α	0.122	0.032	3.813	0.000
sex	β	0.605	0.273	2.216	0.027
			Cure Probability Model		
y-intercept	a_0	1.317	0.382	3.448	0.001
thickness	a	-0.189	0.121	1.562	0.118
sex	b	-0.381	0.435	-0.876	0.381

Table 5.7. The summary of the estimates of covariate effects and change points under the mixture model with change points.

covariate	parameter	estimates	$\widehat{s.d.}$	Wald test-statistics	p-value
			Latency Survival Model		
thickness	α	0.098	0.028	3.500	0.000
sex	β	0.552	0.290	1.903	0.057
sex I(thickness $\leq \zeta$)	γ	-0.532	0.615	-0.865	0.387
threshold	ζ	1.620	0.367	4.414	0.000
			Cure Probability Model		
y-intercept	a_0	1.657	0.247	6.709	0.000
thickness	a	-0.152	0.058	2.621	0.009
sex	b	-0.885	0.394	-2.246	0.025
sex I(thickness $\leq \phi$)	g	1.239	0.576	2.151	0.031
threshold	ϕ	1.290	0.289	4.464	0.000

use tumor thickness for a threshold covariate in both the cure probability model and the latency proportional hazard model. We fit the mixture model with the latency hazard function

$$h(t|Z) = h_0(t) \exp\{\alpha \text{ thickness } + \beta \text{ sex } + \gamma \text{ sex } I(\text{ thickness } \leq \zeta)\}$$

and the cure probability

$$\text{logit}(c(Z, \theta)) = a_0 + a \text{ thickness } + b \text{ sex } + g \text{ sex } I(\text{ thickness } \leq \phi).$$

The estimates of the covariate effects and the change points are given in Table 5.7.

From Table 5.7, the estimated latency hazard function is

$$h(t|Z) = h_0(t) \exp(0.098 \text{ thickness } + 0.020 \text{ sex })$$

for thickness ≤ 1.620 mm and

$$h(t|Z) = h_0(t) \exp(0.098 \text{ thickness } + 0.552 \text{ sex })$$

for thickness > 1.620 mm.

The log relative risk of sex for the thickness less than 1.620 is 0.020. The log relative risk in sex for the thickness higher than 1.620 is 0.552. The risk of death for male increases significantly for those with larger tumor (thickness > 1.620 mm) while it is not much different from female for those with smaller tumor (thickness ≤ 1.620 mm).

The estimated cure probability is

$$\text{logit}(c(Z, \theta)) = 1.657 - 0.152 \text{ thickness } + 0.354 \text{ sex}$$

for thickness ≤ 1.290 mm and

$$\text{logit}(c(Z, \theta)) = 1.657 - 0.152 \text{ thickness } - 0.885 \text{ sex}$$

for thickness > 1.290 mm.

When the thickness is greater than 1.290 mm the estimated odds of cure for male equal 0.41 times the estimated odds for female. The estimated odds of cure were 59% lower for the male group. When the thickness is less than 1.290, the estimated odds of cure for male equal 1.42 times the estimated odds for female. The estimated odds of cure were 42% higher for the male group. The result of higher odds of cure for male group is not consistent with the result of higher risk for male group in the estimated latency hazard function for smaller tumor (thickness $\leq$ 1.290 mm). We found that the 95% confidence interval for $\beta + \gamma$ is $(-1.174, 1.214)$ and the one for $b + g$ is $(-0.765, 1.473)$. This indicates that there is no evidence that there exists significant gender effect for smaller tumor. The inconsistency between the estimated cure probability and the estimated latency hazard function for smaller tumor are due to variation caused by noises. However, the 95% confidence interval for b is $(-1.657, -0.113)$. It shows that there is evidence of significant gender effect for larger tumor (thickness > 1.290 mm) in the cure probability.

5.7. Discussion

This paper studies the failure time mixture models with change points according to thresholds of a covariate. The semiparametric maximum likelihood procedure is proposed and the estimation is carried out through the EM algorithm. Several hypotheses testing procedures are formulated to test the existence of a change point in the covariate effects and its location for the latency survival model as well for the cure probability. The standard errors of the estimated parameters are not directly available under the EM algorithm. The bootstrap method is used to estimate the standard errors of the estimators. We stress that the theoretical properties of the bootstrap is not established, and in this non-standard setting it may be worthwhile to explore the use of the subsampler bootstrap, Politis *et al.* (1999).

We notice a few elevated values for the coverage probability in Table 5.1, which may be caused by the constrains imposed for model identifiability in addition to sample sizes. There is an inhered identifiability issue with the semiparametric mixture model. The cure model is for the ideal situation where a group of subjects are nonsusceptible and the follow-up is infinity. But in practice, all the follow-up periods are limited. The real cure group can't be truly identified. For identifiability we have considered the subjects who survived beyond the longest observed failure time as cured or nonsusceptible. Although it follows the same probability law, the longest observed failure time varies from sample to sample, which causes different model constrains from sample to sample. This additional variability may have caused more variations in the estimation of the parameters and their variances, and thus the coverage probabilities. This issue requires further investigation.

The proposed EM algorithm is used to analyze the failure time data from a malignant melanoma survival study. The failure time of interest is the time to death from malignant melanoma since operation. The time to death without relapse is considered as the censoring time. We simplify the problem in the application by assuming independent censoring. This is in fact a competing risks model where death with relapse and death without relapse can be considered as two competing risks. The censoring caused by death without relapse may be dependent in which case our independent censoring assumption is violated and the proposed estimators may be biased. However, the difficulty is that the independent censoring assumption cannot be tested under the current setting unless additional assumptions are made. It would be interesting to investigate the cure model under the competing risks set-up.

5.8. Acknowledgments

The authors thank Ram Tiwari for helpful discussions. The authors would also like to thank the valuable comments from the referee. This research was partially supported by NSF grants DMS-0604576 and DMS-0905777 and NIH grant R37 AI054165-09.

References

Andersen, P., Borgan, Ø., Keiding, N. and Gill, R. (1993). *Statistical Models Based on Counting Processes* (Springer, New York).

Berkson, J. and Gage, R. (1952). Survival curve for cancer patients following treatment, *Journal of the American Statistical Association* **47**, 501–515.

Davison, A. and Hinkley, D. (1997). *Bootstrap Methods and their Application* (Cambridge University Press, England).

Fang, H., Li, G. and Sun, J. (2005). Maximum likelihood estimation in a semi-paramtric logistic/proportional hazards mixture model, *The Scandinavian Journal of Statistics* **32**, 59–75.

Farewell, V. (1982). The use of mixture models for the analysis of survival data with long-term survivors, *Biometrics* **38**, 1041–1046.

Jespersen, N. (1986). Dichotomizing a continuous covariate in the cox model, Research Report 2, Statistical Research Unit, University of Copenhagen.

Kuk, A. and Chen, C. (1992). A mixture model combining logistic regression with proportional hazards regression, *Biometrika* **79**, 531–541.

Larson, M. and Dinse, G. (1985). A mixture model for the regression analysis of competing risk data, *Applied Statistics* **34**, 201–211.

Lo, Y., Taylor, J., McBride, W. and Withers, H. (1993). The effect of fractionated doses of radiation on mouse spinal cord, *International Journal of Radiation Oncology Biology Physics* **27**, 309–317.

Lu, W. and Ying, Z. (2004). On semiparametric transformation cure models, *Biometrika* **91**, 331–343.

Luo, X. and Boyett, J. (1997). Estimation of a threshold parameter in cox regression, *Communications in Statistics, Theory and Methods* **26**, 2329–2346.

Peng, Y. and Dear, K. (2000). A nonparametric mixture model for cure rate estimation, *Biometrics* **56**, 237–243.

Politis, D., Romano, J. and Wolf, M. (1999). *Subsampling* (Springer, New York).

Pons, O. (2003). Estimation in a cox regression model with a change-point according to a threshold in a covariate, *The Annals of Statistics* **31**, 442–463.

Sy, J. and Taylor, J. (2000). Estimation in a cox proportional hazards cure model, *Biometrics* **56**, 227–237.

Taylor, J. (1995). Semi-parametric estimation in failure time mixture models, *Biometrics* **51**, 899–907.

Tiwari, R., Cronin, K., Davis, W. and Feuer, E. (2005). Bayesian model selection approach to jointpoint regression model, *Journal of the Royal Statistical Society Series B* **54**, 919–939.

Yamaguchi, K. (1992). Acceleration failure-time regression models with a regression model of surviving fraction: An application to the analysis of permanent employment in japan, *Journal of the American Statistical Association* **87**, 284–292.

Yu, B. and Tiwari, R. (2005). Multiple imputation methods for modelling relative survival data, *Statistics in Medicine* **25**, 2946–2955.

Yu, B., Tiwari, R., Cronin, K. and Feuer, E. (2003). Cure fraction estimation from the mixture cure models for grouped survival data, *Statistics in Medicine* **23**, 1733–1747.

Chapter 6

Modeling Survival Data Using the Piecewise Exponential Model with Random Time Grid

Fabio N. Demarqui

Departamento de Estatística
Universidade Federal de Minas Gerais, Brazil
fndemarqui@ufmg.br

Dipak K. Dey

Departament of Statistics
University of Connecticut, USA
dey@stat.uconn.edu

Rosangela H. Loschi and Enrico A. Colosimo

Departamento de Estatística
Universidade Federal de Minas Gerais, Brazil
loschi, enricoc@est.ufmg.br

In this paper we present a fully Bayesian approach to model survival data using the piecewise exponential model (PEM) with random time grid. We assume a joint noninformative improper prior distribution for the time grid and the failure rates of the PEM, and show how the clustering structure of the product partition model can be adapted to accommodate improper prior distributions in the framework of the PEM. Properties of the model are discussed and the use of the proposed methodology is exemplified through the the analysis of a real data set. For comparison purposes, the results obtained are compared with those provided by other methods existing in the literature.

6.1. Introduction

In many practical situations, specially those including medical data, it is often not possible to control a significant part of the sources of variation of an experiment. These uncontrolled sources of variation, when present, may compromise considerably one or more assumptions of a given parametric model assumed to fit the data, which may lead to misleading conclusions.

The Piecewise Exponential Model (PEM) arises as a quite attractive alternative to parametric models for the analysis of time to event data. Although parametric in a strict sense, the PEM can be thought as a non-parametric model as far as it does not have a closed form for its hazard function. This nice characteristic of the PEM allows us to use this model to approximate satisfactorily hazard functions of several shapes. For this reason, the PEM has been widely used to model time to event data in different contexts, such as clinical situations including kidney infection [1], heart transplant data [2], hospital mortality data [3], and cancer studies including leukemia [4], gastric cancer [5], breast cancer [6] (see also [7] for an application to interval-censored data), melanoma [8] and nasopharynx cancer [9], among others. The PEM has also been used in reliability engineering [10, 11], and economics problems [5, 12].

In order to construct the PEM, we need to specify a time grid, say $\tau = \{s_0, s_1, ..., s_J\}$, which divides the time axis into a finite number (J) of intervals. Then, for each interval induced by that time grid, we assume a constant failure rate. Thus, we have a discrete version, in the form of a step function, of the true and unknown hazard function.

The time grid $\tau = \{s_0, s_1, ..., s_J\}$ plays a central role in the goodness of fit of the PEM. It is well know that a time grid having a too large number of intervals might provide unstable estimates for the failure rates, whereas a time grid based on just few intervals might produce a poor approximation for the true survival function. In practice, we desire a time grid which provides a balance between good approximations for both the hazard and survival functions. This issue has been one of the greatest challenges of working with the PEM. Although there exist a vast literature related to the PEM, the time grid $\tau = \{s_0, s_1, ..., s_J\}$ has been arbitrarily chosen in most of those works. According to [13] the selection of the time grid $\tau = \{s_0, s_1, ..., s_J\}$ should be made independent of the data, but they do not provide any procedure to do such. Later, [4] proposes defining the endpoints s_j of the intervals $I_j = (s_{j-1}, s_j]$ as the observed failure times. We shall refer the PEM constructed based on such time grid to as nonparametric PEM. Other heuristic discussions regarding adequate choices for the time grid of the PEM can be found in [5], [1] and [14], to cite few.

The problem of specifying a suitable time grid to fit the PEM can be overcome by assuming that $\tau = \{s_0, s_1, ..., s_J\}$ is itself a random quantity to be estimated using the data information. The first effective effort in such direction is due to [15]. In that work it is assumed that the endpoints of the intervals $I_j = (s_{j-1}, s_j]$ are defined according to a jump process following a martingale structure which is included into the model through the prior distributions. Similar approaches to modeling the time grid of the PEM are considered by [9] and [8]. Independently from those works, [16] also propose an approach which considers a random time grid for the PEM. Based on the usual assumptions for the time grid and assuming independent Gamma prior distributions for the failure rates, they prove that the prior distribution for the time grid has a product form, and use the structure of the Product Partition Model (PPM) (see [17]) to handle the problem. By considering such approach, the use of the reversible jump algorithm to sample from the posteriors is avoided although the dimension of the parametric space is not fixed.

In this paper we extend the approach proposed by [16] by deriving a noninformative joint prior distribution for $(\boldsymbol{\lambda}, \tau)$. Specifically, we assume a discrete uniform prior distribution for the random time grids of the PEM and then, conditionally on those random time grids, we build the joint Jeffreys's prior for the failure rates. Conditions regarding the properties of the joint posterior distribution of (λ, τ) are discussed. Finally, we illustrate the usefulness of the proposed methodology by analyzing the survival time of patients diagnosed with brain cancer in Windham-CT, USA, obtained from the SEER (Surveillance, Epidemiology and End Results) database [18]. For comparison purposes, the results are compared with those provided by other methods existing in the literature.

This paper is organized as follows: the proposed model is introduced in Sec. 6.2. The new methodology is illustrated with the analysis of a real data set in Sec. 6.3. Finally, in Sec. 6.4 some conclusion about the proposed model are draw.

6.2. Model Construction

In this section we introduce a piecewise exponential model which time grid is a random variable. We start our model presentation reviewing the piecewise exponential distribution.

6.2.1. *Piecewise exponential distribution and the likelihood*

Let T be a non-negative random variable. Assume, for instance, that T denotes the time to the event of interest. In order to obtain the probability density function of the PEM we need first to specify a time grid $\tau = \{s_0, s_1, ..., s_J\}$, such that $0 = s_0 < s_1 < s_2 < ... < s_J = \infty$, which induces a partition of the time axis into J intervals I_1, ..., I_J, where $I_j = (s_{j-1}, s_j]$, for $j = 1, ..., J$. Then, we assume a constant failure rate for each interval induced by τ, that is,

$$h(t) = \begin{cases} \lambda_1, \text{if } t \in I_1; \\ \lambda_2, \text{if } t \in I_2; \\ \vdots \\ \lambda_J, \text{if } t \in I_J. \end{cases} \tag{6.1}$$

To conveniently define the cumulative hazard function and the survival and density functions as well, we define

$$t_j = \begin{cases} s_{j-1}, & \text{if } t < s_{j-1}, \\ t, & \text{if } t \in I_j, \\ s_j, & \text{if } t > s_j, \end{cases} \tag{6.2}$$

where $I_j = (s_{j-1}, s_j]$, $j = 1, ..., J$.

The cumulative hazard function of the PEM is computed from (6.1) and (6.2), yielding

$$H(t|\boldsymbol{\lambda}, \tau) = \sum_{j=1}^{J} \lambda_j (t_j - s_{j-1}). \tag{6.3}$$

Consequently, it follows from the identity $S(t) = \exp\{-H(t)\}$ that the survival function of the PEM is given by

$$S(t|\boldsymbol{\lambda}, \tau) = \exp\left\{ -\sum_{j=1}^{J} \lambda_j (t_j - s_{j-1}) \right\}. \tag{6.4}$$

The density function of T is obtained by taking minus the derivative of (6.4). Thus, we say that the random variable T follows a piecewise exponential model with time grid τ and vector parameter $\boldsymbol{\lambda} = (\lambda_1, ..., \lambda_J)'$, denoted by $T \sim PED(\tau, \boldsymbol{\lambda})$, if its probability density function is given by

$$f(t|\boldsymbol{\lambda}, \tau) = \lambda_j \exp\left\{ -\sum_{j=1}^{J} \lambda_j (t_j - s_{j-1}) \right\}, \tag{6.5}$$

for $t \in I_j = (s_{j-1}, s_j]$ and $\lambda_j > 0$, $j = 1, ..., J$.

Let us assume that n individuals were observed independently. Let X_i be the survival time under study for the i-th element, $i = 1., ..., n$. Also, assume that there is a right-censoring scheme working independently of the failure process. Denote by C_i the censored time for the the i-th element, and assume that $C_i \sim G$, for some continuous distribution G defined on the semi-positive real line. Then, the complete information associated to the process is (T_i, δ_i), where $T_i = min\{X_i, C_i\}$ and $\delta_i = I(X_i \leq C_i)$ are, respectively, the observable survival time, and the failure indicator function for the i-th element. Suppose that $(T_i|\tau, \boldsymbol{\lambda}) \sim PEM(\tau, \boldsymbol{\lambda})$, with τ and $\boldsymbol{\lambda}$ as defined before.

In order to properly construct the likelihood function over the J intervals induced by $\tau = \{s_0, s_1, ..., s_J\}$, assume t_{ij} as defined in (6.2). Further, define $\delta_{ij} = \delta_i \nu_j^{(i)}$, where $\nu_j^{(i)}$ is the indicator function assuming value 1 if the survival time of the i-th element falls in the j-th interval, and 0 otherwise. It follows that the contribution of the survival time $t_i \in I_j = (s_{j-1}, s_j]$ for the likelihood function of the PEM is $\lambda_j^{\delta_{ij}} \exp\{-\sum_{j=1}^{J} \lambda_j (t_{ij} - s_{j-1})\}$. Then, given a time grid $\tau = \{s_0, s_1, ..., s_J\}$, we have that the complete likelihood function is given by

$$\begin{aligned} L(D|\lambda, \tau) &= \prod_{i=1}^{n} \left(\prod_{j=1}^{J} \lambda_j^{\delta_{ij}} \exp\{-\lambda_j (t_{ij} - s_{j-1})\} \right) \\ &= \prod_{j=1}^{J} \lambda_j^{\nu_j} \exp\{-\lambda_j \xi_j\}, \end{aligned} \tag{6.6}$$

where the number of failures, $\nu_j = \sum_{l=1}^{n} \delta_{ij}$, and the total time under test, $\xi_j = \sum_{i=1}^{n} (t_{ij} - s_{j-1})$, observed at each interval I_j, are sufficient statistics for λ_j, $j = 1, ..., J$.

It is noticeable that, given τ, the likelihood function given in (6.6) naturally factors into a product of kernels of gamma distributions. As we shall see in the following, along with mild conditions on the joint distribution of the time grid and failure rates, it allows us to use the structure of the PPM proposed by [17] to model the randomness of the time grid of the PEM.

6.2.2. *Priors and the clustering structure*

Following [16], we start our model formulation by imposing some constrains on the set of possible time grids for the PEM. Specifically, we assume the time grid associated to the nonparametric approach as the finest possible time grid for the PEM. We further assume that only time grids whose endpoints are equal to distinct observed failure times are possible. These assumptions guarantee that at least one failure time falls at each interval induced by the random time grid of the PEM.

The randomness of the time grid of the PEM is modeled through the clustering structure of the PPM as follows. Let $\mathscr{F} = \{0, y_1, ..., y_m\}$ be the set formed by the origin and the m distinct ordered observed failure times from a sample of size n. Then, $\mathscr{F}$ defines a partition of time time into disjoint intervals I_j, $j = 1, \ldots, m$, as defined previously. Further, denote by $\mathcal{I} = \{1, ..., m\}$ the set of indexes related to such intervals. Let $\rho = \{i_0, i_1, ..., i_b\}$, $0 = i_0 < i_1 < ... < i_b = m$, be a random partition of $\mathcal{I}$, which divides the m initial intervals into $B = b$ new disjoint intervals. The random variable B denotes the number of clustered intervals related to the random partition ρ. Finally, let $\tau = \tau(\rho) = \{s_0, s_1, ..., s_b\}$ be the time grid induced by the random partition ρ, where

$$s_j = \begin{cases} 0, & \text{if } j = 0, \\ y_{i_j}, & \text{if } j = 1, \ldots, b, \end{cases} \tag{6.7}$$

for $b = 1, ..., m$. Then, it follows that the clustered intervals induced by $\rho = \{i_0, i_1, ..., i_b\}$ are given by

$$I_\rho^{(j)} = \cup_{r=i_{j-1}+1}^{i_j} I_r = (s_{j-1}, s_j], \quad j = 1, ..., b. \tag{6.8}$$

Conditionally on $\rho = \{i_0, i_1, ..., i_b\}$, we assume that

$$h(t) = \lambda_r \equiv \lambda_\rho^{(j)}, \tag{6.9}$$

where $\lambda_\rho^{(j)}$ denotes the common failure rate related to the clustered interval $I_\rho^{(j)}$, for $i_{j-1} < r \leq i_j$, $r = 1, ..., m$ and $j = 1, ..., b$.

In order to complete the model specification, we need to specify the joint prior distribution for $(\boldsymbol{\lambda}_\rho, \rho)$. This is done hierarchically by first specifying a prior distribution for the random partition ρ, and then eliciting prior distributions for $\boldsymbol{\lambda}_\rho$, conditioning on ρ.

Under the assumption that there is no prior information available regarding the time grid, we elicit the Bayes-Laplace prior for $\rho = \{i_0, i_1, ..., i_b\}$, that is,

$$\pi(\rho = \{i_0, i_1, ..., i_b\}) = \frac{1}{2^{m-1}}. \tag{6.10}$$

This prior distribution puts equal mass onto the 2^{m-1} possible partitions associated with the time grids formed by time-points belonging to $\mathscr{F}$, reflecting our lack of information about the time grid. Observe that, if we set $P(\rho = \{i_0, i_1, ..., i_b\}) = 1$ for a particular partition, we return to the usual model that assumes a fixed time grid for the PEM.

Remember that we are defining a random time grid for the PEM in terms of a random partition of the intervals I_j. Furthermore, we are considering that only contiguous intervals are possible, and that the endpoint i_j of each clustered interval $I_\rho^{(j)}$ depends only upon the previous endpoint i_{j-1}. Thus, it follows that the prior distribution (6.10) can be written as the product prior distribution proposed by [17], that is,

$$\pi(\rho = \{i_0, i_1, ..., i_b\}) = \frac{1}{K} \prod_{i=1}^{b} c_{I_\rho^{(j)}}, \tag{6.11}$$

with prior cohesions $c_{I_\rho^{(j)}} = 1$, $\forall\ (i_{j-1}, i_j) \in \mathcal{I}$ and $K = \sum_{\mathscr{C}} \prod_{j=1}^{b} c_{I_\rho^{(j)}} = 2^{m-1}$, where $\mathscr{C}$ denotes the set of all possible partitions of the set $\mathcal{I}$ into b disjoint clustered intervals with endpoints $i_1, ..., i_b$, satisfying the condition $0 = i_0 < i_1 < ... < i_b = m$, for all $b \in \mathcal{I}$.

Conditionally on ρ, we assume the Jeffreys's prior distribution as a joint noninformative prior distribution for $\boldsymbol{\lambda}_\rho$. Let $I(\boldsymbol{\lambda}_\rho)$ denote the Fisher information matrix for $\boldsymbol{\lambda}_\rho$. Then, the joint prior distribution for $\boldsymbol{\lambda}_\rho$, given ρ, is defined as

$$\begin{aligned} \pi(\boldsymbol{\lambda}_\rho|\rho) &\propto |I(\boldsymbol{\lambda}_\rho)|^{\frac{1}{2}} \\ &\propto \prod_{j=1}^{b} \left(\lambda_\rho^{(j)}\right)^{-1}. \end{aligned} \tag{6.12}$$

One attractive characteristic of the Jeffreys's prior is that, regardless of the nature of vector of parameters of model under consideration, this non-informative prior distribution is invariant under one-to-one transformations of those parameter, *i.e.*, the Jeffreys's prior is invariant to parameterizations. In particular, the product form of (6.12) also induces independence among the failure rates into different intervals.

It follows from (6.11) and (6.12) that the (improper) joint prior distribution for $(\boldsymbol{\lambda}_\rho, \rho)$ is given by

$$\begin{aligned} \pi(\boldsymbol{\lambda}_\rho, \rho) &\propto \pi(\boldsymbol{\lambda}_\rho|\rho)\pi(\rho) \\ &\propto \prod_{j=1}^{b} \left(\lambda_\rho^{(j)}\right)^{-1}. \end{aligned} \tag{6.13}$$

Hence, conditionally on $\rho = \{i_0, i_1, ..., i_b\}$, from the product form of (6.6) and (6.13) we have that the joint distribution of the observations has also a product form, given by

$$f(\mathbf{D}|\rho) = \prod_{j=1}^{b} \int \left(\lambda_\rho^{(j)}\right)^{\eta_j - 1} \exp\left\{-\xi_j \lambda_\rho^{(j)}\right\} d\lambda_\rho^{(j)}$$
$$= \prod_{j=1}^{b} \frac{\Gamma(\eta_j)}{\xi_j^{\eta_j}}, \qquad (6.14)$$

where $\Gamma(\cdot)$ denotes the gamma function.

Thus, the joint distribution of the observations given in (6.14) satisfies the product condition required for applying the clustering structure of the PPM to model the randomness of the time grid of the PEM. Bayes inference under noninformative priors for the baseline hazard distribution was also considered by [19], but from a different modeling perspective.

6.2.3. *Posterior distributions and related inference*

Assuming the prior specification in (6.13), the joint posterior distribution of $(\boldsymbol{\lambda}_\rho, \rho)$ becomes

$$\pi(\boldsymbol{\lambda}_\rho, \rho|\mathbf{D}) = L(\mathbf{D}|\boldsymbol{\lambda}_\rho, \rho)\pi(\boldsymbol{\lambda}_\rho|\rho)\pi(\rho)$$
$$\propto \prod_{j=1}^{b} \left(\lambda_\rho^{(j)}\right)^{\nu_j - 1} \exp\left\{-\lambda_\rho^{(j)} \xi_j\right\}. \qquad (6.15)$$

It is noteworthy that the posterior in (6.15) is proper. This is an immediate result of the model formulation we are proposing. Notice that (6.15) corresponds to the product of kernels of gamma distributions, since we can always verify that $\nu_j > 0$ and $\xi_j > 0$, for all j, regardless of the random time grid of the PEM.

The posterior distribution of $\rho = \{i_0, i_1, ..., i_b\}$ is obtained after integrating out (6.15) with respect to $\boldsymbol{\lambda}_\rho$, that is,

$$\pi(\rho|\mathbf{D}) = \int_{\boldsymbol{\lambda}_\rho} L(\mathbf{D}|\boldsymbol{\lambda}_\rho, \rho)\pi(\boldsymbol{\lambda}_\rho|\rho)\pi(\rho) d\boldsymbol{\lambda}_\rho$$
$$= \frac{1}{K^*} \prod_{i=1}^{b} c^*_{I_\rho^{(j)}}, \qquad (6.16)$$

where $c^*_{I_\rho^{(j)}} = (\Gamma(\eta_j)/\xi_j^{\eta_j})$ denotes the posterior cohesion associated with the j-th clustered interval $I_\rho^{(j)}$, and $K^* = \sum_{\mathscr{C}} \prod_{j=1}^{b} c^*_{I_\rho^{(j)}}$.

From the structure of the PPM, we have that the posterior distribution for λ_k, $k = 1, ..., m$, is the following discrete mixture of distributions

$$\pi(\lambda_k|\mathbf{D}) = \sum_{i_{j-1}<k\leq i_j} \pi(\lambda_\rho^{(j)}|\mathbf{D})R(I_\rho^{(j)}|\mathbf{D}), \tag{6.17}$$

where $R(I_\rho^{(j)}|\mathbf{D})$ is named as posterior relevance, and which denotes the probability of each clustered interval $I_\rho^{(j)}$ to belong to the random partition ρ, and $\pi(\lambda_\rho^{(j)}|\mathbf{D})$ denotes the posterior distribution of the common parameter $\lambda_\rho^{(j)}$, $j = 1, ..., b$.

Assuming the squared-error loss function, we have that the product estimate for the failure rate (PEFR) λ_k is given by

$$\hat{\lambda}_k = \sum_{i_{j-1}<k\leq i_j} E(\lambda_\rho^{(j)}|\mathbf{D})R(I_\rho^{(j)}|\mathbf{D}), \tag{6.18}$$

for $j = 1, ..., b$ and $k = 1, ..., m$.

Finally, the posterior survival function for a new element, assumed to be independent of the observed data set, is obtained by averaging the conditional survival function $S(y|\mathbf{D}, \rho)$ over all random partitions $\rho = \{i_0, i_1, ..., i_b\}$, that is,

$$S(y|\mathbf{D}) = \sum_\rho S(y|\mathbf{D}, \rho)\pi(\rho|\mathbf{D}), \tag{6.19}$$

where

$$\begin{aligned} S(y|\mathbf{D}, \rho) &= \int S(y|\lambda_\rho^{(j)}, \rho)\pi(\lambda_\rho^{(j)}|\mathbf{D})d\lambda_\rho^{(j)} \\ &= \left(1 + \frac{y - s_{j-1}}{\gamma_j}\right)^{-\alpha_j} \prod_{r=1}^{j-1} \left(1 + \frac{s_r - s_{r-1}}{\gamma_r}\right)^{-\alpha_r}, \end{aligned} \tag{6.20}$$

for $\lambda_\rho^{(j)} = (\lambda_\rho^{(1)}, ..., \lambda_\rho^{(b)})'$ and $y \in I_\rho^{(j)}$, $j = 1, ..., b$. Throughout this paper we shall refer to (6.20) as the product estimate for the survival function (PESF).

6.3. Numerical Illustration

In this section we use the proposed model to analyze the survival times (in months) of 231 individuals diagnosed with brain cancer in Windham-CT, USA, during 1995 to 2004. This data set was obtained from SEER database. Our interest relies on investigate the performance of our model in estimating both the hazard and survival function. The computational procedures needed to fit the proposed model can be found in [16].

From the 231 patients diagnosed with brain cancer, we observed 134 failures and 97 censored times, resulting in a percentage of failures of the order of 58%. It is also noteworthy that, as the survival times were measured in months, only 32 of the 134 observed failure times correspond to distinct failure times. Thus, under the setup we are proposing, these 32 distinct failure times compose the finest possible time grid for the PEM.

We first examined the performance of the proposed model in estimating the hazard function by comparing the PEFR with the estimates provided by the competing piecewise exponential estimators (PEXE) for the failure rates ([10]), namely, the estimates associated with the nonparametric PEM, obtained via the maximum likelihood approach.

When there are no ties among the observed survival times, the PEXE for the failure rates are not consistent, once we have only one failure time at each interval, regardless of the sample size n. On the other hand, in the presence of ties, asymptotic results could hold only under the assumption that the number of ties in each interval increases without bound, which is not a realistic assumption in practice. Therefore, asymptotic confidence intervals for the failure rates should not be computed in these cases. Furthermore, in the finite sample size scenario, little is known theorically about the PEXE ([10]). These drawbacks of the maximum likelihood approach are easily overcome in the setting we are proposing, since HPD intervals do not rely on the sample size n, and can be obtained straightforwardly.

In Fig. 6.1 we present the PEFR and the PEXE for the failure rates along with the 95% HPD intervals provided by the proposed model. Notice that PEFR and the PEXE for the failure rates are quite similar, and yield essentially the same estimated hazard function for the current data set. From the clinical point of view, the estimated failure rates displayed in Fig. 6.1 indicate a decreasing hazard function, suggesting that the risk of death by brain cancer decreases through time.

The well known Kaplan-Meyer product limiting estimator (KME) arises as a standard estimator for the survival function. In practice, the PEXE yields a smoothing version of the KME for the survival function. Moreover, as shown in [20], the KME and PEXE for the survival function are asymptotically equivalent. Thus, for the sake of simplicity, we compare the PESF with the KME.

Figure 6.2 displays the estimated survival functions provided by the PESF and the KME, along with their corresponding 95% confidence and HPD intervals. The similar performance of the two competing estimators in both point and interval estimation for the survival function is also evident.

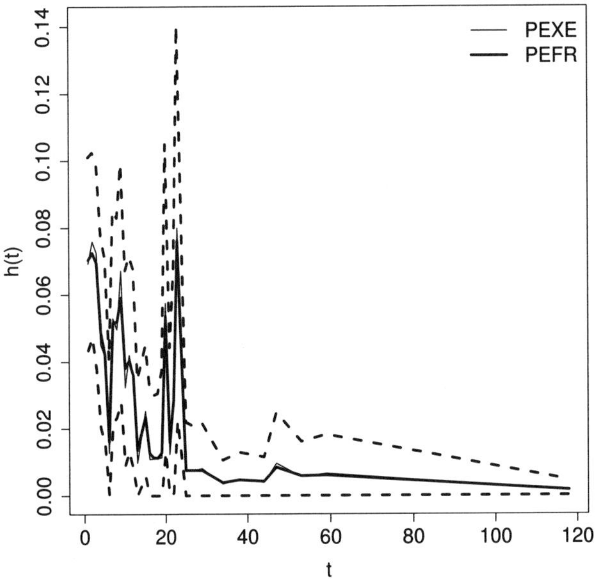

Fig. 6.1. Estimated hazard function (solid lines) for patients with brain cancer and the 95% HPD interval (dashed lines) provided by the PEM with random time grid.

Table 6.1. Posterior summaries for the number of intervals.

Mean	St. Dev.	Min	Max	$P_{2.5}$	$P_{25.0}$	$P_{50.0}$	$P_{75.0}$	$P_{97.5}$
24.779	2.292	15	32	20	23	25	26	29

Another advantage of the proposed model is that it allows us to enrich the analysis by making inferences about the time grid and the number of intervals used to fit the PEM. For instance, the posterior most probable number of intervals is $B = 25$, with probability 0.174. The estimated 95% HPD intervals for B is $[21, 29]$. Other characteristics of the posterior sample of the number of intervals of the PEM are given in Table 6.1 and Fig. 6.3.

6.4. Conclusions

In this paper we presented a fully Bayesian approach to model time to event data using the piecewise exponential model with random time grid.

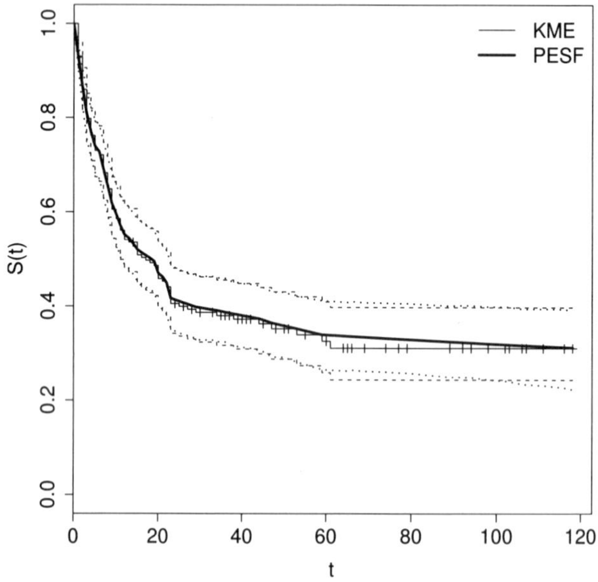

Fig. 6.2. Estimated survival function (solid lines) for patients with brain cancer and their corresponding 95% confidence and HPD intervals (dashed lines) provided by the PEM with random time grid.

Extending the previous work due to [16], we elicited noninformative priors for both the time grid and the failure rates. Fixed the time grid, the Jeffreys' prior for the failure rates is a product distribution favoring the use of the structure of the PPM. It also induces independence among the failure rates in different intervals. Finally, we conducted the analysis of a real data set to illustrate the performance of the proposed model.

The results obtained from the analysis of the brain cancer data set suggest that the estimates provided by the proposed model are comparable with the those yielded by other estimators established in the literature such as the PEXE and the KME. However, interval estimation is straightforward under the framework we are proposing, and it does not rely on asymptotic approximations such as the PEXE and the KME do. Other advantage of the proposed model is that it enriches the analysis by enabling inferences for the time grid of the PEM. Furthermore, the assumption of a joint noninformative prior distribution for the failure rates and the time grids is quite attractive in situations where there is no prior information available.

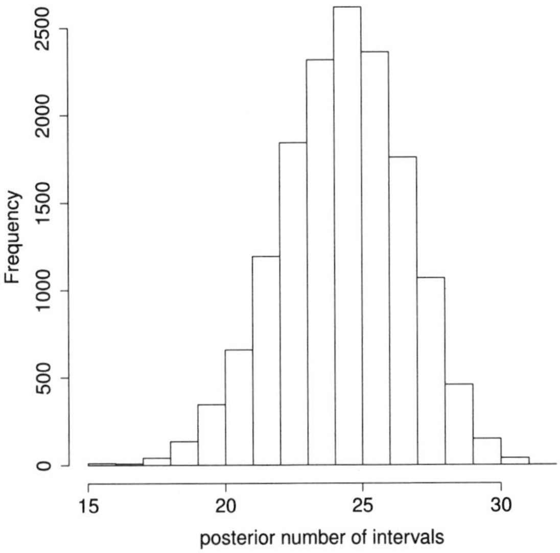

Fig. 6.3. Posterior distribution of the number of intervals.

Acknowledgments

The authors would like to express their gratitude to the editor and the referees for the careful refereeing of the paper. Fabio N. Demarqui's research was sponsored by CAPES(*Coordenação de Aperfeiçoamento de Pessoal de Ensino Superior*) of the Ministry for Science and Technology of Brazil. Rosangela H. Loschi's research has been partially funded by CNPq (*Conselho Nacional de Desenvolvimento Científico e Tecnológico*) of the Ministry for Science and Technology of Brazil, grants 306085/2009-7, 304505/2006-4. Enrico A. Colosimo's research has been partially funded by the CNPq, grants 150472/2008-0 and 306652/2008-0.

References

[1] S. K. Sahu, D. K. Dey, H. Aslanidu, and D. Sinha, A Weibull regression model with gamma frailties for multivariate survival data, *Lifetime Data Analysis.* **3**, 123–137, (1997).

[2] M. Aitkin, N. Laird, and B. Francis, A reanalysis of the Stanford heart transplant data (with discussion), *Journal of the American Statistical Association.* **78**, 264–292, (1983).

[3] D. E. Clark and L. M. Ryan, Concurrent prediction of hospital mortality and length of stay from risk factors on admission, *Health Services Research.* **37**, 631–645, (2002).
[4] N. E. Breslow, Covariance analysis of censored survival data, *Biometrics.* **30**, 89–99, (1974).
[5] D. Gamerman, Dynamic Bayesian models for survival data, *Journal of the Royal Statistical Society: Series C (Applied Statistics).* **40**, 63–79, (1991).
[6] J. G. Ibrahim, M. H. Chen, and D. Sinha, *Bayesian survival analysis.* (Springer-Verlag, New York, 2001).
[7] D. Sinha, M. H. Chen, and S. K. Gosh, Bayesian analysis and model selection for interval-censored survival data., *Biometrics.* **55**, 585–590, (1999).
[8] S. Kim, M. H. Chen, D. K. Dey, and D. Gamerman, Bayesian dynamic models for survival data with a cure fraction, *Lifetime Data Analysis.* **13**, 17–35, (2006).
[9] I. W. McKeague and M. Tighiouart, Bayesian estimators for conditional hazard functions, *Biometrics.* **56**, 1007–1015, (2000).
[10] J. S. Kim and F. Proschan, Piecewise exponential estimator of the survival function, *IEEE Transactions on Reliability.* **40**, 134–139, (1991).
[11] D. Gamerman, Bayes estimation of the piece-wise exponential distribution, *IEEE Transactions on Reliability.* **43**, 128–131, (1994).
[12] L. S. Bastos and D. Gamerman, Dynamic survival models with spatial frailty, *Lifetime Data Analysis.* **12**, 441–460, (2006).
[13] J. D. Kalbfleisch and R. L. Prentice, Marginal likelihoods based on Cox's regression and life models, *Biometrika.* **60**, 267–278, (1973).
[14] Z. Qiou, N. Ravishanker, and D. K. Dey, Multivariate survival analysis with positive stable frailties, *Biometrics.* **55**, 637–644, (1999).
[15] E. Arjas and D. Gasbarra, Nonparametric Bayesian inference from right censored survival data, using the Gibbs sampler, *Statistica Sinica.* **4**, 505–524, (1994).
[16] F. N. Demarqui, R. H. Loschi, and E. A. Colosimo, Estimating the grid of time-points for the piecewise exponential model, *Lifetime Data Analysis.* **14**, 333–356, (2008).
[17] D. Barry and J. A. Hartigan, Product partition models for change point problems, *The Annals of Statistics.* **20**, 260–279, (1992).
[18] Surveillance and Epidemiology and End Results (SEER). *Surveillance Research Program, National Cancer Institute SEER*Stat software (www.seer.cancer.gov/seerstat) version 6.5.1. Database: Incidence - SEER 17 Regs Limited-Use Nov 2006 Sub (1973-2004 varying).*
[19] D. Sinha, J. G. Ibrahim, and M. H. Chen, A bayesian justification of cox's partial likelihood, *Biometrika.* **90**, 629–641, (2003).
[20] J. Kitchin, N. A. Langberg, and F. Proschan. A new method for estimating life distributions from incomplete data. Technical report, Department of Statistics, Florida State University, (1980).

Chapter 7

Proportional Rate Models for Recurrent Time Event Data Under Dependent Censoring: A Comparative Study

Leila D. A. F. Amorim

Federal University of Bahia, Department of Statistics
Av.Adhemar de Barros s/n, Salvador-BA, 40170-115 Brazil
leiladen@ufba.br

Jianwen Cai and Donglin Zeng

University of North Carolina at Chapel Hill
Department of Biostatistics
3101 McGavran-Greenberg, CB 7420, Chapel Hill-NC, 27599 USA

In many biomedical studies, each patient can experience events that can occur repeatedly over time. Modeling techniques have been developed to analyze recurrent time-to-event data assuming independent censoring, i.e., the censoring process is unrelated to the event failure process conditional on the covariates. However, this assumption may not hold in general. This would happen if, for instance, the subjects who are at higher risks of recurrent events tend to withdraw from the study earlier. Another form of dependent censoring could occur due to terminal events, such as death. Two methods were recently proposed to account for dependent censoring with the marginal rate models for analysis of recurrent event data. This paper overviews these approaches and critically compare them through extensive simulation studies. The simulation results showed that the approaches are effective for handling dependent censoring when the source of informative censoring is correctly specified. Further research is still needed for more complex situations.

7.1. Introduction

Many studies involve the occurrence of recurrent events and much attention has been given for the development of modelling techniques that take into account the dependence structure of multiple event data in the last few decades.[1–5] Particularly the marginal hazards models, such as WLW[3] and LWA,[6] and conditional hazards models[1,2] have been used more frequently, since they have been incorporated into many statistical software packages. Methods for estimation of means/rates of recurrent events have also been developed in recent years. Pepe and Cai[4] studied methods to display and estimate rate functions for the analysis of multiple time-to-event data. Later, Lawless and Nadeau[7] presented robust methods for nonparametric estimation of the cumulative mean/rate function as well as proposed a marginal model for the mean/rate estimation. More recently, Lin *et al.*[5] provided rigorous justification for the use of semiparametric regression for the mean and rate functions in the analysis of recurrent time-to-event data.

Almost all of the approaches assume independent censoring, i.e., the censoring process is unrelated to the event failure process. However, the recurrent event times in a typical medical study are often subject to both independent and dependent censoring. The dependent or informative censoring arises when the censoring time depends on the observed or unobserved recurrent event times. This would happen if, for instance, the subjects who are at higher risk of recurrent events tend to withdraw from the study earlier. In such scenario the subjects can potentially experience further events after the censoring time, but these events will not be observed by the investigators. Another form of dependent censoring would occur because of terminal events, such as death. For chronic and complex diseases, it has been observed that some patients who are diagnosed with a certain disease die from the related cause.[8] In these cases, assumption of independence between the main outcome and censoring causes seems implausible. When the informative censoring exists, the *ad hoc* estimation procedures from the observed data will result in inconsistent estimators.[9]

Analysis methods have been proposed to accommodate informative censoring for various models.[8,10–16] In this paper we focus on proportional rate model and compare two recently proposed methods: a modeling strategy that allows the observation times to be correlated with the event process through their connections with an unobserved frailty[17] and an approach that requires modeling of the censoring mechanism.[9]

Wang, Qin and Chiang (WQC)[17] proposed to model the occurrence of recurrent events by a subject-specific nonstationary Poisson process via a latent variable, allowing the censoring mechanism be possibly informative. Their approach adopted a multiplicative intensity function as the underlying model. WQC showed that, under regularity conditions, the resulting estimators are consistent. On the other hand, Miloslavsky, Keles, van der Laan and Butler (MKLB)[9] proposed inverse probability of censoring weighted (IPCW) estimators for the regression parameters in the Andersen-Gill model. They also extended their approach for the proportional rate model. MKLB proposed estimating equations based on IPCW mapping[18] and showed that their estimators are also consistent.

Although the finite sample properties of the estimators proposed by WQC and MKLB had been studied in limited setups for each of those methods and the advantages of MKLB approach over the corresponding method assuming independent censoring have been established, there has been no systematic studies to compare these two fairly recent methods. Since each method investigates dependent censoring through distinct mechanism, it would be of interest to evaluate the relative performance of the two estimators under various scenarios for dependent censoring. Due to the complexity of the data structure and the estimation approaches, an analytic comparison in general seems very difficult, if not impossible, to obtain. Hence, we conducted simulation studies to compare these two approaches for the estimation of covariate effects from recurrent time-to-event data in the presence of dependent censoring. The major contribution of this work is to provide a better understanding of the mechanisms through which the dependent censoring is modeled by each approach and to discuss the advantages/limitations of each of them.

The article is organized as follows. We present an overview of the proportional rate model for recurrent event data as well as the WQC and MKLB approaches in Sec. 7.2. In Sec. 7.3, we outline the simulation framework and in Sec. 7.4 we present results from the simulation studies. In Sec. 7.5 we provide an illustration with recurrent diarrheal data from a clinical trial conducted in Brazil. The conclusions appear in Sec. 7.6.

7.2. Modeling Recurrent Event Data

7.2.1. *Notation and proportional rate model*

Suppose that the data consist of n subjects. Let $X_{i,k} = T_{i,k} \wedge C_i$ be the observed time for the ith subject with respect to the kth event, where C_i denotes censoring time and $T_{i,1}, T_{i,2}, ..., T_{i,k}$ represent the event times. These event times are called total times and represent, for instance, the time since randomization to treatment until the occurrence of the kth event for the ith subject. Using counting process notation, let $N_i(t) = \int_0^t dN_i(s)$ be the number of events in $[0, t]$ for subject i, where $dN_i(s)$ denotes the number of events in the small time interval $[s, s + ds]$. Suppose that the occurrence rate of recurrent events in the interval $[0, \tau]$ is of interest, where τ refers to the end time of the study. Let $Y(t) = I(C \geq t)$ be the at-risk indicator.

An interesting approach for analyzing recurrent event data is to model marginal means and rates functions. As Cai and Schaubel[19] pointed out, one of the main appeals of using such approach is that the mean number of events is an intuitive quantity and is usually of direct interest to the investigators. Using counting process notation, the rate function is defined as $d\mu_i(s) = E[dN_i(s)|\mathbf{Z}_i(s)]$. In order to explore the association between the covariates $\mathbf{Z}(t)$ and $N(.)$, consider the marginal proportional rate model

$$E\{dN(t)|\mathbf{Z}(t)\} = Y(t)d\mu_0(t)\exp\{\beta^{'}\mathbf{Z}(t)\}$$

where $dN(t)$ denotes the number of events in the time interval $[t, t+dt]$, β is a $p \times 1$ vector of fixed regression parameters and $d\mu_0(t)$ is an unspecified baseline rate function.

We summarize and compare two approaches for the marginal rate model in the presence of dependent censoring. These approaches are presented in the following subsections.

7.2.2. *WQC method*

WQC only considers time-independent covariates and proposes to model the occurrence of recurrent events by a subject-specific nonstationary Poisson process via a latent variable, allowing the censoring mechanism be possibly informative. The distribution of both the censoring and latent variables are treated as nuisance parameters. They assume that there exists a non-negative valued latent variable u so that, conditioning on $(\mathbf{Z}, u)$, $N(t)$ is a nonstationary Poisson process with intensity function $u\lambda_0(t)\exp\{\beta^{'}\mathbf{Z}\}$, where the baseline intensity function $\lambda_0(t)$ is an unspecified continuous function. The latent variable satisfies $E[u|\mathbf{Z}] = 1$. The marginal proportional rate model defined in Sec. 7.2 holds under this assumption. The second assumption is that conditioning on $(\mathbf{Z}, u)$, $N(.)$ is independent of C.

Because of the Poisson property, $N_i(C_i \wedge \tau)$ is a sufficient statistic for each individual frailty u_i. Thus, using a conditional likelihood approach by conditioning on this sufficient statistic, WQC constructed a class of estimating equations independent of frailties for $(\alpha, \beta^{'})$ as

$$n^{-1}\sum_{i=1}^{n} w_i(1, \mathbf{Z}_i)^{'}(N_i(C_i \wedge \tau)\hat{F}^{-1}(C_i) - e^{\alpha+\beta^{'}\mathbf{Z}_i}) = 0,$$

where w_i is a weight function depending on $\mathbf{Z}_i$ and $(\beta, \hat{F})$, $\alpha = \log\Lambda_0(\tau) = \log\int_0^{\tau}\lambda_0(s)ds$, and $\hat{F}(t) = \prod_{s_{(l)}>t}(1 - d_{(l)}/N_{(l)})$, where $\{s_{(l)}\}$ are the ordered and distinct values of the recurrent event times, $\{d_{(l)}\}$ is the number of recurrent events occurring at $s_{(l)}$, and $\{N_{(l)}\}$ is the total number of recurrent events less than or equal to $s_{(l)}$ for the subjects whose C_i satisfies $s_{(l)} \leq C_i$.

Wang, Qin and Chiang[17] showed that the solution of this class of estimating equations has the property that $\sqrt{n}(\hat{\beta} - \beta)$ converges weakly to a multivariate normal distribution with mean zero and covariance matrix which can be consistently estimated if the marginal rate model is correctly specified.

7.2.3. *MKLB method*

The MKLB method allows time-dependent covariates and uses inverse probability of censoring weighted(IPCW)[18] estimators for the regression parameters in the proportional rates model. The main interest is modeling the rate of recurrent events conditional on covariates. The proposed estimator requires modeling of the censoring mechanism, producing the weights to be used through IPCW estimating functions. The authors considered the *coarsening at random* (CAR) assumption, which states that given $\mathbf{Z}(t)$, the censoring event depends only on the observed part of the data.

Let $\mathbf{V}$ stands for everything that can be observed on a randomly selected subject in the interval $(0, \tau]$ and let $\bar{G}(t|\mathbf{V}) = P(C > t|\mathbf{V})$, i.e, the survival function for the censoring time given $\mathbf{V}$. This methodology requires a model for the censoring mechanism, which can be given by

$$\lambda_c(t|\mathbf{V}) = Y_c(t)\lambda_{0,c}(t)\exp\{\beta_c^{'}\xi_c(t)\},$$

where $Y_c(t)$ is the at-risk indicator for censoring, $\lambda_{0,c}(t)$ is an unspecified baseline hazard function and $\xi_c(t)$ is a known function of observed $\mathbf{V}$ up to time t. The assumptions of this approach implies that $N(.)\perp C|\mathbf{V}$.

A choice of IPCW estimating function for β is then given by

$$\sum_{i=1}^{n}\int_0^{\tau}\left(\mathbf{Z}_i(t) - \frac{\sum_{j=1}^{n}[\hat{G}(t|\mathbf{V}_j)^{-1}\mathbf{Z}_j(t)\hat{G}\{t|\mathbf{Z}_j(t)\}Y_j(t)\exp\{\beta'\mathbf{Z}_j(t)\}]}{\sum_{j=1}^{n}[\hat{G}(t|\mathbf{V}_j)^{-1}\hat{G}\{t|\mathbf{Z}_j(t)\}Y_j(t)\exp\{\beta'\mathbf{Z}_j(t)\}]}\right)$$

$$\times\frac{\hat{G}\{t|\mathbf{Z}_i(t)\}Y_i(t)dN_i(t)}{\hat{G}(t|\mathbf{V}_i)} = 0,$$

where $\hat{G}$ is estimated by fitting the proportional hazards model for the censoring process. The left-hand side is an estimating function because at the true parameters, the left-hand side approximates

$$E\left[\int_0^{\tau}\left(\mathbf{Z}_i(t)-\frac{E[G(t|\mathbf{V}_j)^{-1}\mathbf{Z}_j(t)G\{t|\mathbf{Z}_j(t)\}Y_j(t)\exp\{\beta'\mathbf{Z}_j(t)\}]}{E[G(t|\mathbf{V}_j)^{-1}G\{t|\mathbf{Z}_j(t)\}Y_j(t)\exp\{\beta'\mathbf{Z}_j(t)\}]}\right)\right.$$

$$\left.\times\frac{G\{t|\mathbf{Z}_i(t)\}Y_i(t)dN_i(t)}{G(t|\mathbf{V}_i)}\right]$$

$$=E\left[\int_0^{\tau}\left(\mathbf{Z}_i(t)-\frac{E[\mathbf{Z}_j(t)G\{t|\mathbf{Z}_j(t)\}\exp\{\beta'\mathbf{Z}_j(t)\}]}{E[G\{t|\mathbf{Z}_j(t)\}\exp\{\beta'\mathbf{Z}_j(t)\}]}\right)\right.$$

$$\left.\times G\{t|\mathbf{Z}_i(t)\}e^{\beta'\mathbf{Z}_i(t)}d\mu_0(t)\right]=0.$$

Furthermore, Miloslavsky and colleagues[9] showed that the obtained estimator for β is asymptotically consistent and normal. They also pointed out that the proposed estimating function is biased if the model for the censoring mechanism does not include all relevant covariates.

7.2.4. *WQC vs MKLB method*

Besides the theoretical derivation of their estimation approaches, WQC and MKLB have also conducted simulation studies to evaluate the finite sample properties of their estimators. WQC used 500 samples of size 400 to estimate the effect of a Bernoulli variable on the occurrence of recurrent events. They computed bias, standard errors and 95% bootstrap confidence intervals for their estimator, concluding for its validity. MKLB, on the other hand, considered 2,000 samples of size 200, with fixed levels of censoring (10%, 20%, 50%), to estimate the parameter of interest and to compare the proposed method with the corresponding method that assumes independent censoring. They concluded that their weighted estimator outperformed the 'naive' unweighted estimator. With an example dataset, WQC and MKLB compared the results of the application of their method with the corresponding 'naive' method.

Both approaches characterizes the rate of the counting process under the marginal rate model, allowing arbitrary dependence structures among recurrent events. However, the two approaches differ in their ways to adjusting for dependent censoring. WQC introduces a latent variable to handle the informative or dependent censoring, while MKLB deals with the problem of informative censoring by modelling the censoring time using observable covariate information.

Computer programs for general purpose are not available to model recurrent time-to-event data using these approaches. A library for software R (version 1.9.1.) is available upon request to the WQC authors for fitting the WQC model. The MKLB approach can be implemented by adapting standard routines available in statistical software packages. In the next section, we describe the set up for the simulation studies designed to compare these two methods.

7.3. Simulation Framework

Consider a clinical trial where each subject is randomly assigned to a treatment arm of interest. Let $N(t) = \sum_k I(T_k \leq t)$ be the recurrent events counting process of interest while Z denotes the treatment variable and W a baseline covariate. Suppose that the goal of this study is to estimate the effect of the treatment. Assuming a proportional rate model, the parameter of interest is the regression coefficient β in the model

$$d\mu(t|Z) = d\mu_0(t)\exp(\beta Z).$$

We generated T following distribution with the intensity function $\lambda_T(t|Z,W,u) = u\lambda_{0,T}(t)\exp(\beta_0 Z + \gamma_0 W)$. Conditioning on (Z,W,u), the censoring time C is generated from $\lambda_C(t|Z,W,u) = u\lambda_{0,C}(t)\exp(\tilde{\beta}_0 Z + \tilde{\gamma}_0 W)$. We generated independent Z from Bernoulli distribution (Z $\sim$ Bernoulli (0.5)) and W from various distributions including Uniform, Bernoulli and Normal (W $\sim$ Bernoulli (0.5), W $\sim$ Uniform (0,1), W $\sim$ Normal (0, 1), truncated at (-1,1)). The failure indicator Δ_i is defined as $\Delta_i = I(T_i \leq C_i)$. We generated independent u_i $(i = 1, ..., n)$ from gamma distribution, with mean 1 and variance σ^2. Large values of σ^2 reflect greater heterogeneity between subjects and a stronger association between events from the same subject. Large values of σ^2 also indicates stronger association between the censoring time and recurrent events. We considered a constant baseline hazard function for all configurations. The values for $\lambda_{o,T}(t)$ and $\lambda_{o,C}(t)$ varied for the different simulation set-ups, such that an average of approximately 3.6 events per subject was observed.

The focus of this simulation study is on the performance of WQC and MKLB approaches in the estimation of β for various combinations of σ^2 (σ^2=0, 1, 4), sample size ($n = 200, 500$), treatment effect ($\beta_0 = -1.2, \tilde{\beta}_0 = 1$) and baseline covariate effect ($\gamma_0 = 0, 8; \tilde{\gamma}_0 = 0, 5$). The simulation set-up only meets the assumptions of WQC when $\gamma_o = 0$, $\tilde{\gamma}_o = 0$ and $\sigma^2 > 0$, while it meets the assumptions of MKLB when $\gamma_o \neq 0$, $\tilde{\gamma}_o \neq 0$ and

$\sigma^2 = 0$. We generated 1,000 samples for each configuration of simulation parameters. We use the sample bias and sample variances to measure, respectively, the accuracy and efficiency of regression parameter estimates from the two approaches. The mean squared errors are also computed using the sample bias and variances. The sample bias and sample variance are defined, respectively, as the average bias and the variance from the 1,000 random samples.

Note that the parameter of interest β, the covariate effect in the working proportional rate model, may not be the same as the β_0 which is used to generate the data through a conditional model. According to Miloslavsky *et al.*,[9] we obtain a good estimate of the true parameter β by generating a large number of observations (e.g. N=100,000) from the data-generating distribution and fitting the marginal model using the full data(T,Z,W). This estimate corresponds to the minimizer of the Kullback-Leibler projection of the true data-generating distribution of the model of interest. β is the covariate effects in the working model (proportional rate models) and it may be different from the covariate effect in the true model for data generation, especially when the true survival function depends on other covariates or latent frailty.

For the results summarized in Table 7.1, the parameters were set as follows: $\beta_0 = -1.2$, $\gamma_0 = 0$ and 8, $\tilde{\beta}_0 = 1$, $\tilde{\gamma}_0 = 0$ and 5, $\sigma^2 = 0$, 1 and 4, $\tau = 4$ months, $n = 200$ and 500, $Z \sim$ Bernoulli(0.5) and $W \sim$ Uniform(3,4).

The simulation study was conducted in R v.1.9.1 software. The standard coxph( ) routine in R, providing the appropriate weights, was used for the MKLB approach. We used the crf R library, developed by WQC, in order to fit their model.

7.4. Simulation Results

Under the scenario of independent censoring ($\sigma^2 = 0, \gamma_0 = 0, \tilde{\gamma}_0 = 0$), the use of both methods leads to approximately unbiased estimates. In addition, the results obtained from the MKLB method are similar to those for proportional rate model without dependent censoring.[5] When censoring is dependent through covariates ($\sigma^2 = 0, \gamma_0 \neq 0, \tilde{\gamma}_0 \neq 0$), modeling the censoring mechanism with proper covariate information using IPCW estimators in the MKLB method is approximately unbiased while the estimate from the proportional rate model ignoring dependent censoring is biased (bias=0.1912 for $n = 500$). In this set up, the MKLB estimator is less biased and more precise than the WQC estimator (Table 7.1). However WQC estimator is also approximately unbiased.

Table 7.1. Simulation results on bias, empirical standard error and mean squared error of the three estimators for the regression parameter β based on 1,000 replicates (with $\beta_0 = -1.2$, $\tilde{\beta}_0 = 1$) and $W \sim$ Uniform(3,4).

			MKLB Method			WQC Method			Indep.cens. Method		
σ^2	$(\gamma_0, \tilde{\gamma}_0)$	n	Bias	ESE	MSE	Bias	ESE	MSE	Bias	ESE	MSE
0	(0,0)	500	0.0038	0.0642	0.0041	0.0064	0.1326	0.0176	0.0031	0.0641	0.0041
		200	0.0064	0.1040	0.0109	0.0165	0.1969	0.0390	0.0091	0.1037	0.0108
	(8,5)	500	0.0046	0.1733	0.0301	0.0079	0.1908	0.0365	0.1912	0.1659	0.0641
		200	0.0580	0.2819	0.0828	0.0748	0.2977	0.0942	0.1388	0.2689	0.0916
1	(0,0)	500	0.2347	0.1103	0.0673	0.0215	0.1654	0.0278	0.2339	0.1101	0.0668
		200	0.2357	0.1786	0.0875	0.0422	0.2256	0.0527	0.2345	0.1776	0.0865
	(8,5)	500	0.1052	0.2398	0.0686	0.0641	0.2621	0.0728	0.1730	0.2288	0.0896
		200	0.0498	0.3317	0.1125	0.1344	0.3595	0.1473	0.1356	0.3182	0.1196
4	(0,0)	500	0.4613	0.1585	0.2379	0.0811	0.1942	0.0443	0.4612	0.1583	0.2378
		200	0.3554	0.2460	0.1868	0.1683	0.2509	0.0913	0.3559	0.2459	0.1871
	(8,5)	500	0.2362	0.2958	0.1433	0.1743	0.3312	0.1401	0.2692	0.2843	0.1533
		200	0.1600	0.4334	0.2134	0.2776	0.5020	0.3291	0.1972	0.4155	0.2115

Table 7.2. Simulation results on bias, empirical standard errors and mean squared errors of the three estimators for the regression parameter β based on 1,000 replicates (with $\beta_0 = -1.2$, $\tilde{\beta}_0 = 1$) and $W \sim$ Bernoulli(0.5).

			MKLB Method			WQC Method			Indep.cens. Method		
σ^2	$(\gamma_0, \tilde{\gamma}_0)$	n	Bias	ESE	MSE	Bias	ESE	MSE	Bias	ESE	MSE
0	(0,0)	500	0.0027	0.0640	0.0041	0.0104	0.1160	0.0136	0.0017	0.0639	0.0041
		200	0.0011	0.0979	0.0096	0.0033	0.1698	0.0288	0.0021	0.0972	0.0094
	(8,5)	500	0.0398	0.1152	0.0149	0.0194	0.1586	0.0255	0.2260	0.1306	0.0681
		200	0.0361	0.1808	0.0340	0.0203	0.2290	0.0529	0.2211	0.2053	0.0911
1	(0,0)	500	0.0398	0.1106	0.0138	0.0057	0.1248	0.0156	0.0397	0.1105	0.0138
		200	0.0496	0.1628	0.0280	0.0001	0.1888	0.0356	0.0496	0.1628	0.0290
	(8,5)	500	0.1087	0.1709	0.0410	0.0371	0.1876	0.0366	0.1454	0.1701	0.0501
		200	0.0947	0.2590	0.0760	0.0342	0.2767	0.0777	0.1303	0.2600	0.0846
4	(0,0)	500	0.1117	0.1926	0.0496	0.0047	0.2119	0.0449	0.1119	0.1924	0.0495
		200	0.0814	0.2809	0.0856	0.0299	0.2996	0.0907	0.0819	0.2808	0.0856
	(8,5)	500	0.1334	0.2720	0.0918	0.0252	0.2881	0.0837	0.1459	0.2699	0.0941
		200	0.1055	0.3910	0.1641	0.0042	0.4084	0.1668	0.1168	0.3884	0.1645

The WQC method uses a latent variable to characterize the heterogeneity among subjects and assume that the latent variable u_i is the only factor that explains the heterogeneity from different subjects (besides Z_i). It is evident from Table 7.1 that in this case ($\sigma^2 = 4, \gamma_0 = 0, \tilde{\gamma}_0 = 0$) the estimator from WQC method (bias=0.0811 for $n = 500$) are much less biased than that obtained from MKLB method (bias=0.4613 for $n = 500$). Similar patterns are observed when the variability of the latent variable is reduced ($\sigma^2 = 1, \gamma_0 = 0, \tilde{\gamma}_0 = 0$). The bias for the MKLB method is smaller for smaller σ^2. Same is observed for WQC method. At the same time, when considering that the censoring mechanism depends on both the unmeasured factors (u) and on the observed baseline covariates (W), both estimators become biased (bias=0.2362 and 0.1743 for MKLB and WQC methods, respectively, for $n = 500$). The empirical standard errors (ESE) of MKLB method are consistently smaller than those obtained using WQC method. We also compared these methods through the use of the mean squared error (MSE) as the comparison criterion. Note that for the results presented in Table 7.1, the smallest MSE is generally observed for the corresponding method with the smallest bias.

When $W \sim$ Bernoulli(0.5) and $W \sim$ Normal(0,1) (Tables 7.2 and 7.3, respectively), the magnitude of the bias is generally reduced for all scenarios compared to the results presented in Table 7.1. Regardless of the distribution associated with W, the smallest bias and ESE for all methods are generally observed under the scenario of independent censoring ($\sigma^2 = 0, \gamma = 0, \tilde{\gamma} = 0$) and the MKLB method has the smallest MSE when censoring is dependent through covariate W ($\sigma^2 = 0, \gamma \neq 0, \tilde{\gamma} \neq 0$). The WQC method outperforms the MKLB method in terms of bias when the dependence between event and censoring times is introduced only through a latent variable ($\sigma^2 = 1$ or 4, $\gamma_0 = 0, \tilde{\gamma}_0 = 0$). Due to the general reduced magnitude of the bias when $W \sim$ Bernoulli(0.5) and $W \sim$ Normal(0,1), the values of MSE in Tables 7.2 and Table 7.3 are strongly influenced by ESE, which are consistently smaller for MKLB method. Hence, in those scenarios the MSE is mostly driven by the efficiency instead of by the bias of the estimates.

The worst performance is observed when the censoring mechanism depends on both the observed baseline covariate (W) and on unmeasured factors (u) ($\sigma^2 = 1$ or 4, $\gamma \neq 0, \tilde{\gamma} \neq 0$). For all parameter configurations considered in these simulation studies, the sampling variances increase as the sample size decreases from 500 to 200.

Table 7.3. Simulation results on bias, empirical standard errors and mean squared errors of the three estimators for the regression parameter β based on 1,000 replicates (with $\beta_0 = -1.2$, $\tilde{\beta}_0 = 1$) and $W \sim$ Normal(0,1) truncated at (-1,1).

			MKLB Method			WQC Method			Indep.cens. Method		
σ^2	$(\gamma_0, \tilde{\gamma}_0)$	n	Bias	ESE	MSE	Bias	ESE	MSE	Bias	ESE	MSE
0	(0,0)	500	0.0049	0.0634	0.0040	0.0088	0.1408	0.0199	0.0060	0.0633	0.0040
		200	0.0048	0.1008	0.0102	0.0011	0.1834	0.0336	0.0065	0.1005	0.0102
	(8,5)	500	0.0046	0.1694	0.0287	0.0180	0.1757	0.0312	0.1041	0.1780	0.0425
		200	0.0373	0.2737	0.0763	0.0255	0.2890	0.0842	0.0542	0.2794	0.0810
1	(0,0)	500	0.0502	0.1059	0.0137	0.0046	0.1275	0.0163	0.0502	0.1058	0.0137
		200	0.0355	0.1634	0.0280	0.0098	0.1809	0.0328	0.0355	0.1633	0.0279
	(8,5)	500	0.0233	0.4797	0.2307	0.1398	0.4817	0.2516	0.0788	0.4582	0.2162
		200	0.1323	0.6479	0.4373	0.2557	0.6383	0.4728	0.0629	0.6293	0.3999
4	(0,0)	500	0.0647	0.1814	0.0371	0.0433	0.2015	0.0425	0.0647	0.1815	0.0371
		200	0.0916	0.2879	0.0913	0.0173	0.3036	0.0925	0.0937	0.2877	0.0915
	(8,5)	500	0.0956	0.4106	0.1777	0.1013	0.3943	0.1657	0.1266	0.4054	0.1803
		200	0.1055	0.3911	0.1641	0.0042	0.4084	0.1668	0.1168	0.3884	0.1645

To compare the effect of the relative magnitude of W on the estimation process, we display in Table 7.4 the results of a simulation study considering $W \sim$ Uniform(0,1). The bias magnitudes are generally reduced compared to when $W \sim$ Uniform(3,4). The WQC approach again has the least bias in the presence of a latent variable and without any effect of W on the event occurrence ($\sigma^2 = 1$ or 4, $\gamma = 0, \tilde{\gamma} = 0$). However, when introducing the dependent censoring through covariate W ($\gamma \neq 0, \tilde{\gamma} \neq 0$), the bias for the WQC approach is slightly smaller than that for MKLB approach for $\sigma^2 = 0$ while MKLB approach outperforms WQC approach in terms of the bias measure when $\sigma^2 \neq 0$. Such results are different from those obtained when $W \sim$ Uniform(3,4) and somewhat similar to those obtained when $W \sim$ Normal(0,1).

Computationally, the WQC method is more demanding than the MKLB. However, in order to properly consider the MKLB approach, the investigator has to define a model for the censoring mechanism first for obtaining the appropriate weights for modeling the rate of recurrent events.

In summary, all methods are approximately unbiased for the scenario of independent censoring ($\sigma^2 = 0, \gamma = 0, \tilde{\gamma} = 0$) and the MKLB approach produces the most efficient estimates. When the only source of dependent or informative censoring is known to be due to the covariates ($\sigma^2 = 0, \gamma \neq 0, \tilde{\gamma} \neq 0$), which can be properly modelled through the censoring mechanism, the MKLB method generally yields the more accurate and efficient estimates compared to WQC method. Particularly in such scenario the MKLB estimates are much less biased than those obtained by fitting a marginal rates model under the assumption of independent censoring regardless of the relative magnitude and distribution of W. If the censoring is independent of the covariates and other sources of dependence, the MKLB estimator is expected to be more efficient than the naive estimator.[9] Nevertheless, when the heterogeneity among subjects is introduced only through a latent variable ($\sigma^2 \neq 0, \gamma = 0, \tilde{\gamma} = 0$), WQC approach always outperforms MKLB approach in terms of accuracy for the configurations studied here. On the other hand, when both covariate (W) and latent variable (u) are used to introduce dependent censoring on the event occurrence ($\sigma^2 \neq 0, \gamma \neq 0, \tilde{\gamma} \neq 0$), the results are not consistent across the parameter configurations considered in this paper. In those situations, the accuracy and efficiency of the estimates seem to vary for distinct relative magnitudes as well as probability distributions associated with W.

Table 7.4. Simulation results on bias, empirical standard errors and mean squared errors of the three estimators for the regression parameter β based on 1,000 replicates (with $\beta_0 = -1.2$, $\tilde{\beta}_0 = 1$) and $W \sim$ Uniform(0,1).

			MKLB Method			WQC Method			Indep.cens. Method		
σ^2	$(\gamma_0, \tilde{\gamma}_0)$	n	Bias	ESE	MSE	Bias	ESE	MSE	Bias	ESE	MSE
0	(0,0)	500	0.0035	0.0649	0.0042	0.0073	0.1246	0.0156	0.0001	0.0976	0.0095
	(8,5)	500	0.0173	0.1764	0.0314	0.0136	0.1836	0.0339	0.0383	0.1736	0.0316
1	(0,0)	500	0.0361	0.1058	0.0125	0.0131	0.1278	0.0165	0.0361	0.1059	0.0125
	(8,5)	500	0.0379	0.2365	0.0574	0.0593	0.2609	0.0716	0.0987	0.2283	0.0619
4	(0,0)	500	0.1178	0.1923	0.0509	0.0099	0.2067	0.0428	0.1180	0.1922	0.0509
	(8,5)	500	0.0266	0.3391	0.1157	0.1284	0.3366	0.1509	0.0471	0.3346	0.1142

7.5. An Example: Modeling Times to Recurrent Diarrhea in Children

In this section we apply the aforementioned methods to recurrent diarrhea data to illustrate the modelling process. We use data from 1,191 children aged 6-48 months at baseline, who participated in a randomized community trial conducted in Brazil between 1990 and 1991 to evaluate the effect of high dosages of vitamin A supplementation on the occurrence of recurrent diarrheal episodes. The complete study was described elsewhere.[20] For the analysis presented here, we consider data available from the first treatment cycle, i.e, between the first and second dosages of vitamin A. During this period, the mean number of episodes of diarrhea is 2.53 (sd=2.41, range=0-15). The covariates include demographic, economic and health indicators. We consider the following covariates for modelling diarrhea occurrence: age (in months, at baseline), sex, treatment group (TRT=placebo or vitamin A) and an indicator of existence of toilet (TOILET) in the household. To capture their health status we consider as covariates weight-for-age Z-score (WAZ) and previous occurrence of measles. Among these children, 26.4% lived in houses that do not have toilets and 89.3% had measles previously. Censoring time is defined as the time when the participant withdrew from the study or died, or the study ended.

Table 7.5. Estimated coefficients for the marginal rates model of diarrhea occurrence considering independent censoring, WQC and MKLB approaches.

Variables	Indep. censoring		MKLB		WQC	
	Param estimate	Std error	Param estimate	Std error	Param estimate	Std error
TRT	-0.136	0.0556	-0.137	0.0568	-0.131	0.0516
AGE	-0.030	0.0024	-0.030	0.0027	-0.025	0.0021
TOILET	0.254	0.0622	0.253	0.0630	0.210	0.0558
WAZ	-0.050	0.0181	-0.051	0.0289	-0.042	0.0166

We applied the WQC method, MKLB method, and the 'naive' method assuming independent censoring to analyze this dataset. In order to implement the MKLB approach, we first obtained a 'good' model for the censoring mechanism. For the selection of such model, we considered all available covariates. The only important covariate for the censoring mechanism was WAZ ($\hat{\beta} = -0.059(0.017)$). The weights were then estimated by using the estimated censoring survival probability obtained from the model for the censoring mechanism. The estimates of the regression coefficients from these three methods are given in Table 7.5. The standard errors for the WQC and MKLB methods were estimated using bootstrap.

The estimated coefficients and standard errors do not change noticeably when we employ WQC and MKLB approaches. The dependent censoring could have been introduced in this study if children who were at higher risk of having recurrent diarrheal episodes withdrawn from the study earlier. Another form of dependent censoring could have been introduced due to terminal events, such as death. However, the few death cases that occurred during this study were equally distributed among the two treatment groups and were not associated to diarrhea occurrence. The results from WQC approach do not indicate any other source of dependent censoring in these data.

Both models lead to the same clinical conclusions. TRT has a strong effect on the rate of diarrhea occurrence (RR=0.87). Based on these results, the rate of diarrhea occurrence in children receiving vitamin A supplementation is 13% lower than the corresponding rate in children in the placebo group. The increase in age and in weight-for-age Z-score also contributes for a significant reduction on the rate of diarrhea occurrence. On the contrary, the existence of toilet in the house leads to an increase of 29% on the rate of diarrhea, which could be associated to poor hygiene practices in this community.

7.6. Concluding Remarks

Several other approaches have been proposed in the literature to handle dependent censoring.[10–12,14,15] Some of this work has focused on joint parametric modeling of recurrent events and survival data using panel data,[10,14] others have accounted for dependent censoring in the cure modeling framework[8] and also on the joint modeling of both repeated measures and recurrent events in the presence of a terminal event.[15] Ghosh and Lin[11] proposed a semiparametric joint model that accounts for dependent censoring, considering the use of accelerated failure time models. In the context of randomized clinical trials with repeated events, Matsui[12] proposed the use of structural failure time models for both times to repeated events and to dependent censoring, leaving unspecified the dependence structure between those times. Hsu and Taylor[16] adjusted for dependent censoring via multiple imputation to allow comparison of two survival distributions. Other recent developments accounting the dependent censoring were also related to nonparametric methods.[21] More recently, Bandyopadhyay and Sen[22] considered a Bayesian framework to model association between the intensity of a recurrent event and the hazard of a terminating event. However, none of these approaches had used the proportional rate model for analysis of recurrent event data.

In this paper we compared two approaches (WQC and MKLB) using the proportional rate model under dependent censoring for the estimation of covariate effects for recurrent time-to-event data. We found that they produce approximately unbiased estimates when the dependent or informative censoring is not present. The variances of the parameter estimates from the two approaches increase with decreasing sample size, as expected. When dependent censoring is present, the WQC method performs better when the dependent censoring is introduced through a latent variable and the MKLB method performs better when the dependent censoring is introduced through covariates. Generally, the empirical standard errors from WQC approach are larger than those from MKLB approach. According to Wang *et al.*,[17] approaches that model the censoring time using observable covariates are expected to achieve optimal estimation efficiency at the price of modelling the censoring mechanism with proper covariate information. It is important to emphasize that the MKLB approach requires observation of a larger number of covariates, under which the censoring is independent of the counting process, as well as the correct modelling of the censoring time. This factor may contribute to its improved efficiency over WQC approach. Biased results were found when the informative or dependent censoring was introduced simultaneously by a covariate and a latent variable. Further research is still needed under such situations.

Our results show that the assumptions on how censoring is related to the recurrent process are important for these two approaches. These methods are not robust when the assumptions are violated. Researchers should think carefully about how censoring is related to the recurrent event process before adopting an approach in the study. As a practical recommendation, we suggest taking the following steps: (1) conduct the analysis using both methods. If the results are similar, then one can conclude that the censoring is not informative about the recurrent event process either through covariates or shared frailty. In this case, either method can be employed. (2) If the results are different, then one should conduct a sensitivity analysis for the censoring process to verify whether any of the collected covariates are associated with the censoring process. If no covariates are associated with the censoring process, then one should proceed with the WQC method. On the other hand, if there are some covariates which are associated with the censoring process, then one should carefully verify which of the collected covariates are associated with the censoring process and proceed with the

MKLB approach. An alternative approach that also can be considered to evaluate presence of dependent censoring in a longitudinal study was recently proposed by Sun and Lee.[23]

Acknowledgments

The authors would like to thank Dr.Barreto and colleagues for providing the Vitamin A data. We also would like to thank the editors and referees for the helpful comments and suggestions. Dr.Amorim was supported by a scholarship from Brazilian Agency CAPES during this work. This work is partially supported by US National Institute of Health grants RO1 HL-57444 and PO1CA-142538.

References

1. R. L. Prentice, B.J. Williams and A.V. Peterson, On the Regression Analysis of Multivariate Failure Time Data, *Biometrika*, **68**, 373–379, (1981).
2. P. K. Andersen and Gill, Cox's Regression Model for Counting Processes: A Large Sample Study, *The Annals of Statistics*, **10**, 110–1120, (1982).
3. L. J. Wei, D.Y. Lin and L. Weissfeld, Regression Analysis of Multivariate Incomplete Failure Time Data by Modeling Marginal Distributions, *Journal of the American Statistical Association*, **84**, 1065–1073, (1989).
4. M. S. Pepe and J. Cai, Some graphical displays and Marginal Regression analysis for Recurrent Failure Times and Time Dependent Covariates, *Journal of American Statistical Association*, **88**, 811–820, (1993).
5. D.Y. Lin, L.J. Wei, I. Yang and Z. Ying, Semiparametric regression for the mean and rate functions of recurrent events, *Journal of Royal Statistics Society - Series B*, **62** (4), 711–730, (2000).
6. E.W. Lee, L.J. Wei and D.A. Amato, Cox-Typed Regression Analysis for Large Numbers of Small Groups of Correlated Failure Time Observations, In: *Survival Analysis: State of the Art*, 237–247, (1992).
7. J.F. Lawless and C. Nadeau, Some Simple Robust Methods for the Analysis of Recurrent Events, *Technometrics*, **37**, 158–168, (1995).
8. Y. Li, R.C. Tiwari and S. Guha, Mixture cure survival models with dependent censoring, *Journal of Royal Statistics Society - Series B*, **69** (3), 285–306, (2007).
9. M. Miloslavsky, S. Keles, M.J. van der Laan and S. Butler, Recurrent events analysis in the presence of time-dependent covariates and dependent censoring, *Journal of Royal Statistics Society - Series B*, **66** (1), 239–257, (2004).
10. A. Lancaster and O. Intrator, Panel data with survival: hospitalization of HIV-positive patients, *Journal of American Statistical Association*, **93**, 46–53, (1998).

11. D. Ghosh and D.Y. Lin, Semiparametric Analysis of Recurrent Event Data in the Presence of Dependent Censoring, *Biometrics*, **59**, 877–885, (2003).
12. S. Matsui, Analysis of times to repeated events in two-arm randomized trials with noncompliance and dependent censoring, *Biometrics*, **60**, 965–976, (2004).
13. D. Zeng, Estimating marginal survival function by adjusting for dependent censoring using many covariates, *The Annals of Statistics*, **32**, 1533–1555, (2004).
14. C-Y. Huang, M-C. Wang and Y. Zhang, Analysing panel count data with informative observation times, *Biometrika*, **93**, 763–775, (2006).
15. L. Liu and X. Huang, Joint analysis of correlated repeated measures and recurrent events processes in the presence of death, with application to a study on acquired immune deficiency syndrome, *Journal of Royal Statistics Society - Series C*, **58** (1), 65–81, (2009).
16. C-H. Hsu and J.M.G. Taylor, Nonparametric comparison of two survival functions with dependent censoring via nonparametric multiple imputation, *Statistics in Medicine*, **28**, 462–475, (2009).
17. M-C. Wang, J. Qin and C-T. Chiang, Analyzing Recurrent Event Data With Informative Censoring, *Journal of the American Statistical Association*, **96**, 1057–1065, (2001).
18. J. Robbins and A. Rotnitzky, Recovery of information and adjustment for dependent censoring using surrogate markers, In: *AIDS Epidemiology, Methodological Issues. Boston: Birkhäuser*, 297–331, (1992).
19. J. Cai and E. Schaubel, Analysis of Recurrent Event Data, *Handbook of Statistics*, **23**, 603–623, (2004).
20. M.L. Barreto, L.M.P. Santos, A.M.O. Assis, M.P.N. Araujo, G.G. Farenzena, P.A.B. Santos and R.L. Fiaccone, Effect of vitamin A supplementation on diarrhoea and acute lower-respiratory-tract infections in young children in Brazil, *Lancet*, **344**, 228–231, (1994).
21. B. Pradhan and A. Dewanji, On induced dependent censoring for quality adjusted lifetime (QAL) data in a simple illnessdeath model, *Statistics & Probability Letters*, **79**, 2170–2176, (2009).
22. N. Bandyopadhyay and A. Sen, Bayesian Modeling of Recurrent Event Data with Dependent Censoring, *Communications in Statistics - Simulation and Computation*, **39**, 641–654, (2010).
23. Y. Sun and J. Lee, Testing independent censoring for longitudinal data, *Statistica Sinica*, to appear, (2010).

Chapter 8

Efficient Algorithms for Bayesian Binary Regression Model with Skew-Probit Link

Rafael B. A. Farias* and Marcia D. Branco†

Department of Statistics, University of Sao Paulo
Sao Paulo, SP, Brazil
**rfarias@ime.usp.br*
†mbranco@ime.usp.br

We here propose different Gibbs sampling algorithms in the context of the binary response regression models under the skew-probit link. We use latent variables, which is a technique widely used to attainment of the full posterior conditional distributions. Analytical expressions for the full conditional posterior distributions are obtained and presented. These algorithms are compared through two measures of efficiency, the average Euclidean update distance between interactions and the effective sample size. We conclude that the new algorithms are more efficient than the usual Gibbs sampling, where one of them leads to around 160% of improvement in the effective sample size measure. The developed procedures are illustrated with an application to a medical data set.

8.1. Introduction

Bayesian inference is getting more and more dependent on stochastic simulation methods and of their efficiency. The introduction of latent variables is a technique widely used to the attainment of the full posterior conditional distributions, which allows the implementation of the Gibbs sampling algorithm. However, the introduction of these latent variables many times provides algorithms for which the values are highly correlated, which harms the convergence rate. The grouping of the unknown quantities in blocks in the joint simulation of these quantities is feasible and can be an alternative to reduce the autocorrelation. Liu (1994) used the idea of blocking and collapsing in Gibbs sampling and showed a method that can lead to good results. Chib and Carlin (1999) developed new approaches for MCMC simulation of longitudinal models based on the use of blocking that provided significant improvements.

A standard way to deal with the inferential problem in binary or binomial response regression model is using the maximum likelihood approach and the related asymptotic theory. Thus, the accuracy of this inference is questionable for small sample sizes. On the other hand, Bayesian inference which is based on the posterior distribution performs well for small samples. Albert and Chib (1993) introduced the use of latent variables for binary probit regression models and showed that under suitable prior distributions their approach renders known conditional posterior distributions. In this case, conjugate prior distributions are available and the Gibbs sampling algorithm is implemented easily. Nevertheless, given the strong posterior correlation between the regression coefficients and the latent variables, this algorithm is not very efficient. In view of that, Holmes and Held (2006) use the ideas of using Gibbs block sampler from Liu (1994) and Chib and Carlin (1999) for reduction of the correlation in the simulated sample in order to find a more efficient simulation framework. The construction of these new algorithms depend on obtaining explicit form for marginal distributions of some parameters instead the full conditional distributions. The main difference of these algorithms is the fact that the first algorithm simulates from the posterior conditional distribution of the latent variables given the regression parameters of the model, while the second simulates from the posterior marginal distribution of the latent variables. The last permits the joint updating of the regression coefficients and auxiliary variables. The original algorithm will be denoted by *Conditional* algorithm and the second by *Marginal* algorithm.

Binary response data are usually fitted with symmetric link functions, namely, probit and logit links. However, skew link functions are more flexible to modeling binary data, and very important in situations where the success probability approaches to zero in a different rate than approaches to one. Chen (2004) and Bazán *et al.* (2005) presented several reasons for using skewed link functions. Chen *et al.* (1999) define a general class of skewed link functions which includes a skew-probit link. This class contains the probit link as a particular case. Later on, Bazán *et al.* (2005) presented an alternative skew-probit link. In Bazán *et al.* (2010) the relationship between these two skew-probit links is discussed in detail.

The main goal of this paper is to propose and compare new Gibbs sampler algorithms for the skew-probit regression model given by Chen *et al.* (1999). The motivation comes from Holmes and Held (2006) paper.

The rest of this paper is organized as follows. First, we review symmetric models, with special attention for the simulation algorithms under the probit link. In Sec. 8.3, we present the skew-probit model with latent variables and obtain full conditional distributions. Section 8.4 develops some analytical results which will be helpful for the proposal of the new algorithms. In the next section, four joint updating algorithms using latent variable and blocks are presented. Section 8.5 compares the algorithms using two efficiency measures. An application to a real data set is presented and discussed in Sec. 8.6. Finally, some conclusions are presented in Sec. 8.7. The proofs of the propositions and pseudo-codes are presented in the appendices.

8.2. Symmetric Models and the Use of Latent Variables

The models commonly used in binary regression are obtained using symmetric cumulative distribution functions. The most popular are the probit and the logistic models. They are adequate when there is no evidence that the probability of success increases in a different rate than the one it decreases.

Let $\boldsymbol{y} = (y_1, \ldots, y_n)^T$ be a set of binary variables (0/1), where $y_1, \ldots, y_n$ are independent random variables. Consider also that $\boldsymbol{x}_i = (x_{i1}, \ldots, x_{ip})^T$ a set of previous fixed quantities associated with y_i, where x_{i1} can be equal 1 (that corresponds to intercept). The binary regression model with independent responses is given by

$$p_i = p(y_i = 1) = F(\boldsymbol{x}_i^T \boldsymbol{\beta}), \tag{8.1}$$

where F is a function that linearizes the relationship between the probability of success and the covariates and $\boldsymbol{\beta}$ is a p-dimensional vector of the regression coefficients. The function F^{-1} is called link function under generalized linear models (GLM) theory. The inverse of the link function is a monotone and differentiable function. Typically, F is a cumulative distribution function (cdf) of a random variable with support on real set. Sometimes, the link function depends on additional parameters denoted here by λ. The Bayesian binary regression model is given by

$$\begin{aligned} y_i &\sim Bernoulli\,(F(\eta_i))\,, \\ \eta_i &= \boldsymbol{x}_i^T \boldsymbol{\beta}, \\ \boldsymbol{\beta} &\sim \pi_1(\boldsymbol{\beta}), \\ \lambda &\sim \pi_2(\lambda), \end{aligned} \tag{8.2}$$

where $\pi_1(\boldsymbol{\beta})$ and $\pi_2(\lambda)$ are suitable prior distributions associated with the parameters of the model. Since the posterior distributions do not have a standard form, MCMC methods to sample from the posterior distribution will be considered.

8.2.1. *Probit regression*

The probit models are widely used in several fields, mainly in clinical trials. It is obtained when we assume $F(u) = \Phi(u)$ in (8.2), where $\Phi(\cdot)$ denotes a cdf of a standard normal distribution. Alternatively, we can represent a Bayesian probit regression model using latent variables such as

$$\begin{aligned} y_i &= \begin{cases} 1 & \text{if } z_i > 0, \\ 0 & \text{otherwise,} \end{cases} \\ z_i &= \boldsymbol{x}_i^T \boldsymbol{\beta} + \epsilon_i, \\ \epsilon_i &\sim \mathcal{N}(0,1), \\ \boldsymbol{\beta} &\sim \pi(\boldsymbol{\beta}), \end{aligned} \tag{8.3}$$

where y_i is now a deterministic conditional on the sign of the stochastic latent variable z_i. The advantage of working with representation (8.3) is that, for a good choice of the prior distribution for $\boldsymbol{\beta}$, it is straightforward to obtain closed forms for the conditional posterior distributions. Albert and Chib (1993) have obtained the conditional distributions $\pi(\boldsymbol{z}|\boldsymbol{\beta}, \boldsymbol{y})$ and $\pi(\boldsymbol{\beta}|\boldsymbol{z}, \boldsymbol{y})$ for the model (8.3). The inclusion of auxiliary variables offers a convenient framework to use the Gibbs algorithm. However, a potential problem is the strong posterior dependence between $\boldsymbol{\beta}$ and $\boldsymbol{z}$, as indicated in model (8.3). This dependence provides a slow mixing in the Markov chain. The marginal algorithm permits joint updating of regression coefficients and of the auxiliary variables using the following factorization $\pi(\boldsymbol{z}, \boldsymbol{\beta}|\boldsymbol{y}) = \pi(\boldsymbol{z}|\boldsymbol{y})\pi(\boldsymbol{\beta}|\boldsymbol{z}, \boldsymbol{y})$. Holmes and Held (2006) assume that $\boldsymbol{\beta}$ has a p-variate prior normal distribution with mean vector zero and showed through an empirical study that the joint updating algorithm (marginal) in the probit and logit models is more efficient than the conditional algorithm.

8.3. Skew-Probit Regression

8.3.1. *A general class of skewed links*

Chen (2004) carried out a simulation study to investigate the importance of the choice of a link function in prediction of binary response variables.

He considered two simulation schemes: (i) the data set was generated according to probit model; and (ii) the data was generated according to C log-log model. In both situations the probit, logit and C log-log models were fitted. The author observed that, when the true link function is probit, there are few differences between the probit and logit models; while the C log-log model is inadequate. On the other hand, when the true link function is C log-log, the symmetric models are clearly inadequate. The author has concluded from his empirical study that the choice of the link function is very important, and if it is not very well specified it can provide poor predictions.

We consider the following class of distributions to parametric asymmetric link functions

$$\mathcal{F}_{\mathcal{A}} = \left\{ F_\lambda(z) = \int_{[0,\infty)} F(z + \lambda w) dG(w) \right\}, \tag{8.4}$$

where $\lambda \in \mathbb{R}$, F is a cdf of a symmetrical distribution around zero with support on the real line, and G is a cdf of an asymmetrical distribution on $[0, \infty)$. The model defined in (8.4) has some attractive proprieties, namely: (a) when $\lambda = 0$ or G is a degenerate distribution, then the model reduces to a model with symmetrical link; (b) the skewness of link function can be characterized by λ and G; and (c) heavy and light tails for F_λ can be obtained according to choice of F.

The binary regression model presented in (8.1) with inverse of link function given by $F_\lambda \in \mathcal{F}_{\mathcal{A}}$, is characterized by

$$p_i = p(y_i = 1) = F_\lambda(\boldsymbol{x}_i^T \boldsymbol{\beta}) = \int_{[0,\infty)} F(\boldsymbol{x}_i^T \boldsymbol{\beta} + \lambda w) dG(w), \tag{8.5}$$

where F and G are defined in (8.4).

8.3.2. *Bayesian regression with skew-probit link*

A particular case of the general skewed model presented in the last subsection is obtained when F is the cdf of a normal distribution and G is the cdf of a half-normal distribution. This skewed regression problem was extended to the class of elliptical distribution by Sahu *et al.* (2003). Note that, the standard skewed normal distribution given in Sahu *et al.* (2003) is not the same given by Azzalini (1985). However, there is an easy relationship between both, as showed by Bazán *et al.* (2010).

Considering that the prior distributions of $\boldsymbol{w} = (w_1, \ldots, w_n)^T$, $\boldsymbol{\epsilon} = (\epsilon_1, \ldots, \epsilon_n)^T$, $\boldsymbol{\beta} = (\beta_1, \ldots, \beta_p)$ and λ are independent, the Bayesian skew-probit model can be represented as

$$y_i = \begin{cases} 1 & \text{if } z_i > 0, \\ 0 & \text{otherwise}, \end{cases}$$
$$z_i = \boldsymbol{x}_i^T \boldsymbol{\beta} + \lambda w_i + \epsilon_i, \tag{8.6}$$

where

$$\epsilon_i \sim \mathcal{N}(0,1), \;\; w_i \sim \mathcal{N}(0,1) I(w_i > 0), \;\; \boldsymbol{\beta} \sim \pi_1(\boldsymbol{\beta}) \text{ and } \lambda \sim \pi_2(\lambda). \tag{8.7}$$

We have considered the notations $\mathcal{N}(\mu, \sigma^2)I(A)$ and $\mathcal{SN}(\mu, \sigma^2, \lambda)I(A)$ to indicate a truncated normal and skew-normal distribution in a region A, respectively. Moreover, $\mathcal{SN}(\mu, \sigma^2, \lambda)$ denotes a skew-normal distribution with location parameter μ, scale parameter σ^2 and shape parameter λ (Azzalini, 1985). which the density probability function is as follows

$$f_{\mathcal{SN}}(x) = \left(\frac{2}{\sigma}\right) \phi\left(\frac{x-\mu}{\sigma}\right) \phi\left(\lambda \frac{x-\mu}{\sigma}\right). \tag{8.8}$$

Although the skew-probit link was proposed by Chen *et al.* (1999) and discussed in more details by Bazán *et al.* (2010), neither of them showed the complete conditional distributions that should be used for the Gibbs algorithm. For all of them, the prior distributions considered for $\boldsymbol{\beta}$ and λ are, respectively,

$$\pi_1(\boldsymbol{\beta}) = \mathcal{N}_p(\boldsymbol{b}, \boldsymbol{\nu}) \quad \text{and} \quad \pi_2(\lambda) = \mathcal{N}(\alpha, \tau). \tag{8.9}$$

Proposition 8.1. *Under the skew-probit model (8.6), and assuming the prior specification (8.9), the full conditional distributions of* $\pi(\boldsymbol{\beta}, \boldsymbol{z}, \boldsymbol{w}, \lambda | \boldsymbol{y})$ *are*

$$a) \quad \boldsymbol{\beta} | \boldsymbol{z}, \boldsymbol{w}, \lambda \sim \mathcal{N}_p(\boldsymbol{B}, \boldsymbol{V}), \tag{8.10}$$

with $\boldsymbol{B} = \boldsymbol{V}\left[\boldsymbol{\nu}^{-1}\boldsymbol{b} + \boldsymbol{X}^T(\boldsymbol{z} - \lambda \boldsymbol{w})\right]$ *and* $\boldsymbol{V} = (\boldsymbol{\nu}^{-1} + \boldsymbol{X}^T\boldsymbol{X})^{-1}$, *where* $\boldsymbol{X} = (\boldsymbol{x}_1, \boldsymbol{x}_2, \ldots, \boldsymbol{x}_n)^T$.

$$b) \quad z_i | y_i, \boldsymbol{\beta}, w_i, \lambda \sim \begin{cases} \mathcal{N}(\boldsymbol{x}_i^T\boldsymbol{\beta} + \lambda w_i, 1) I(z_i > 0) & \text{if } y_i = 1, \\ \mathcal{N}(\boldsymbol{x}_i^T\boldsymbol{\beta} + \lambda w_i, 1) I(z_i \leq 0) & \text{if } y_i = 0, \end{cases} \tag{8.11}$$

where z_i, $i = 1, \ldots, n$, *are conditional independent random variables.*

$$c) \quad w_i | z_i, \boldsymbol{\beta}, \lambda \propto \mathcal{N}\left(\frac{\lambda}{1+\lambda^2}(z_i - \boldsymbol{x}_i^T\boldsymbol{\beta}), \frac{1}{1+\lambda^2}\right) I(w_i > 0), \tag{8.12}$$

where w_i, $i = 1, \ldots, n$, *are conditional independent random variables.*

$$d)\quad \lambda|\boldsymbol{z}, \boldsymbol{\beta}, \boldsymbol{w} \sim \mathcal{N}(m, \nu), \tag{8.13}$$

where $m = \nu\left[\tau^{-1}\alpha - \boldsymbol{w}^T(\boldsymbol{z} - \boldsymbol{X\beta})\right]$ *and* $\nu = (\tau^{-1} + \boldsymbol{w}^T\boldsymbol{w})^{-1}$.

Note that the conditional posterior distributions given in (8.10), (8.12) and (8.13) do not depend on $\boldsymbol{y}$. Moreover, like in the probit model, there is a strong posterior dependence between $\boldsymbol{\beta}$ and $\boldsymbol{z}$. Furthermore, there is also a strong correlation between λ and $\boldsymbol{w}$, clearly indicated in (8.6). The grouping of these quantities in blocks in such a way that the joint simulation of this quantities is feasible, is an alternative to reduce the autocorrelation, and therefore, it should improve the efficiency on the simulation procedure.

8.4. New Simulation Algorithms

The use of a multivariate vector as a block usually provides an improvement of the convergence speed of the Markov chain. Specially when these variables are highly correlated. This happens because the block incorporates the correlation structure between its components. Despite there is no general rule to choose an excellent block formation, blocks which are easy to sample from are natural blocks. For more discussion about that, see for example, Gamerman and Lopes (2006). Some of these blocks schemes will be proposed in this section, after we present some useful analytical results. Pseudo-codes and proposition's proof are presented in the appendices.

8.4.1. *Analytical results*

The following propositions will be helpful to define the blocks used for the simulation procedure.

Proposition 8.2. *Considering the skew-probit model (8.6) and the prior specification (8.9), we have that*

$$a)\quad \boldsymbol{z}|\boldsymbol{y}, \boldsymbol{w}, \lambda \sim \mathcal{N}_n(\boldsymbol{Xb} + \lambda\boldsymbol{w}, I_n + \boldsymbol{X\nu X}^T)Ind(\boldsymbol{y}, \boldsymbol{z}), \tag{8.14}$$

where $Ind(\boldsymbol{y}, \boldsymbol{z}) = \prod_{i=1}^{n}\{I(y_1 = 1)I(z_i > 0) + I(y_1 = 0)I(z_i \leq 0)\}$.

$$b)\quad z_i|y_i, \boldsymbol{\beta}, \lambda \sim \begin{cases} \mathcal{SN}(\boldsymbol{x}_i^T\boldsymbol{\beta}, 1 + \lambda^2, \lambda)I(z_i > 0) & \text{if } y_i = 1, \\ \mathcal{SN}(\boldsymbol{x}_i^T\boldsymbol{\beta}, 1 + \lambda^2, \lambda)I(z_i \leq 0) & \text{if } y_i = 0, \end{cases} \tag{8.15}$$

where z_i, $i = 1\ldots,n$, are independent random variables.

c) $$\boldsymbol{z}|\boldsymbol{y},\boldsymbol{w},\boldsymbol{\beta} \sim \mathcal{N}_n(\boldsymbol{X\beta}+\alpha\boldsymbol{w}, \boldsymbol{I}_n+\tau^{-1}\boldsymbol{w}\boldsymbol{w}^T)Ind(\boldsymbol{y},\boldsymbol{z}), \tag{8.16}$$

where $Ind(\boldsymbol{y},\boldsymbol{z})$ is the same indicator function presented in (8.14).

Since it is not easy to simulate efficiently from a multivariate truncated normal distribution directly, we suggest to construct a new Gibbs sampling to simulate from (8.14) and (8.16).

Proposition 8.3. *Considering the skew-probit model in (8.6) and the prior specification (8.9), we have that*

a) $$z_i|\boldsymbol{z}_{-i}, y_i, \boldsymbol{w}, \lambda \propto \begin{cases} \mathcal{N}(m_i,\nu_i)I(z_i>0) & \text{if } y_i=1, \\ \mathcal{N}(m_i,\nu_i)I(z_i\le 0) & \text{if } y_i=0, \end{cases} \tag{8.17}$$

with $\boldsymbol{z}_{-i}$ denoting the vector $\boldsymbol{z}$ with the ith variable removed, and

$$m_i = \boldsymbol{x}_i^T\boldsymbol{b}+\lambda w_i + \frac{1}{1-h_{ii}}\sum_{k=1,k\neq i}^{n} h_{ik}(z_k-\boldsymbol{x}_k^T\boldsymbol{b}-\lambda w_k) \quad \text{and} \quad \nu_i = \frac{1}{1-h_{ii}},$$

where h_{ik} denotes the ith element of the kth column of the matrix $\boldsymbol{H} = \boldsymbol{XVX}^T$, with $\boldsymbol{V}$ defined in (8.10).

b) $$z_i|\boldsymbol{z}_{-i}, y_i, \boldsymbol{w}, \boldsymbol{\beta} \propto \begin{cases} \mathcal{N}(m_i,\nu_i)I(z_i>0) & \text{if } y_i=1, \\ \mathcal{N}(m_i,\nu_i)I(z_i\le 0) & \text{if } y_i=0, \end{cases} \tag{8.18}$$

where $\boldsymbol{z}_{-i}$ denotes the vector $\boldsymbol{z}$ with the ith variable removed, and

$$m_i = \boldsymbol{x}_i^T\boldsymbol{\beta} - w_i m - \frac{h_i}{1-h_i}\left[z_i - \left(\boldsymbol{x}_i^T\boldsymbol{\beta}-w_i m\right)\right] \quad \text{and} \quad \nu_i = \frac{1}{1-h_i},$$

with m given in (8.13) and $h_i = \nu w_i^2$.

We notice that the skew-probit regression model (8.6) can be represented in a similar way to probit model (8.3) since λ is considered a kind of regression coefficient associated to latent variable $\boldsymbol{w}$. Considering that $\boldsymbol{a}_i^T = \left[\boldsymbol{x}_i^T, w_i\right]$ and $\boldsymbol{\theta} = (\boldsymbol{\beta},\lambda)^T$, the model can be represented as

$$\begin{aligned} y_i &= \begin{cases} 1 & \text{if } z_i>0, \\ 0 & \text{otherwise}, \end{cases} \\ z_i &= \boldsymbol{a}_i^T\boldsymbol{\theta}+\epsilon_i, \\ \epsilon_i &\sim \mathcal{N}(0,1), \\ \boldsymbol{\theta} &\sim \mathcal{N}_{p+1}(\boldsymbol{b},\boldsymbol{\nu}). \end{aligned} \tag{8.19}$$

Proposition 8.4. *Considering the skew-probit model (8.19), it follows that*

$$\boldsymbol{z}|\boldsymbol{y},\boldsymbol{w} \sim \mathcal{N}_n(\boldsymbol{X}\boldsymbol{b}, \boldsymbol{I}_n + \boldsymbol{A}\boldsymbol{\nu}\boldsymbol{A}^T) Ind(\boldsymbol{y},\boldsymbol{z}) \tag{8.20}$$

and

$$z_i|\boldsymbol{z}_{-i},\boldsymbol{y},\boldsymbol{w},\lambda \sim \begin{cases} \mathcal{N}(m_i,\nu_i)I(z_i>0) & \text{if } y_i = 1, \\ \mathcal{N}(m_i,\nu_i)I(z_i\leq 0) & \text{if } y_i = 0, \end{cases} \tag{8.21}$$

where m_i and ν_i are given by

$$m_i = \boldsymbol{a}_i^T\boldsymbol{B} - \frac{h_i}{1-h_i}(z_i - \boldsymbol{a}_i^T\boldsymbol{B}) \quad \text{and} \quad \nu_i = \frac{1}{1-h_i}, \tag{8.22}$$

with $\boldsymbol{B} = \boldsymbol{V}\left(\boldsymbol{\nu}^{-1}\boldsymbol{b} + \boldsymbol{A}^T\boldsymbol{z}\right)$ and h_i being the ith element of the diagonal of the matrix $\boldsymbol{H} = \boldsymbol{A}\boldsymbol{V}\boldsymbol{A}^T$, where $\boldsymbol{V} = (\boldsymbol{\nu}^{-1} + \boldsymbol{A}^T\boldsymbol{A})^{-1}$ and $\boldsymbol{A} = (\boldsymbol{a}_1^T, \boldsymbol{a}_2^T, \ldots, \boldsymbol{a}_n^T)$.

8.4.2. *Joint updating of $\{\boldsymbol{z},\boldsymbol{\beta}\}$*

The joint updating method of the auxiliary variables $\boldsymbol{z}$ and of the regression coefficients $\boldsymbol{\beta}$ for skew-probit model is an extension of what was presented in Holmes and Held (2006) for the probit link. We update here $\{\boldsymbol{z},\boldsymbol{\beta}\}$ jointly based in the following factorization

$$\pi(\boldsymbol{\beta},\boldsymbol{z}|\boldsymbol{y},\boldsymbol{w},\lambda) = \pi(\boldsymbol{z}|\boldsymbol{y},\boldsymbol{w},\lambda)\pi(\boldsymbol{\beta}|\boldsymbol{z},\boldsymbol{w},\lambda).$$

The distribution $\pi(\boldsymbol{\beta}|\boldsymbol{z},\boldsymbol{w},\lambda)$ is given in (8.10) and the distribution $\pi(\boldsymbol{z}|\boldsymbol{y},\boldsymbol{w},\lambda)$ is a multivariate truncated normal given in (8.14). Since it is difficult to sample directly from a n-variate truncated normal distribution, we have considered (8.17) in the Proposition 8.3 to sample from $\pi(\boldsymbol{z}|\boldsymbol{y},\boldsymbol{w},\lambda)$. An efficient alternative to obtain the location parameter m_i given in (8.17) using matrix operations is

$$m_i = \boldsymbol{x}_i^T\boldsymbol{B} + \lambda w_i - \frac{h_i}{1-h_i}\left[z_i - \left(\boldsymbol{x}_i^T\boldsymbol{B} + \lambda w_i\right)\right],$$

where z_i is the current value of the ith component of the vector $\boldsymbol{z}$, h_i denotes the ith component of the diagonal of the matrix $\boldsymbol{H}$, and $\boldsymbol{B}$ is given in (8.10). Since $\boldsymbol{B}$ is a function of auxiliary variables z_i, we can update it using the relationship

$$\boldsymbol{B} = \boldsymbol{B}^{\text{old}} + \boldsymbol{s}_i\left(z_i - z_i^{\text{old}}\right),$$

where $\boldsymbol{B}^{\text{old}}$ and z_i^{old} denote, respectively, the values of $\boldsymbol{B}$ and z_i prior to the update of z_i, and $\boldsymbol{s}_i$ is the ith column vector of matrix $\boldsymbol{S} = \boldsymbol{V}\boldsymbol{X}^T$. This algorithm will be denoted here by Marginal($\boldsymbol{z},\boldsymbol{\beta}$). Note that, when $\lambda = 0$

we recover all the expression given by Holmes and Held (2006) for probit model.

8.4.3. *Joint updating of* $\{\boldsymbol{z}, \boldsymbol{w}\}$

The posterior distribution of $\{\boldsymbol{z}, \boldsymbol{w}\}$ given $\{\boldsymbol{\beta}, \lambda\}$ can be factorized as

$$\pi(\boldsymbol{z}, \boldsymbol{w}|\boldsymbol{y}, \boldsymbol{\beta}, \lambda) = \pi(\boldsymbol{z}|\boldsymbol{y}, \boldsymbol{\beta}, \lambda)\pi(\boldsymbol{w}|\boldsymbol{z}, \boldsymbol{\beta}, \lambda),$$

where $\pi(\boldsymbol{w}|\boldsymbol{z}, \boldsymbol{\beta}, \lambda)$ and $\pi(\boldsymbol{z}|\boldsymbol{y}, \boldsymbol{\beta}, \lambda)$ are given in (8.12) and (8.15), respectively. We can sample from an univariate truncated skew-normal distribution using the procedure described in Devroye (1986) to sample from truncated distributions. For that, it is only necessary to evaluate the cdf of the skew-normal distribution and its inverse function. Those functions are implemented in *sn* package available in statistical software R (R Development Core Team, 2009). This algorithm will be denoted by Marginal($\boldsymbol{z}, \boldsymbol{w}$).

8.4.4. *Joint updating of* $\{\boldsymbol{z}, \boldsymbol{\lambda}\}$

An alternative for skew-probit model is to consider the block $\{\boldsymbol{z}, \lambda\}$. The conditional posterior distribution of $\{\boldsymbol{z}, \lambda\}$ can be factorized as follows

$$\pi(\boldsymbol{z}, \lambda|\boldsymbol{y}, \boldsymbol{\beta}, \boldsymbol{w}) = \pi(\boldsymbol{z}|\boldsymbol{y}, \boldsymbol{\beta}, \boldsymbol{w})\pi(\lambda|\boldsymbol{z}, \boldsymbol{\beta}, \boldsymbol{w}),$$

where $\pi(\lambda|\boldsymbol{z}, \boldsymbol{\beta}, \boldsymbol{w})$ is presented in (8.13) and $\pi(\boldsymbol{z}|\boldsymbol{y}, \boldsymbol{\beta}, \boldsymbol{w})$ is given in Equ. (8.16) of the Proposition 8.2. However, we suggest to use (8.18) from Proposition 8.3 to sample from $\pi(\boldsymbol{z}|\boldsymbol{y}, \boldsymbol{\beta}, \boldsymbol{w})$. Note that, the value m_i in (8.18) must be updated for each new value of z_i using the relationship $m = m_i^{\text{old}} + s_i\left(z_i - z_i^{\text{old}}\right)$, where m_i^{old} and z_i^{old} denote the values of m_i and z_i before the update, respectively, and $s_i = \nu w_i$. This algorithm will be denoted by Marginal($\boldsymbol{z}, \lambda$).

8.4.5. *Joint updating of* $\{\boldsymbol{z}, \boldsymbol{\beta}, \boldsymbol{\lambda}\}$

The last algorithm proposed here considers the joint updating of a larger block of the parameters. It is based on (8.19) and in the following factorization

$$\pi(\boldsymbol{z}, \boldsymbol{\theta}|\boldsymbol{y}, \boldsymbol{w}) = \pi(\boldsymbol{z}|\boldsymbol{y}, \boldsymbol{w})\pi(\boldsymbol{\theta}|\boldsymbol{z}, \boldsymbol{w}),$$

where $\boldsymbol{\theta} = (\boldsymbol{\beta}, \lambda)$. The conditional distribution of $\boldsymbol{\theta}$ given $\{\boldsymbol{z}, \boldsymbol{w}\}$ is still normal and it is given by

$$\boldsymbol{\theta}|\boldsymbol{z}, \boldsymbol{w} \sim \mathcal{N}_{p+1}(\boldsymbol{B}, \boldsymbol{V}), \tag{8.23}$$

where $\boldsymbol{B}$ and $\boldsymbol{V}$ are given in (8.22). The distribution $\pi(\boldsymbol{z}|\boldsymbol{y}, \boldsymbol{w})$ is given in (8.20) from the Proposition 8.4. An efficient way to sample from $\pi(\boldsymbol{z}|\boldsymbol{y}, \boldsymbol{w})$ is given in (8.21).

The update of $\boldsymbol{B}$ in (8.21) is carried through each updating of any z_i using the relationship $\boldsymbol{B} = \boldsymbol{B}^{\text{old}} + \boldsymbol{s}_i \left(z_i - z_i^{\text{old}}\right)$, where $\boldsymbol{B}^{\text{old}}$ and z_i^{old} are the same as defined previously, and $\boldsymbol{s}_i$ denotes the ith column vector of the matrix $\boldsymbol{S} = \boldsymbol{V}\boldsymbol{A}^T$. Since the matrix $\boldsymbol{A} = [\boldsymbol{X}, \boldsymbol{w}]$ is a function of the auxiliary variables, we must update it for each updating of this vector of the variables. The update of $\boldsymbol{w}$ is made using the distribution $\pi(\boldsymbol{w}|\boldsymbol{z}, \boldsymbol{\beta}, \lambda)$ presented in (8.12). This algorithm will be denoted by Marginal$(\boldsymbol{z}, \boldsymbol{\beta}, \lambda)$.

8.5. Comparison between Algorithms

In this section we consider the beetle mortality data set presented in Bliss (1935). This data set records the number of the adult flour beetles killed after five hours of exposure to gaseous carbon disulfide at various concentration levels. For more details about this data set see, for example, Bazán *et al.* (2010). Our objective is to analyze the efficiency of our proposed algorithms over simple Gibbs sampling algorithms after fitting the model.

We consider two efficiency measures to illustrate the gain with the use of the joint updating algorithms in comparison to the simple Gibbs sampling algorithm. The first measure is the average Euclidean update distance between interactions of vector of parameters, which is defined as

$$\text{DIS} = \frac{1}{M-1}\sum_{i=1}^{M-1} ||\boldsymbol{\theta}^{(i)} - \boldsymbol{\theta}^{(i+1)}||,$$

where $||\cdot||$ denote the Euclidean norm and $\boldsymbol{\theta}^{(i)}$ denotes the ith vector of a sample with size M from the posterior distribution of $\boldsymbol{\theta}$ obtained by MCMC method. This distance shows us how the Markov chain is mixing itself. Large values of DIS indicate a large mixture in the chain.

The second measure is the effective sample size (ESS), given by

$$\text{ESS} = \frac{M}{1 + 2\sum_{s=1}^{\infty}\rho(s)},$$

where $\rho(s)$ is the sth serial autocorrelation (Ripley, 1987). The ESS can be interpreted as the number of observations belonging to a simple random sample that estimates the interest parameter with the same precision of correlated sample of size M obtained by MCMC method. Large values of ESS indicate a better precision to estimates the interest parameter.

The programs are written using the programming language S, implemented in statistical program R and run on a desktop PC. The choice of this program is suggested because it has open source and it has several useful statistical packages implemented for this work. We start considering known skewness parameter, and then study the case where the skewness parameter is unknown.

8.5.1. *Efficiency analysis with known skewness*

The skewness parameter can be considered known when we use results from previous studies or when we want to estimate it by using a grid of points. In that case, we can only use the Conditional, Marginal($\boldsymbol{z}, \boldsymbol{\beta}$) and Marginal($\boldsymbol{z}, \boldsymbol{w}$) algorithms to obtain a posterior sample from the regression parameters.

For each one of these algorithms it is considered eight parallel chains. For each chain, it is monitored the graphs of the ergodic averages and considered Gelman-Rubin and Geweke tests [see, for example Gamerman and Lopes (2006)]. A sample size of 20000 and a burn-in of 20000 interactions are considered.

The linear predictor of the model is defined by $\eta_i = \beta_0^* + \beta_1(x_i - \bar{x})$, with $\beta_0 = \beta_0^* - \beta_1\bar{x}$, where x_i is the dosage received by ith beetle, and $\bar{x}$ is the average dosage. It is considered that β_0^* and β_1 have independent normal prior distributions with mean 0 and variance 1000. It is also assumed that $\lambda = 4$, which is close to the posterior median obtained by Bazán *et al.* (2005).

Table 8.1 presents the performance of the three algorithms according to ESS and DIS measures. The second column records the CPU run time, in seconds, to generate a sample of size 1000. The third and forth columns list the average of the values of ESS and DIS obtained over eight chains. The fifth and sixth show the ratio between the marginal and conditional ESS and DIS means. The last two columns show the relative efficiency of the new approach with joint updating according to conditional algorithm. The values of the standard deviation are all smaller than 0.01 and the standard deviation of the ESS is approximately 5% of the averages for all algorithms.

We note from Table 8.1 that Marginal($\boldsymbol{z}, \boldsymbol{w}$) and Conditional algorithms are similar according to our proposed measures. Nevertheless, we have that the Marginal($\boldsymbol{z}, \boldsymbol{\beta}$) algorithm is much more efficient than both of them, obtaining a gain of efficiency larger than 120% according to ESS measure and improvement larger than 50% according to DIS measure. These results

Table 8.1. Values of CPU run time, in seconds, and the ESS and DIS measures for different algorithms.

Algorithm	CPU(s)	Mean		Ratio	
		ESS	DIS	ESS	DIS
Conditional	38	422.87	1.68	—	—
Marginal($\boldsymbol{z}, \boldsymbol{\beta}$)	41	945.39	2.67	2.26	1.55
Marginal($\boldsymbol{z}, \boldsymbol{w}$)	45	395.94	1.68	0.94	1.00

show that the chain obtained by Marginal($\boldsymbol{z}, \boldsymbol{\beta}$) is mixing itself more quickly and it needs a smaller sample size than the other algorithms. Therefore, we recommend using Marginal($\boldsymbol{z}, \boldsymbol{\beta}$) algorithm when the skewness parameter is considered fixed or known.

8.5.2. *Efficiency analysis with unknown skewness parameter*

Next we consider that the skewness parameter λ is unknown. It is considered the following flat prior distribution, $\mathcal{N}(0, 1000)$ for the skewness parameter. The convergence monitoring using the graphs of the ergodic averages and the tests of Gelman-Rubin and Geweke showed that the simulated values can be considered a sample from posterior distribution. Furthermore, the analysis of this same data set carried out by Bazán *et al.* (2005) using the Conditional algorithm in WinBugs provided results very close to the one obtained for each one of the 40 chain simulated in this study.

The results of the performance of all five algorithms according to DIS and ESS measures are presented in Tables 8.2 and 8.3. Table 8.3 presents the CPU run time, in seconds, to generate a sample of size 1000 in the second column.

Table 8.2. Values of DIS for different algorithms.

Algorithm	DIS	Ratio
Conditional	1.68	–
Marginal($\boldsymbol{z}, \boldsymbol{\beta}$)	2.24	1.33
Marginal($\boldsymbol{z}, \boldsymbol{w}$)	1.68	1.00
Marginal($\boldsymbol{z}, \lambda$)	1.68	1.00
Marginal($\boldsymbol{z}, \boldsymbol{\beta}, \lambda$)	2.63	1.56

Table 8.3. Values of CPU run time and ESS for different algorithms.

Algorithm	CPU(s)	ESS				
		β_0	β_1	λ	Mean	Ratio
Conditional	64	47.53	49.23	34.16	43.64	–
Marginal($\boldsymbol{z}, \boldsymbol{\beta}$)	69	63.58	66.04	42.77	57.46	1.32
Marginal($\boldsymbol{z}, \boldsymbol{w}$)	72	60.13	62.54	46.19	56.28	1.29
Marginal($\boldsymbol{z}, \lambda$)	92	79.13	82.60	40.58	67.44	1.54
Marginal($\boldsymbol{z}, \boldsymbol{\beta}, \lambda$)	87	118.89	119.14	104.98	114.34	2.62

From Tables 8.2 and 8.3, it is observed that there is a gain with the use of new approach of joint updating, in particularly, with the use of Marginal($\boldsymbol{z}, \boldsymbol{\beta}, \lambda$). Table 8.3 shows that the chains provided by Marginal($\boldsymbol{z}, \boldsymbol{\beta}$) and Marginal($\boldsymbol{z}, \boldsymbol{\beta}, \lambda$) algorithms are mixing more quickly than the others. The relative gain of the Marginal($\boldsymbol{z}, \boldsymbol{\beta}, \lambda$) algorithm is larger than 160%. This is expected because is updated jointly a larger number of variables. We recommend the use of Marginal($\boldsymbol{z}, \boldsymbol{\beta}, \lambda$) algorithm to obtain a sample of posterior distribution when the skewness parameter is unknown.

8.6. Application

We consider as a motivating example the data set presented in Christensen (1997). It consists of a randomly selected subset of 300 patients admitted to the University of New Mexico Trauma Center between the years 1991 and 1994. Of these, 22 died. One of the objectives of this study was to explain the probability of the patient eventually died due to the injuries by using binary regression model and considering the following explanatory variables: injury severity score (ISS), revised trauma score (RTS), patient's age (AGE) and the type of injuries (TI), that is, whether they were blunt (TI=0) or penetrating (TI=1). The response variable is 1 if the patient died and 0 if the patient survives. The ISS is an overall index of injury based on the (approximately) 1300 injuries catalogued in the Abbreviated Injury Scale. It takes values from 0 for patient with no injury to 75 for patient with severe injuries. The RTS is an index of physiologic injury and is constructed as a weighted average of an incoming patient's systolic blood pressure, respiratory rate and Glasgow Coma Scale. It can takes values from 0 for patient with no vital signs to 7.84 for patient with normal vital signs.

The data considered here has been analyzed by means of binary regression model assuming different link functions, such as logistic, probit and complementary log-log models, see Christensen (1997). He compared these models through of Bayes Factor (Kass and Raftery, 1995) and suggest against complementary log-log model, but there are not a serious preference between logistic and probit models. However, there exists significant difference between the observed number of 0's (278 survivors) and 1's (22 fatalities) on the data set, that indicates a skewed link. Thus, based on this argument, we propose a skew-probit link to analyze the data for this motivating example. The skew-probit link is able to fit positively and negatively skewed data. Christensen (1997) fitted a model using an intercept and the predictors ISS, RTS, AGE, TI and the interaction AGE and TI. However, we verify that the intercept is not significant and a model with null intercept fits better the data. Thus, the intercept is assumed to be null on our analysis. Furthermore, we compare three models through of several Bayesian criteria in order to verify which is most appropriate for the data: logistic ($\mathcal{M}_1$), probit ($\mathcal{M}_2$) or skew-probit ($\mathcal{M}_3$).

First, we consider independent normal diffuse prior with mean 0 and variance 1000 for each parameter for all models. We check the convergence of the MCMC using several diagnostic procedures, such as the graphs of the ergodic averages and the Geweke statistic. These diagnostic procedures showed that convergence had been achieved. Finally, the Monte Carlo sample size was taken to be M = 3000 in all calculations.

To compare the models we obtained the values of the Deviance information criterion (DIC), the Bayes factor (Kass and Raftery, 1995) and the pseudo-Bayes factor (Geisser and Eddy, 1979). These values are presented in Table 8.4. Notice that skew-probit model outperforms logistic and probit models for all criteria used.

Based on the analysis reported in Table 8.4, the skew-probit model should be prefered than logistic and probit models. Therefore, we can

Table 8.4. Values of DIC, Bayes factor and pseudo-Bayes factor for Trauma data.

		Bayes factor			Pseudo-Bayes factor		
	DIC	$\mathcal{M}_1$	$\mathcal{M}_2$	$\mathcal{M}_3$	$\mathcal{M}_1$	$\mathcal{M}_2$	$\mathcal{M}_3$
$\mathcal{M}_1$	112.998	—	0.269	0.159	—	0.551	0.191
$\mathcal{M}_2$	112.032	1.689	—	0.591	1.813	—	0.347
$\mathcal{M}_3$	109.781	6.275	3.715	—	5.224	2.881	—

conclude that the skew-probit model seems to be more appropriate to fit this data set than logistic and probit models. The final selected model in our analysis is given by

$$P(Y_i = 1|\boldsymbol{\beta}, \boldsymbol{X}) = F_\lambda(\beta_1 \text{ISS}_i + \beta_2 \text{RTS}_i + \beta_3 \text{AGE}_i + \beta_3 \text{TI}_i + \beta_3 \text{AGE}_i \times \text{TI}_i),$$

where $i = 1, \ldots, 300$ and F_λ is a cdf of a skew-normal distribution given in Chen *et al.* (1999). That is, the skew-normal distribution presented in (8.8) with $\mu = 0$ and $\sigma^2 = 1 + \lambda^2$. Table 8.5 lists posterior summaries of the parameters for skew-probit model, where SD and HDP represent the standard deviation from the posterior distributions and the 95% highest posterior density interval, respectively.

Recall that large values of ISS and low values of RTS are bad for the patient, so the tendency of ISS and RTS coefficients to be positive and negative, respectively, is reasonable. Moreover, as indicated in the estimate HPD for λ, a negative skewed link fits better this data.

8.7. Conclusion

We conclude that the new algorithms are more efficient than the conventional one (without blocks). The Marginal($\boldsymbol{z}, \boldsymbol{\beta}$) algorithm showed best performance when the skewness parameter is fixed. Its proportional gains are more than 120% for the effective sample size. However, when the skewness parameter is unknown, the Marginal($\boldsymbol{z}, \boldsymbol{\beta}$) and Marginal($\boldsymbol{z}, \boldsymbol{\beta}, \lambda$) algorithms are most efficient, where the Marginal($\boldsymbol{z}, \boldsymbol{\beta}, \lambda$) algorithm provided around 160% improvement in the effective sample size measure when comparing to Conditional algorithm. We believe that this performance could be improved even more using a more efficient way to sample from the multivariate truncated normal. These results show that the chains obtained by

Table 8.5. Inference summaries for Trauma data.

Parameter	Posterior mean	Posterior SD	95% HPD interval	
			Lower	Upper
β_1	0.088	0.032	0.029	0.151
β_2	−0.553	0.137	−0.823	−0.303
β_3	0.053	0.021	0.0183	0.098
β_4	0.722	1.272	−1.771	3.263
β_5	0.006	0.034	−0.063	0.074
λ	-4.652	1.988	−8.924	−1.204

the proposed algorithms mixing itself more quickly. Therefore, we need a smaller sample size in the Conditional algorithm to obtain good estimates for the parameters. Finally, we present an application to a medical data set. The skew-probit model seems to be more appropriate to fit the Trauma data set than the logistic and probit models.

Acknowledgment

We gratefully acknowledge grant from FAPESP (grant no. 2007/03598-3) and CNPq (Brazil). The authors are also grateful to a referee for helpful comments and suggestions.

Appendix.

A.1. Proofs of Propositions

Proof. [Proof of Proposition 8.1]

a) Note that

$$\pi(\boldsymbol{\beta}|\boldsymbol{z},\boldsymbol{w},\lambda) = \pi(\boldsymbol{z}|\boldsymbol{\beta},\boldsymbol{w},\lambda)\pi(\boldsymbol{\beta})/\pi(\boldsymbol{z}|\boldsymbol{w},\lambda) \tag{A.1}$$

and

$$\pi(\boldsymbol{z}|\boldsymbol{\beta},\boldsymbol{w},\lambda) = \phi_n\left(\boldsymbol{z};\boldsymbol{X\beta}-\lambda\boldsymbol{w},I_n\right), \tag{A.2}$$

where $\phi_n(\cdot;\boldsymbol{\mu},\boldsymbol{\Sigma})$ denote the pdf of a n-variate normal distribution with means vector $\boldsymbol{\mu}$ and covariance matrix $\boldsymbol{\Sigma}$. Using well known matrix operations, it follows that

$$\pi(\boldsymbol{z},\boldsymbol{\beta}|\boldsymbol{w},\lambda) = K\exp\left\{-\frac{1}{2}(\boldsymbol{\beta}-\boldsymbol{B})^T\boldsymbol{V}^{-1}(\boldsymbol{\beta}-\boldsymbol{B})\right\}, \tag{A.3}$$

where $\boldsymbol{B}=\boldsymbol{V}[\boldsymbol{\nu}^{-1}\boldsymbol{b}+\boldsymbol{X}^T(\boldsymbol{z}-\lambda\boldsymbol{w})]$ and $\boldsymbol{V}=(\boldsymbol{\nu}^{-1}+\boldsymbol{X}^T\boldsymbol{X})^{-1}$, with $K=(2\pi)^{\frac{-(n+p)}{2}}|\boldsymbol{\nu}|^{-\frac{1}{2}}\exp\{-\frac{1}{2}[\boldsymbol{z}-(\boldsymbol{Xb}+\lambda\boldsymbol{w})]^T(\boldsymbol{I}_n+\boldsymbol{X\nu X}^T)^{-1}[\boldsymbol{z}-(\boldsymbol{Xb}+\lambda\boldsymbol{w})]\}$. Therefore,

$$\pi(\boldsymbol{\beta}|\boldsymbol{z},\boldsymbol{w},\lambda)\propto\exp\left\{-\frac{1}{2}(\boldsymbol{\beta}-\boldsymbol{B})^T\boldsymbol{V}^{-1}(\boldsymbol{\beta}-\boldsymbol{B})\right\}. \tag{A.4}$$

It is the kernel of a p-variate normal distribution and $\pi(\boldsymbol{z}|\boldsymbol{w},\lambda)$ in (A.1) is a pdf of a n-variate normal distribution given by

$$\pi(\boldsymbol{z}|\boldsymbol{w},\lambda) = \phi_n\left(\boldsymbol{z};\boldsymbol{Xb}+\lambda\boldsymbol{w},\boldsymbol{I}_n+\boldsymbol{X\nu X}^T\right). \tag{A.5}$$

Then, the full conditional distribution of $\boldsymbol{\beta}$ is given by $\boldsymbol{\beta}|\boldsymbol{z},\boldsymbol{w},\lambda \sim \mathcal{N}_p(\boldsymbol{B},\boldsymbol{V})$, where $\boldsymbol{B}$ and $\boldsymbol{V}$ are given in (A.3).

b) Note that

$$\pi(\boldsymbol{z}|\boldsymbol{y},\boldsymbol{\beta},\boldsymbol{w},\lambda) \propto \pi(\boldsymbol{z}|\boldsymbol{\beta},\boldsymbol{w},\lambda)\pi(\boldsymbol{y}|\boldsymbol{z},\boldsymbol{\beta},\boldsymbol{w},\lambda), \tag{A.6}$$

where $\pi(\boldsymbol{z}|\boldsymbol{\beta},\boldsymbol{w},\lambda)$ is given in (A.2) and

$$\pi(\boldsymbol{y}|\boldsymbol{z},\boldsymbol{\beta},\boldsymbol{w},\lambda) = \pi(\boldsymbol{y}|\boldsymbol{z}) = \operatorname{Ind}(\boldsymbol{y},\boldsymbol{z}) = \prod_{i=1}^{n} \operatorname{Ind}(y_i, z_i), \tag{A.7}$$

with $\operatorname{Ind}(y_i,z_i) = I(z_i > 0)I(y_i = 1) + I(z_i \le 0)I(y_i = 0)$, where $I(A)$ denotes the indicator function in the set A. Then, replacing (A.7) and (A.2) in (A.6), we have $\pi(\boldsymbol{z}|\boldsymbol{y},\boldsymbol{\beta},\boldsymbol{w},\lambda) \propto \phi_n(\boldsymbol{z};\boldsymbol{X}\boldsymbol{\beta} - \lambda\boldsymbol{w},\boldsymbol{I}_n)\operatorname{Ind}(\boldsymbol{y},\boldsymbol{z})$, which is the kernel of a n-variate truncated normal distribution. Then, the components z_i, $i = 1\ldots,n$ are independent random variables with truncated normal distributions given by

$$z_i|y_i,w_i,\boldsymbol{\beta},\lambda \propto \begin{cases} \mathcal{N}(\boldsymbol{x}_i^T\boldsymbol{\beta} + \lambda w_i, 1)I(z_i > 0) & \text{if } y_i = 1, \\ \mathcal{N}(\boldsymbol{x}_i^T\boldsymbol{\beta} + \lambda w_i, 1)I(z_i \le 0) & \text{if } y_i = 0. \end{cases}$$

c) From the independence between the prior distributions of $\boldsymbol{\beta},\boldsymbol{w}$ and λ, follows that

$$\begin{aligned}\pi(\boldsymbol{z},\boldsymbol{w}|\boldsymbol{\beta},\lambda) &= \exp\left\{-\frac{1}{2}\left[\boldsymbol{z} - (\boldsymbol{X}\boldsymbol{\beta} + \lambda\boldsymbol{w})\right]^T\left[\boldsymbol{z} - (\boldsymbol{X}\boldsymbol{\beta} - \lambda\boldsymbol{w})\right]\right\} \\ &\quad \times\pi^{-n}\exp\left\{-\frac{1}{2}\boldsymbol{w}^T\boldsymbol{w}\right\} I(\boldsymbol{w} > \boldsymbol{0}) \\ \pi(\boldsymbol{z},\boldsymbol{w}|\boldsymbol{\beta},\lambda) &= K\exp\left\{-\frac{1+\lambda^2}{2}\left(\boldsymbol{w} - \boldsymbol{m}\right)^T\left(\boldsymbol{w} - \boldsymbol{m}\right)\right\},\end{aligned} \tag{A.8}$$

where $K = \pi^{-n}\exp\left\{-\frac{1}{2}\left(\boldsymbol{z} - \boldsymbol{X}\boldsymbol{\beta}\right)^T\left[2(1+\lambda^2)\right]^{-1}\left(\boldsymbol{z} - \boldsymbol{X}\boldsymbol{\beta}\right)\right\}$ and $m = \lambda(1+\lambda^2)^{-1}\left(\boldsymbol{z} - \boldsymbol{X}\boldsymbol{\beta}\right)$. Therefore, $\boldsymbol{w}|\boldsymbol{z},\boldsymbol{\beta},\lambda \propto \mathcal{N}_n\left(\frac{\lambda}{1+\lambda^2}(\boldsymbol{z} - \boldsymbol{X}\boldsymbol{\beta}), \frac{1}{1+\lambda^2}\boldsymbol{I}_n\right) I(\boldsymbol{w} > \boldsymbol{0})$. As the covariance matrix $(1+\lambda^2)^{-1}\boldsymbol{I}_n$ is diagonal, then w_i, $i = 1,\ldots,n$ are independent random variables with truncated normal distribution given by

$$w_i|z_i,\boldsymbol{\beta},\lambda \propto \mathcal{N}\left(\frac{\lambda}{1+\lambda^2}(z_i - \boldsymbol{x}_i^T\boldsymbol{\beta}), \frac{1}{1+\lambda^2}\right) I(w_i > 0).$$

d) Note that

$$\pi(\lambda|\boldsymbol{z},\boldsymbol{\beta},\boldsymbol{w}) = \pi(\boldsymbol{z},\lambda|\boldsymbol{\beta},\boldsymbol{w})/\pi(\boldsymbol{z}|\boldsymbol{\beta},\boldsymbol{w}). \tag{A.9}$$

From the independence between the prior of $\boldsymbol{\beta}, \boldsymbol{w}$ and λ,

$$\pi(\boldsymbol{z},\lambda|\boldsymbol{\beta},\boldsymbol{w}) = \exp\left\{-\frac{1}{2}\left[\boldsymbol{z} - (\boldsymbol{X\beta} - \lambda\boldsymbol{w})\right]^T\left[\boldsymbol{z} - (\boldsymbol{X\beta} - \lambda\boldsymbol{w})\right]\right\}$$
$$\times(2\pi)^{-n/2}(2\pi\tau)^{-1/2}\exp\left\{-\frac{1}{2}\frac{(\lambda-\alpha)^2}{\tau}\right\}$$
$$\pi(\boldsymbol{z},\lambda|\boldsymbol{\beta},\boldsymbol{w}) = K\exp\left\{-\frac{1}{2}\frac{(\lambda-m)^2}{\nu}\right\}, \tag{A.10}$$

where $m = \nu\left[\tau^{-1}\alpha - \boldsymbol{w}^T(\boldsymbol{z} - \boldsymbol{X\beta})\right]$, $\nu = \left(\boldsymbol{w}^T\boldsymbol{w} + \tau^2\right)^{-1}$ and $K = (2\pi)^{-\frac{n+1}{2}}\,\tau^{-1/2}\exp\left\{-\frac{1}{2}\left[\boldsymbol{z} - (\boldsymbol{X\beta} + \alpha\boldsymbol{w})\right]^T(\boldsymbol{I}_n + \tau\boldsymbol{w}\boldsymbol{w}^T)^{-1}\right.$ $\left.\left[\boldsymbol{z} - (\boldsymbol{X\beta} + \alpha\boldsymbol{w})\right]\right\}$. Then, $\pi(\lambda|\boldsymbol{z},\boldsymbol{\beta},\boldsymbol{w}) \propto \exp\left\{-(\lambda-m)^2/(2\nu)\right\}$ is the kernel of a normal distribution and $\pi(\boldsymbol{z}|\boldsymbol{\beta},\boldsymbol{w})$ in (A.9) is a pdf of a n-variate normal distribution given by

$$\pi(\boldsymbol{z}|\boldsymbol{\beta},\boldsymbol{w}) = \phi_n\left(\boldsymbol{z};\boldsymbol{X\beta} + \alpha\boldsymbol{w}, \boldsymbol{I}_n + \tau\boldsymbol{w}\boldsymbol{w}^T\right), \tag{A.11}$$

Then, the full conditional distribution of λ is given by $\lambda|\boldsymbol{z},\boldsymbol{w},\boldsymbol{\beta} \sim \mathcal{N}(m,\nu)$, where m and ν are given in (A.10).

□

Proof. [Proof of Proposition 8.2]

a) From (A.5) and (A.7) follows that

$$\pi(\boldsymbol{z}|\boldsymbol{y},\boldsymbol{w},\lambda) = \frac{\phi_n\left(\boldsymbol{z};\boldsymbol{Xb} + \lambda\boldsymbol{w}, \boldsymbol{I}_n + \boldsymbol{X}\boldsymbol{\nu}\boldsymbol{X}^T\right)}{\bar{\boldsymbol{\Phi}}_n(R(\boldsymbol{y});\boldsymbol{Xb} - \lambda\boldsymbol{w}, \boldsymbol{I}_n + \boldsymbol{X}\boldsymbol{\nu}\boldsymbol{X}^T)}\text{Ind}(\boldsymbol{y},\boldsymbol{z}),$$

where $\bar{\boldsymbol{\Phi}}_n(R(\boldsymbol{y},\boldsymbol{z});\boldsymbol{\mu},\boldsymbol{\Sigma})$ denote the cdf in the region $R(\boldsymbol{y},\boldsymbol{z}) = \{\boldsymbol{z} = (z_1, z_2, \ldots, z_n)^T; z_i > 0 \text{ if } y_i = 1 \text{ or } z_i \leq 0 \text{ if } y_i = 0\}$. Hence, the conditional distribution of $\boldsymbol{z}$ is given by

$$\boldsymbol{z}|\boldsymbol{y},\boldsymbol{w},\lambda \propto \mathcal{N}_n(\boldsymbol{Xb} + \lambda\boldsymbol{w}, \boldsymbol{I}_n + \boldsymbol{X}\boldsymbol{\nu}\boldsymbol{X}^T)\text{Ind}(\boldsymbol{y},\boldsymbol{z}), \tag{A.12}$$

where $\text{Ind}(\boldsymbol{y},\boldsymbol{z})$ is the indicator function given in (A.7). Note that, for $\lambda = 0$ the distribution (A.12) reduces to posterior marginal distribution of $\boldsymbol{z}$ presented in Holmes and Held (2006) for the probit model.

b) Considering the model in (8.6), we have that $z_i|w_i, \boldsymbol{\beta}, \lambda \sim \mathcal{N}(\boldsymbol{x}_i^T\boldsymbol{\beta} + \lambda w_i, 1)$ and $w_i \sim \mathcal{N}(0,1)I(w_i > 0)$. Then, using the properties of the skew-normal distribution (Azzalini, 1985) follows that $z_i|\boldsymbol{\beta}, \lambda \sim \mathcal{SN}(\boldsymbol{x}_i^T\boldsymbol{\beta}, 1+\lambda^2, \lambda)$. Hence, the conditional distribution of the vector $\boldsymbol{z}$ is given by

$$\pi(\boldsymbol{z}|\boldsymbol{\beta}, \lambda) = \prod_{i=1}^{n} \phi\left(\frac{z_i - \boldsymbol{x}_i^T\boldsymbol{\beta}}{\sqrt{1+\lambda^2}}\right) \Phi\left[\frac{\lambda}{\sqrt{1+\lambda^2}}(z_i - \boldsymbol{x}_i^T\boldsymbol{\beta})\right]. \quad \text{(A.13)}$$

On the other hand, we have that $\pi(\boldsymbol{z}, \boldsymbol{y}|\boldsymbol{\beta}, \lambda) = \pi(\boldsymbol{z}|\boldsymbol{\beta}, \lambda)\pi(\boldsymbol{y}|\boldsymbol{z})$, where $\pi(\boldsymbol{z}|\boldsymbol{\beta}, \lambda)$ and $\pi(\boldsymbol{y}|\boldsymbol{z})$ are given in (A.13) and (A.7), respectively. Then,

$$\pi(\boldsymbol{z}, \boldsymbol{y}|\boldsymbol{\beta}, \lambda) = \prod_{i=1}^{n} \phi\left(\frac{z_i - \boldsymbol{x}_i^T\boldsymbol{\beta}}{\sqrt{1+\lambda^2}}\right) \Phi\left[\frac{\lambda(z_i - \boldsymbol{x}_i^T\boldsymbol{\beta})}{\sqrt{1+\lambda^2}}\right] \prod_{i=1}^{n} \mathrm{Ind}(y_i, z_i)$$

and $\pi(\boldsymbol{z}|\boldsymbol{y}, \boldsymbol{\beta}, \lambda) \propto \prod_{i=1}^{n} \left\{\phi\left(\frac{z_i - \boldsymbol{x}_i^T\boldsymbol{\beta}}{\sqrt{1+\lambda^2}}\right) \Phi\left[\frac{\lambda(z_i - \boldsymbol{x}_i^T\boldsymbol{\beta})}{\sqrt{1+\lambda^2}}\right] \mathrm{Ind}(y_i, z_i)\right\}$. Therefore, z_i, $i = 1\ldots, n$ are independent random variables with truncated skew-normal distribution given by

$$z_i|y_i, \boldsymbol{\beta}, \lambda \propto \begin{cases} \mathcal{SN}(\boldsymbol{x}_i^T\boldsymbol{\beta}, 1+\lambda^2, \lambda)I(z_i > 0) & \text{if } y_i = 1, \\ \mathcal{SN}(\boldsymbol{x}_i^T\boldsymbol{\beta}, 1+\lambda^2, \lambda)I(z_i \leq 0) & \text{if } y_i = 0. \end{cases}$$

c) From (A.7) and (A.11) follows

$$\pi(\boldsymbol{z}, \boldsymbol{y}|\boldsymbol{w}, \boldsymbol{\beta}) = \phi_n\left(\boldsymbol{z}; \boldsymbol{X\beta} + \alpha\boldsymbol{w}, \boldsymbol{I}_n + \boldsymbol{w}\nu\boldsymbol{w}^T\right) \mathrm{Ind}(\boldsymbol{y}, \boldsymbol{z}). \quad \text{(A.14)}$$

Therefore,

$$\pi(\boldsymbol{z}|\boldsymbol{y}, \boldsymbol{w}, \boldsymbol{\beta}) = C^{-1}\phi_n\left(\boldsymbol{z}; \boldsymbol{X\beta} + \alpha\boldsymbol{w}, \boldsymbol{I}_n + \boldsymbol{w}\nu\boldsymbol{w}^T\right) \mathrm{Ind}(\boldsymbol{y}, \boldsymbol{z}). \quad \text{(A.15)}$$

Notice that (A.15) is a pdf since $C = \bar{\boldsymbol{\Phi}}_n(R(\boldsymbol{y}); \boldsymbol{Xb}, \boldsymbol{I}_n + \boldsymbol{w}\nu\boldsymbol{w}^T)$, where $\bar{\boldsymbol{\Phi}}_n(\cdot)$ and $R(\boldsymbol{y}, \boldsymbol{z})$ are given in (A.12). Therefore, the distribution of $\boldsymbol{z}|\boldsymbol{y}, \boldsymbol{w}, \boldsymbol{\beta}$ is given by

$$\boldsymbol{z}|\boldsymbol{y}, \boldsymbol{w}, \boldsymbol{\beta} \propto \mathcal{N}_n(\boldsymbol{X\beta} + \alpha\boldsymbol{w}, \boldsymbol{I}_n + \boldsymbol{w}\nu\boldsymbol{w}^T)\mathrm{Ind}(\boldsymbol{y}, \boldsymbol{z}), \quad \text{(A.16)}$$

where $\mathrm{Ind}(\boldsymbol{y}, \boldsymbol{z})$ is the indicator function given in (A.7).

□

Proof. [Proof of Proposition 8.3]

a) Note that

$$\pi(z_i|\boldsymbol{z}_{-i}, \boldsymbol{y}, \boldsymbol{w}, \lambda) = \pi(\boldsymbol{z}|\boldsymbol{z}_{-i}, \boldsymbol{y}, \boldsymbol{w}, \lambda)/\pi(\boldsymbol{z}_{-i}|\boldsymbol{y}, \boldsymbol{w}, \lambda), \quad \text{(A.17)}$$

where $\boldsymbol{z}_{-i}$ denotes the vector $\boldsymbol{z}$ with the ith variable removed. On the other hand, we can write (A.12) as

$$\pi(\boldsymbol{z}|\boldsymbol{y},\boldsymbol{w},\lambda) \propto \exp\left\{\frac{h_{ii}-1}{2}\left[z_i - \left(\sum_{j=1,j\neq i}^{n}\frac{h_{ij}(z_j-\mu_j)}{1-h_{ii}}+\mu_i\right)\right]^2\right\}$$

where $\boldsymbol{z} \in R(\boldsymbol{y},\boldsymbol{z})$, with $R(\boldsymbol{y},\boldsymbol{z})$ given in (A.12). Therefore, we can write the full conditional distribution of z_i as

$$z_i|\boldsymbol{z}_{-i}, y_i, \boldsymbol{w}, \lambda \propto \begin{cases} \mathcal{N}(m_i,\nu_i)I(z_i>0) & \text{if } y_i = 0, \\ \mathcal{N}(m_i,\nu_i)I(z_i\leq 0) & \text{otherwise,}\end{cases} \tag{A.18}$$

where $m_i = \boldsymbol{x}_i^T\boldsymbol{b} + \lambda w_i + (1-h_{ii})^{-1}\sum_{k=1,k\neq i}^{n} h_{ik}(z_k - \boldsymbol{x}_k^T\boldsymbol{b} - \lambda w_k)$ and $\nu_i = (1-h_{ii})^{-1}$. On the other hand, the location parameter m_i can be rewritten in function of $\boldsymbol{B} = \boldsymbol{V}[\boldsymbol{\nu}^{-1}\boldsymbol{b} + \boldsymbol{X}^T(\boldsymbol{z}+\lambda\boldsymbol{w})]$, where $\boldsymbol{V} = (\boldsymbol{\nu}^{-1} + \boldsymbol{X}^T\boldsymbol{X})^{-1}$. This reparametrisation provides a suitable structure for using the Gibbs algorithm. Then, writing $h_{ij} = \boldsymbol{x}_i^T\boldsymbol{V}\boldsymbol{x}_j$, follows that

$$m_i = \boldsymbol{x}_i^T\boldsymbol{B} + \lambda w_i - h_i(1-h_{ii})^{-1}\left[z_i - \left(\boldsymbol{x}_i^T\boldsymbol{B} + \lambda w_i\right)\right].$$

b) We have that

$$\pi(z_i|\boldsymbol{z}_{-i},\boldsymbol{y},\boldsymbol{w},\boldsymbol{\beta}) = \pi(\boldsymbol{z}|\boldsymbol{z}_{-i},\boldsymbol{y},\boldsymbol{w},\boldsymbol{\beta})/\pi(\boldsymbol{z}_{-i}|\boldsymbol{y},\boldsymbol{w},\boldsymbol{\beta}), \tag{A.19}$$

where $\boldsymbol{z}_{-i}$ denotes the vector $\boldsymbol{z}$ with the ith variable removed. On the other hand, we can write (A.16) as

$$\pi(\boldsymbol{z}|\boldsymbol{y},\boldsymbol{w},\lambda) \propto \exp\left\{\frac{h_{ii}-1}{2}\left[z_i - \left(\sum_{j=1,j\neq i}^{n}\frac{h_{ij}(z_j-\mu_j)}{1-h_{ii}}+\mu_i\right)\right]^2\right\},$$

where $\boldsymbol{z} \in R(\boldsymbol{y},\boldsymbol{z})$, with $R(\boldsymbol{y},\boldsymbol{z})$ given in (A.12), $h_{ij} = w_i^T(\tau^{-1} + \boldsymbol{w}^T\boldsymbol{w})^{-1}w_j$ and $\mu_i = \boldsymbol{x}_i^T\boldsymbol{\beta} - \alpha w_i$. Then, for each z_i the full conditional distribution is given by

$$z_i|\boldsymbol{z}_{-i}, y_i, \boldsymbol{w}, \lambda \propto \begin{cases} N(m_i,\nu_i)I(z_i>0) & \text{if } y_i = 0, \\ N(m_i,\nu_i)I(z_i\leq 0) & \text{otherwise,}\end{cases} \tag{A.20}$$

where $m_i = \boldsymbol{x}_i^T\boldsymbol{\beta} + \alpha w_i + (1-h_{ii})^{-1}\sum_{k=1,k\neq i}^{n} h_{ik}(z_k - \boldsymbol{x}_k^T\boldsymbol{\beta} - \alpha w_k)$ and $\nu_i = (1-h_{ii})^{-1}$. On the other hand, the location parameter m_i

can be rewritten as a function of the posterior mean of λ, namely, as a function of $m = \nu[\tau^{-1}\alpha - \boldsymbol{w}^T(\boldsymbol{z} - \boldsymbol{X\beta})]$, where $\nu = (\boldsymbol{w}^T\boldsymbol{w} + \tau^2)^{-1}$. This reparametrisation provides a suitable structure for using the Gibbs Algorithm. Then, writing $h_{ij} = w_i \nu w_j$, we have that

$$m_i = w_i m + \alpha w_i - h_i(1 - h_{ii})^{-1}\left[z_i - (w_i m + \alpha w_i)\right].$$

□

Proof. [Proof of Proposition 8.4] Observe that, given $\boldsymbol{w}$, the model (8.19) is reduced to probit model. Then, the proof of this Proposition is obtained in a similar way from the proofs of Propositions 2 and 3 when it is assumed that $\lambda = 0$. □

A.2. Pseudo-codes

It is considered that: $A[i]$ denotes the ith element of a column matrix A; $A[i, j]$ denotes the ith, jth element of matrix A; $A[i,]$ and $A[, j]$ denote, respectively, the ith row and the jth column of A; AB denotes matrix multiplication of A and B; $A[i,]B[, j]$ denotes the row, column inner product of A and B. Commented lines are preceded by ##.

A.2.1. *The pseudo-code for Marginal*$\{z, \beta\}$

```
## it is stored the unaltered constants within MCMC loop
    V ← (X^T X + v^{-1})^{-1} ## V is the covariance matrix of β
    S ← V X^T
    FOR j = 1 to number of observations
          H[j] ← X[j,]S[,j];    T[j] ← H[j]/(1 − H[j]);    Q[j] ← T[j] + 1
    END
    ## It is initialized the latent variables Z and W
    Z ~ N_n(0, I_n)Ind(Y, Z);    W ~ N_n(0, I_n)I(W > 0)
    FOR i = 1 to number of MCMC interactions
          B ← V[v^{-1}b + X^T(Z + λ[i]W)]
          FOR j = 1 to number of observations
                Z_old ← Z[j];    m ← X[j,]B + λ[i]W[j]
                m ← m − T[j](Z[j] − m)
                m ← m − T[j](Z[j] − m)
                Z[j] ~ N(m, Q[j])Ind(Y[j], Z[j])
                B ← B + (Z[j] − Z_old)S[,j] ## Updating B
          END
          β[,i] ~ N_p(B, V)
          s ← 1/(1 + λ[i]^2)
          FOR j = 1 to number of observations
                r ← sλ[i](Z[j] − X[j,]β[,i])
```

```
            W[j] ~ N(r, s)I(W[j] > 0)
        END
        q ← 1/(τ^{-1} + W^T W);     d ← q[τ^{-1}α − W^T(Z − Xβ[, i])]
        λ[i] ~ N(d, q)
    END of MCMC interactions; RETURN β and λ
```

A.2.2. *The pseudo-code for Marginal{z, w}*

```
## it is stored the unaltered constants within MCMC loop
    V ← (X^T X + v^{-1})^{-1};     S ← V X^T
    FOR j = 1 to number of observations
        H[j] ← X[j, ]S[, j];     T[j] ← H[j]/(1 − H[j]);     Q[j] ← T[j] + 1
    END
    Z ~ N_n(0, I_n)Ind(Y, Z);     W ~ N_n(0, I_n)I(W > 0)
    FOR i = 1 to number of MCMC interactions
        B ← V[v^{-1}b + X^T(Z + λ[i]W)]
        β[, i] ~ N_p(B, V)
        FOR j = 1 to number of observations
            m ← X[j, ]β[, i] + λ[i]W[j]
            ## It is drawn Z[j] from truncated skew-normal
            Z[j] ~ SN(m, 1 + λ[i]^2, λ[i])Ind(Y[j], Z[j])
        END
        s ← 1/(1 + λ[i]^2)
        FOR j = 1 to number of observations
            r ← sλ[i](Z[j] − X[j, ]β[, i])
            W[j] ~ N(r, s)I(W[j] > 0)
        END
        q ← 1/(τ^{-1} + W^T W);     d ← q[τ^{-1}α − W^T(Z − Xβ[, i])]
        λ[i] ~ N(d, q)
    END of MCMC interactions; RETURN β and λ
```

A.2.3. *The pseudo-code for Marginal{z, λ}*

```
V ← (X^T X + v^{-1})^{-1}
    Z ~ N_n(0, I_n)Ind(Y, Z);     W ~ N_n(0, I_n)I(W > 0)
    FOR i = 1 to number of interactions
        S ← τW;     q ← 1/(τ^{-1} + W^T W)
        d ← q[τ^{-1}α − W^T(Z − Xβ[, i])]
        FOR j = 1 to number of observations
            H[j] ← W[j]S[j];     T[j] ← H[j]/(1 − H[j])
            Q[j] ← T[j] + 1;     Z_old ← Z[j]
            m ← X[j, ]β[, i] − W[j];     m ← m − T[j] (Z[j] − m)
            Z[j] ~ N(m, Q[j])Ind(Y[j], Z[j])
            ## Update d through the following relationship
            d ← d + (Z[j] − Z_old)S[j]
```

```
        END
        λ[i] ~ N(d, q)
        s ← 1/(1 + λ[i]^2)
        FOR j = 1 to number of observations
            r ← sλ[i](Z[j] − X[j, ]β[, i])
            W[j] ~ N(r, s)I(W[j] > 0)
        END
        B ← V[v^{-1}b + X^T(Z − λ[i]W)]
        β[, i] ~ N_p(B, V)
    END of MCMC interactions; RETURN β and λ
```

A.2.4. *The pseudo-code for Marginal$\{z, \beta, \lambda\}$*

```
## It is initialized the latent variables Z and W
    Z ~ N_n(0, I_n)Ind(Y, Z);    W ~ N_n(0, I_n)I(W > 0)

    A ← [X ⋮ − W] ## concatenates the matrix X and the vector W
    FOR i = 1 to number of MCMC interactions
        V ← (A^T A + v^{-1})^{-1};    S ← V A^T;    B ← V(v^{-1}b + A^T Z)
        FOR j = 1 to number of observations
            H[j] ← A[j, ]S[, j];    T[j] ← H[j]/(1−H[j]);    Q[j] ← T[j]+1
            Z_old ← Z[j];    m ← A[j, ]B − T[j](Z[j] − A[j, ]B)
            Z[j] ~ N(m, Q[j])Ind(Y[j], Z[j])
            B ← B + (Z[j] − Z_old)S[, j]
        END
        β[, i] ~ N_{p+1}(B, V)
        β ← β[1 : p, i];    λ ← β[p + 1, i];    s ← 1/(1 + λ^2)
        FOR j = 1 to number of observations
            r ← sλ(Z[j] − X[j, ]β)
            W[j] ~ N(r, s)I(W[j] > 0)
        END
    END of MCMC interactions; RETURN β and λ.
```

References

Albert, J.H. and Chib, S. (1993), Bayesian analysis of binary and polychotomous response data, *Journal of the American Statistical Association.* 88, pp. 669–679.

Azzalini, A. (1985), A class of distributions which includes the normal ones, *Scandinavian Journal Statistical.* 12, pp. 171–178.

Bliss, C.I. (1935), The calculation of the dose-mortality curve, *Annals Applied Biology.* 22, pp. 134–167.

Bazán, J.L., Branco, M.D. and Bolfanine, H. (2005), A skew item response model, *Bayesian Analysis.* 1, pp 861–892.

Bazán, J.L., Bolfanine, H. and Branco, M.D. (2010), A framework for skew-probit links in binary regression, *Communication in Statistics - Theory and Methods.* 39, pp. 678–697 .

Chen, M-H. (2004), The skewed link models for categorical response data, in *Skew-Elliptical Distributions and Their Applications: A Journey Beyond Normality,* Genton, M.G, ed. Boca Raton: Chapman & Hall/CRC.

Chen, M-H. and Dey, D.K. Bayesian modelling of correlated binary responses via scale mixture of multivariate normal link functions, *Sankhyã, Ser. A.* 60 (1998), pp. 322–343.

Chen, M-H, Dey, D.K. and Shao, Q-M. (1999), A new skewed link model for dichotomous quantal response data, *Journal of the American Statistical Association.* 98, pp. 1172–1186.

Chib, S. and Carlin, B.P. (1999), On MCMC sampling in hierarchical longitudinal models, *Statistics and Computing.* 9, pp. 17–26.

Christensen, R. (1997), *Log-Linear Models and Logistic Regression.* 2nd ed. Springer-Verlag: New York.

Devroye, L. (1986), *Non-Uniform Random Variate Generation,* New York: Springer.

Gamerman, D. and Lopes, H.F. (2006), *Markov Chain Monte Carlo: Stochastic Simulation for Bayesian Inference,* Chapman & Hall/CRC, Boca Raton, USA.

Geisser, S. and Eddy, W. (1979), A predictive approach to model selection. *Journal of the American Statistical Association.* 74, pp. 153-160.

Holmes, C.C. and Held, L. (2006), Bayesian auxiliary variable models for binary and multinomial regression, *Bayesian Analysis.* 1, pp.145–168.

Liu, J.C. (1994), The collapsed Gibbs sampler in Bayesian computations with applications to a gene regulation problem, *Journal of the American Statistical Association.* 89 pp. 958–966.

Kass, R.E. and Raftery,A.E. (1995), Bayes factors. *Journal of the American Statistical Association.* 90, pp. 773–795.

R Development Core Team (2009), *R: A language and environment for statistical computing,* R Foundation for Statistical Computing, Vienna, Austria. ISBN 3-900051-07-0, avaliable at http://www.R-project.org.

Ripley, B.D. (1987), *Stochastic Simulation,* New York, Wiley.

Sahu, S.K., Dey, D.K. and Branco, M.D. (2003), A new class of multivariate skew distributions with applications to Bayesian regression models, *The Canadian Journal of Statistics.* 29, pp. 217–232.

Chapter 9

M-Estimation Methods in Heteroscedastic Nonlinear Regression Models

Changwon Lim[*,§], Pranab K. Sen[†,‡,¶] and Shyamal D. Peddada[*,||]

[*]*Biostatistics Branch, NIEHS, NIH*
111 T. W. Alexander Dr. RTP, NC 27709, USA
[†]*Department of Statistics and Operations Research*
University of North Carolina at Chapel Hill
338 Hanes Hall, CB#3260, Chapel Hill, NC 27599, USA
[‡]*Department of Biostatistics*
University of North Carolina at Chapel Hill
3101 McGavran-Greenberg, CB#7420, Chapel Hill, NC 27599, USA
[§]*limc2@niehs.nih.gov*
[¶]*pksen@bios.unc.edu*
[||]*peddada@niehs.nih.gov*

In many applications it is common to encounter data which depart from various modeling assumptions such as homogeneity of variances. The problem is further complicated by the existence of potential outliers and influential observations. In such situations results based on ordinary least squares (OLS) and model based maximum likelihood (ML) methods may be inappropriate and even misleading. In this study a robust M-estimation based methodology for heteroscedastic nonlinear regression models is considered. An M-estimator is proposed and its asymptotic properties, including asymptotic normality, are studied under suitable regularity conditions. The proposed methodology is illustrated using a real example from toxicology.

9.1. Introduction

In many applications, such as in toxicological sciences, researchers are interested in developing nonlinear statistical models to describe the relationship between a response variable (Y) and an independent variable (X) (Velarde *et al.*, 1999; Avalos *et al.*, 2001; Pounds *et al.*, 2004). The commonly used ordinary least squares (OLS) methodology for model fitting and drawing inferences on the unknown parameters relies on various assumptions such as homoscedasticity of error variances and normality of the residuals (Seber and Wild, 1989). However, in practice, often such assumptions are not satisfied (Morris *et al.*, 2002; Gaylor and Aylward, 2004; Barata *et al.*, 2006). In addition, the presence of outliers and influential observations can affect the performance of inference based on OLS and maximum likelihood (ML) (Huber, 1981; Hampel *et al.*, 1986).

Estimators given by classical methods, such as OLS and ML procedures, are usually nonrobust to outliers/influential observations or departures from the specified distribution of the response variable. Robust methods such as M-estimation methods are prefered in such situations. During the past several decades, the M-estimators have been well studied in the context of linear models (cf. Huber, 1981; Jureĉková and Sen, 1996; Maronna *et al.*, 2006). Huber (1973) proposed the M-estimator of the regression parameters in the univariate linear model and showed that under certain regularity conditions the M-estimator is consistent and asymptotically normal. The asymptotic theory for the M-estimator has been studied by Relles (1968), Huber (1973), Yohai and Maronna (1979), and Klein and Yohai (1981); for multivariate linear models one may refer to Marrona (1976), Singer and Sen (1985), Kent and Tyler (2001), Tyler (2002), and Maronna and Yohai (2008), among others. Asymptotic behavior of the M-estimator under the nonstandard/nonregular conditions has been also considered recently in the literature (Bose and Chatterjee, 2001; Bantli and Hallin, 2001; Bantli, 2004). Under such conditions the limiting distributions of the M-estimator are usually non-Gaussian and the M-estimator is no longer consistent.

Sanhueza (2000) and Sanhueza and Sen (2001, 2004) proposed M-estimators in the context of nonlinear regression models and studied their asymptotic properties. More recently, Sanhueza *et al.* (2009) extended these methods to nonlinear models for repeated measures data.

Assuming the error variance to be a known function of unknown parameters of the regression model, M-estimation methods in heteroscedastic nonlinear models are considered in this paper. The proposed methodology is a variation to one proposed in Sanhueza and Sen (2001). Difference between the two methods is detailed in Sec. 9.2.

An M-estimator for the parameters in the heteroscedastic nonlinear model is proposed in Sec. 9.2, along with notation and needed regularity conditions. In Sec. 9.3, the asymptotic properties of the M-estimator are established. In Sec. 9.4 we illustrate the proposed methodology using a real toxicology data from the US National Toxicology Program (NTP). In Sec. 9.5 the results are summarized and ongoing research discussed.

9.2. Definitions and Regularity Conditions

Let

$$y_i = f(\mathbf{x}_i, \boldsymbol{\theta}) + \sigma_i \epsilon_i, \quad i = 1, \ldots, n \tag{9.1}$$

where y_i are the observable random variables of size n, $\mathbf{x}_i = (x_{1i}, x_{2i}, \ldots, x_{mi})^t$ are known regression constants, $\boldsymbol{\theta} = (\theta_1, \theta_2, \ldots, \theta_p)^t$ is a vector of unknown parameters, $f(\cdot)$ is a nonlinear function of $\boldsymbol{\theta}$ of specified form; and the errors ϵ_{ij} are assumed to be independent random variables with mean 0 and variance 1. It is further assumed that $\sigma_i = \sigma(\mathbf{x}_i, \boldsymbol{\theta})$ for $i = 1, \ldots, n$, where $\sigma(\cdot)$ is a known function.

An M-estimator of $\boldsymbol{\theta}$ is defined as one that solves the following minimization problem:

$$\hat{\boldsymbol{\theta}}_n = Argmin \left\{ \sum_{i=1}^{n} h^2 \left(\frac{y_i - f(\mathbf{x}_i, \boldsymbol{\theta})}{\sigma(\mathbf{x}_i, \boldsymbol{\theta})} \right) : \boldsymbol{\theta} \in \boldsymbol{\Theta} \subseteq \Re^p \right\}, \tag{9.2}$$

where $h(\cdot)$ is a real valued function, and $\boldsymbol{\Theta}$ is a compact subset of $\Re^p$. Let $\psi(z) = (\partial/\partial z)h^2(z)$, then the estimating equation for the minimization problem Eq. (9.2) is given by:

$$\sum_{i=1}^{n} \lambda(\mathbf{x}_i, y_i, \hat{\boldsymbol{\theta}}_n) = \mathbf{0}, \tag{9.3}$$

where

$$\lambda(\mathbf{x}_i, y_i, \boldsymbol{\theta}) = \psi \left(\frac{y_i - f(\mathbf{x}_i, \boldsymbol{\theta})}{\sigma(\mathbf{x}_i, \boldsymbol{\theta})} \right) \left\{ \frac{\mathbf{f}_{\boldsymbol{\theta}}(\mathbf{x}_i, \boldsymbol{\theta})}{\sigma(\mathbf{x}_i, \boldsymbol{\theta})} + \frac{y_i - f(\mathbf{x}_i, \boldsymbol{\theta})}{\sigma(\mathbf{x}_i, \boldsymbol{\theta})} \frac{\boldsymbol{\sigma}_{\boldsymbol{\theta}}(\mathbf{x}_i, \boldsymbol{\theta})}{\sigma(\mathbf{x}_i, \boldsymbol{\theta})} \right\}, \tag{9.4}$$

$\mathbf{f}_{\boldsymbol{\theta}}(\mathbf{x}_i, \boldsymbol{\theta}) = (\partial/\partial\boldsymbol{\theta}) f(\mathbf{x}_i, \boldsymbol{\theta})$, and $\boldsymbol{\sigma}_{\boldsymbol{\theta}}(\mathbf{x}_i, \boldsymbol{\theta}) = (\partial/\partial\boldsymbol{\theta})\sigma(\mathbf{x}_i, \boldsymbol{\theta})$.

The methodology and asymptotic theory of the M-estimation in this paper are similar to those in Sanhueza and Sen (2001). They proposed the following M-estimator for generalized nonlinear models when the distribution of the response variable belongs to an exponential family:

$$\hat{\boldsymbol{\theta}}_n = Argmin \left\{ \sum_{i=1}^{n} \frac{1}{h(v_i(f(\mathbf{x}_i, \boldsymbol{\theta})))} h^2(y_i - f(\mathbf{x}_i, \boldsymbol{\theta})) : \boldsymbol{\theta} \in \boldsymbol{\Theta} \subseteq \Re^p \right\}, \tag{9.5}$$

where $v_i(\cdot)$ is assumed to be a known variance function. The difference between Eq. (9.2) and Eq. (9.5) are summarized as follows:

(a) In Eq. (9.5), the variance term is separated from the residual term, whereas Eq. (9.2) uses standardized residual after scaling the residual by the error standard deviation. Although Eq. (9.2) as well as Eq. (9.5) assume that the error variance is a known function of the unknown regression parameters, the methods differ in how the parameters are estimated. A major difference between the present paper and Sanhueza and Sen (2001) is in the derivation of the asymptotic normality of the M-estimator. While they derive the asymptotic distribution by assuming that the variance used in the estimating equation is known, such an assumption is avoided here.

(b) In the present formulation the standardized residual is robustified rather than individually robustifying the residual and the variance as in Sanhueza and Sen (2001).

The proof of the consistency and the asymptotic normality of the M-estimator derived in this paper relies on the Hájek-Šidak Central Limit Theorem, which in turn requires Noether type conditions (Singer and Sen, 1985; Jurečková and Sen, 1996; Sanhueza and Sen, 2001). To enable this, the following regularity conditions are needed regarding (A) the score function ψ, (B) the function f, and (C) the function σ:

[A1]:
ψ is nonconstant, absolutely continuous and differentiable with respect to $\boldsymbol{\theta}$.

[A2]:
Let $\epsilon = \{y - f(\mathbf{x}, \boldsymbol{\theta})\}/\sigma(\mathbf{x}, \boldsymbol{\theta})$,

(i) $E\psi(\epsilon) = 0$; $E\psi^2(\epsilon) = \sigma^2_{\psi 1} u(\mathbf{x}) < \infty$, and $E\{\psi(\epsilon)\epsilon\} = \gamma_1 (\neq 0)$; $\text{Var}\{\psi(\epsilon)\epsilon\} = \sigma^2_{\psi 2} v(\mathbf{x}) < \infty$.

(ii) $E|\psi'(\epsilon)|^{1+\delta} < \infty$, $E|\psi'(\epsilon)\epsilon|^{1+\delta} < \infty$, $E|\psi'(\epsilon)\epsilon^2|^{1+\delta} < \infty$ for some $0 < \delta \leq 1$, and $E\psi'(\epsilon) = \gamma_2\,(\neq 0)$; $E\{\psi'(\epsilon)\epsilon\} = 0$; $E\{\psi'(\epsilon)\epsilon^2\} = \gamma_3\,(\neq 0)$.

[A3]:
Let $\epsilon(\boldsymbol{\theta}) = \{y - f(\mathbf{x}, \boldsymbol{\theta})\}/\sigma(\mathbf{x}, \boldsymbol{\theta})$,

(i) $\lim_{\delta \to 0} E\left\{\sup_{\|\boldsymbol{\Delta}\| \leq \delta} |\psi(\epsilon(\boldsymbol{\theta} + \boldsymbol{\Delta})) - \psi(\epsilon(\boldsymbol{\theta}))|\right\} = 0$.

(ii) $\lim_{\delta \to 0} E\left\{\sup_{\|\boldsymbol{\Delta}\| \leq \delta} |\psi(\epsilon(\boldsymbol{\theta} + \boldsymbol{\Delta}))\,\epsilon(\boldsymbol{\theta} + \boldsymbol{\Delta}) - \psi(\epsilon(\boldsymbol{\theta}))\,\epsilon(\boldsymbol{\theta})|\right\} = 0$.

(iii) $\lim_{\delta \to 0} E\left\{\sup_{\|\boldsymbol{\Delta}\| \leq \delta} |\psi'(\epsilon(\boldsymbol{\theta} + \boldsymbol{\Delta})) - \psi'(\epsilon(\boldsymbol{\theta}))|\right\} = 0$.

(iv) $\lim_{\delta \to 0} E\left\{\sup_{\|\boldsymbol{\Delta}\| \leq \delta} |\psi'(\epsilon(\boldsymbol{\theta} + \boldsymbol{\Delta}))\,\epsilon(\boldsymbol{\theta} + \boldsymbol{\Delta}) - \psi'(\epsilon(\boldsymbol{\theta}))\,\epsilon(\boldsymbol{\theta})|\right\} = 0$.

(v) $\lim_{\delta\to 0} E\left\{\sup_{\|\boldsymbol{\Delta}\|\le\delta}\left|\psi'\left(\epsilon(\boldsymbol{\theta}+\boldsymbol{\Delta})\right)\epsilon^2(\boldsymbol{\theta}+\boldsymbol{\Delta})-\psi'\left(\epsilon(\boldsymbol{\theta})\right)\epsilon^2(\boldsymbol{\theta})\right|\right\}=0.$

[B1]:
$f(\mathbf{x},\boldsymbol{\theta})$ is continuous and twice differentiable with respect to $\boldsymbol{\theta}\in\boldsymbol{\Theta}$, where $\boldsymbol{\Theta}$ is a compact subset of $\Re^p$.

[B2]:

(i) $\lim_{n\to\infty}\boldsymbol{\Gamma}_{1n}(\boldsymbol{\theta})/n=\boldsymbol{\Gamma}_1(\boldsymbol{\theta})$, where

$$\begin{aligned}\boldsymbol{\Gamma}_{1n}(\boldsymbol{\theta})=\sum_{i=1}^{n}\Bigg\{&\frac{-\gamma_1}{\sigma^2(\mathbf{x}_i,\boldsymbol{\theta})}\boldsymbol{\Sigma}_{\boldsymbol{\theta}}(\mathbf{x}_i,\boldsymbol{\theta})+\frac{\gamma_2}{\sigma^2(\mathbf{x}_i,\boldsymbol{\theta})}\mathbf{f}_{\boldsymbol{\theta}}(\mathbf{x}_i,\boldsymbol{\theta})\mathbf{f}_{\boldsymbol{\theta}}^t(\mathbf{x}_i,\boldsymbol{\theta})\\&+\frac{\gamma_3+2\gamma_1\sigma(\mathbf{x}_i,\boldsymbol{\theta})}{\sigma^4(\mathbf{x}_i,\boldsymbol{\theta})}\boldsymbol{\sigma}_{\boldsymbol{\theta}}(\mathbf{x}_i,\boldsymbol{\theta})\boldsymbol{\sigma}_{\boldsymbol{\theta}}^t(\mathbf{x}_i,\boldsymbol{\theta})\Bigg\},\end{aligned}$$

$\boldsymbol{\Sigma}_{\boldsymbol{\theta}}(\mathbf{x}_i,\boldsymbol{\theta})=(\partial^2/\partial\boldsymbol{\theta}\partial\boldsymbol{\theta}^t)\sigma(\mathbf{x}_i,\boldsymbol{\theta})$, and $\boldsymbol{\Gamma}_1(\boldsymbol{\theta})$ is a positive definite matrix.

(ii) $\lim_{n\to\infty}\boldsymbol{\Gamma}_{2n}(\boldsymbol{\theta})/n=\boldsymbol{\Gamma}_2(\boldsymbol{\theta})$, where

$$\begin{aligned}\boldsymbol{\Gamma}_{2n}(\boldsymbol{\theta})=\sum_{i=1}^{n}\frac{1}{\sigma^2(\mathbf{x}_i,\boldsymbol{\theta})}\{&\sigma_{\psi 1}^2u(\mathbf{x}_i)\mathbf{f}_{\boldsymbol{\theta}}(\mathbf{x}_i,\boldsymbol{\theta})\mathbf{f}_{\boldsymbol{\theta}}^t(\mathbf{x}_i,\boldsymbol{\theta})\\&+\sigma_{\psi 2}^2v(\mathbf{x}_i)\boldsymbol{\sigma}_{\boldsymbol{\theta}}(\mathbf{x}_i,\boldsymbol{\theta})\boldsymbol{\sigma}_{\boldsymbol{\theta}}^t(\mathbf{x}_i,\boldsymbol{\theta})\},\end{aligned}$$

and $\boldsymbol{\Gamma}_2(\boldsymbol{\theta})$ is a positive definite matrix.

(iii) (Noether's condition)
$\max_i\left\{u(\mathbf{x}_i)\mathbf{f}_{\boldsymbol{\theta}}^t(\mathbf{x}_i,\boldsymbol{\theta})\boldsymbol{\Gamma}_{2n}^{-1}(\boldsymbol{\theta})\mathbf{f}_{\boldsymbol{\theta}}(\mathbf{x}_i,\boldsymbol{\theta})/\sigma^2(\mathbf{x}_i,\boldsymbol{\theta})\right\}\longrightarrow 0$, as $n\to\infty$.

(iv) (Noether's condition)
$\max_i\left\{v(\mathbf{x}_i)\boldsymbol{\sigma}_{\boldsymbol{\theta}}^t(\mathbf{x}_i,\boldsymbol{\theta})\boldsymbol{\Gamma}_{2n}^{-1}(\boldsymbol{\theta})\boldsymbol{\sigma}_{\boldsymbol{\theta}}(\mathbf{x}_i,\boldsymbol{\theta})/\sigma^2(\mathbf{x}_i,\boldsymbol{\theta})\right\}\longrightarrow 0$, as $n\to\infty$.

[B3]:
For $j,l=1,\ldots,p$

(i) $$\begin{aligned}\lim_{\delta\to 0}\sup_{\|\boldsymbol{\Delta}\|\le\delta}\Big|&(\partial/\partial\theta_j)f(\mathbf{x},\boldsymbol{\theta}+\boldsymbol{\Delta})(\partial/\partial\theta_l)f(\mathbf{x},\boldsymbol{\theta}+\boldsymbol{\Delta})\\&-(\partial/\partial\theta_j)f(\mathbf{x},\boldsymbol{\theta})(\partial/\partial\theta_l)f(\mathbf{x},\boldsymbol{\theta})\Big|=0.\end{aligned}$$

(ii) $\lim_{\delta\to 0}\sup_{\|\boldsymbol{\Delta}\|\le\delta}\left|(\partial^2/\partial\theta_j\partial\theta_l)f(\mathbf{x},\boldsymbol{\theta}+\boldsymbol{\Delta})-(\partial^2/\partial\theta_j\partial\theta_l)f(\mathbf{x},\boldsymbol{\theta})\right|=0.$

[C1]:
$\sigma(\mathbf{x},\boldsymbol{\theta})$ is continuous and twice differentiable with respect to $\boldsymbol{\theta}\in\boldsymbol{\Theta}$.

[C2]:
For $j, l = 1, \ldots, p$

(i) $\lim_{\delta \to 0} \sup_{\|\boldsymbol{\Delta}\| \le \delta} \Big| (\partial/\partial\theta_j)\sigma(\mathbf{x}, \boldsymbol{\theta} + \boldsymbol{\Delta})(\partial/\partial\theta_l)\sigma(\mathbf{x}, \boldsymbol{\theta} + \boldsymbol{\Delta}) - (\partial/\partial\theta_j)\sigma(\mathbf{x}, \boldsymbol{\theta})(\partial/\partial\theta_l)\sigma(\mathbf{x}, \boldsymbol{\theta}) \Big| = 0.$

(ii) $\lim_{\delta \to 0} \sup_{\|\boldsymbol{\Delta}\| \le \delta} \Big| (\partial^2/\partial\theta_j\partial\theta_l)\sigma(\mathbf{x}, \boldsymbol{\theta} + \boldsymbol{\Delta}) - (\partial^2/\partial\theta_j\partial\theta_l)\sigma(\mathbf{x}, \boldsymbol{\theta}) \Big| = 0.$

While conditions **[A1]**, **[A3]**, **[B1]**, **[B3]**, **[C1]** and **[C2]** are needed for proving the asymptotic linearity of the proposed M-estimator, conditions **[A2]** and **[B2]** are needed to prove its asymptotic normality. Noether type conditions **[B2]** (iii) and (iv) are standard to linear and nonlinear regression models. Condtions, such as **[B1]** and **[B3]**, can be verified for commonly used nonlinear models in toxicology, such as the Hill model. Also, regularity conditions regarding the score function ψ (**[A1]**–**[A3]**) are usually satisfied for commonly used h functions, such as the Huber function.

9.3. Asymptotic Results

The following lemmas (proved in the Appendix) are needed for proving that asymptotically the M-estimator can be linearized (Theorem 9.1). This linearization result is exploited to prove the consistency and the asymptotic normality of the M-estimator (Theorem 9.2 and Corollary 9.1).

Lemma 9.1. *Let the conditions* **[A1]**–**[A3]**, **[B1]**–**[B3]**, *and* **[C1]**–**[C2]** *hold and let* $\lambda_l(\mathbf{x}_i, y_i, \boldsymbol{\theta})$ *be the* l*th element of the vector* $\lambda(\mathbf{x}_i, y_i, \boldsymbol{\theta})$ *for* $l = 1, \ldots, p$*. Then for* $l = 1, \ldots, p$

$$\sup_{\|\mathbf{t}\| \le C} \left| \frac{1}{n} \sum_{i=1}^{n} \sum_{j=1}^{p} t_j \Big\{ (\partial/\partial\theta_j)\lambda_l\Big(\mathbf{x}_i, y_i, \boldsymbol{\theta} + \frac{s\mathbf{t}}{\sqrt{n}}\Big) - (\partial/\partial\theta_j)\lambda_l(\mathbf{x}_i, y_i, \boldsymbol{\theta}) \Big\} \right| = o_p(1), \tag{9.6}$$

where

$$\lambda_l(\mathbf{x}_i, y_i, \boldsymbol{\theta}) = \psi\left(\frac{y_i - f(\mathbf{x}_i, \boldsymbol{\theta})}{\sigma(\mathbf{x}_i, \boldsymbol{\theta})}\right) \left\{ \frac{f_{\theta_l}(\mathbf{x}_i, \boldsymbol{\theta})}{\sigma(\mathbf{x}_i, \boldsymbol{\theta})} + \frac{y_i - f(\mathbf{x}_i, \boldsymbol{\theta})}{\sigma(\mathbf{x}_i, \boldsymbol{\theta})} \frac{\sigma_{\theta_l}(\mathbf{x}_i, \boldsymbol{\theta})}{\sigma(\mathbf{x}_i, \boldsymbol{\theta})} \right\}, \tag{9.7}$$

$f_{\theta_l}(\mathbf{x}_i, \boldsymbol{\theta}) = (\partial/\partial\theta_l)f(\mathbf{x}_i, \boldsymbol{\theta})$, *and* $\sigma_{\theta_l}(\mathbf{x}_i, \boldsymbol{\theta}) = (\partial/\partial\theta_l)f(\mathbf{x}_i, \boldsymbol{\theta})$ *for* $l = 1, \ldots, p$.

Lemma 9.2. *Let the conditions* **[A1]**–**[A3]**, **[B1]**–**[B3]**, *and* **[C1]**–**[C2]** *hold and let* $\lambda_l(\mathbf{x}_i, y_i, \boldsymbol{\theta})$ *be the* l*th element of the vector* $\lambda(\mathbf{x}_i, y_i, \boldsymbol{\theta})$ *for* $l = 1, \ldots, p$. *Then for* $l = 1, \ldots, p$

$$\sup_{\|\mathbf{t}\| \le C} \Bigg| \frac{1}{n} \sum_{i=1}^{n} \sum_{j=1}^{p} t_j (\partial / \partial \theta_j) \lambda_l(\mathbf{x}_i, y_i, \boldsymbol{\theta}) + \frac{1}{n} \sum_{i=1}^{n} \sum_{j=1}^{p} t_j \bigg\{ \frac{-\gamma_1}{\sigma^2(\mathbf{x}_i, \boldsymbol{\theta})} (\partial^2 / \partial \theta_l \partial \theta_j) \sigma(\mathbf{x}_i, \boldsymbol{\theta}) + \frac{\gamma_2}{\sigma^2(\mathbf{x}_i, \boldsymbol{\theta})} f_{\theta_l}(\mathbf{x}_i, \boldsymbol{\theta}) f_{\theta_j}(\mathbf{x}_i, \boldsymbol{\theta}) + \frac{\gamma_3 + 2\gamma_1 \sigma(\mathbf{x}_i, \boldsymbol{\theta})}{\sigma^4(\mathbf{x}_i, \boldsymbol{\theta})} \sigma_{\theta_l}(\mathbf{x}_i, \boldsymbol{\theta}) \sigma_{\theta_j}(\mathbf{x}_i, \boldsymbol{\theta}) \bigg\} \Bigg| = o_p(1), \tag{9.8}$$

where $\lambda_l(\mathbf{x}_i, y_i, \boldsymbol{\theta})$ *is defined in Eq. (9.7).*

Now the uniform asymptotic linearity of M-estimator shall be proved.

Theorem 9.1. *Let the conditions* **[A1]**–**[A3]**, **[B1]**–**[B3]**, *and* **[C1]**–**[C2]** *hold. Then*

$$\sup_{\|\mathbf{t}\| \le C} \left\| \frac{1}{\sqrt{n}} \sum_{i=1}^{n} \{\lambda(\mathbf{x}_i, y_i, \boldsymbol{\theta} + n^{-\frac{1}{2}} \mathbf{t}) - \lambda(\mathbf{x}_i, y_i, \boldsymbol{\theta})\} + \frac{1}{n} \boldsymbol{\Gamma}_{1n}(\boldsymbol{\theta}) \mathbf{t} \right\| = o_p(1) \tag{9.9}$$

as $n \to \infty$, *where* $\lambda(\mathbf{x}_i, y_i, \boldsymbol{\theta})$ *was defined in Eq. (9.4).*

Proof. Consider the lth element of the vector $\lambda(\mathbf{x}_i, y_i, \boldsymbol{\theta})$ denoted as Eq. (9.7). Then, by using the first order term in the Taylor's expansion, for $0 < s < 1$,

$$\begin{aligned} &\lambda_l(\mathbf{x}_i, y_i, \boldsymbol{\theta} + n^{-\frac{1}{2}} \mathbf{t}) - \lambda_l(\mathbf{x}_i, y_i, \boldsymbol{\theta}) \\ &\quad = \frac{1}{\sqrt{n}} \sum_{j=1}^{p} t_j \{(\partial / \partial \theta_j) \lambda_l(\mathbf{x}_i, y_i, \boldsymbol{\theta})\} \\ &\qquad + \frac{1}{\sqrt{n}} \sum_{j=1}^{p} t_j \Big\{ (\partial / \partial \theta_j) \lambda_l \Big(\mathbf{x}_i, y_i, \boldsymbol{\theta} + \frac{s\mathbf{t}}{\sqrt{n}}\Big) - (\partial / \partial \theta_j) \lambda_l(\mathbf{x}_i, y_i, \boldsymbol{\theta}) \Big\}. \end{aligned}$$

And for $l = 1, \ldots, p$

$$\sup_{\|\mathbf{t}\| \le C} \left| \frac{1}{\sqrt{n}} \sum_{i=1}^{n} \Big\{ \lambda_l \Big(\mathbf{x}_i, y_i, \boldsymbol{\theta} + \frac{\mathbf{t}}{\sqrt{n}}\Big) - \lambda_l(\mathbf{x}_i, y_i, \boldsymbol{\theta}) \Big\} \right.$$

$$+\frac{1}{n}\sum_{i=1}^{n}\sum_{j=1}^{p} t_j\left\{\frac{-\gamma_1}{\sigma^2(\mathbf{x}_i,\boldsymbol{\theta})}(\partial^2/\partial\theta_l\partial\theta_j)\sigma(\mathbf{x}_i,\boldsymbol{\theta})+\frac{\gamma_2}{\sigma^2(\mathbf{x}_i,\boldsymbol{\theta})}f_{\theta_l}(\mathbf{x}_i,\boldsymbol{\theta})f_{\theta_j}(\mathbf{x}_i,\boldsymbol{\theta})\right.$$
$$\left.+\frac{\gamma_3+2\gamma_1\sigma(\mathbf{x}_i,\boldsymbol{\theta})}{\sigma^4(\mathbf{x}_i,\boldsymbol{\theta})}\sigma_{\theta_l}(\mathbf{x}_i,\boldsymbol{\theta})\sigma_{\theta_j}(\mathbf{x}_i,\boldsymbol{\theta})\right\}\Bigg|$$
$$\le \sup_{\|\mathbf{t}\|\le C}\left|\frac{1}{\sqrt{n}}\sum_{i=1}^{n}\sum_{j=1}^{p} t_j\left\{(\partial/\partial\theta_j)\lambda_l(\mathbf{x}_i,y_i,\boldsymbol{\theta}+\frac{s\mathbf{t}}{\sqrt{n}})-(\partial/\partial\theta_j)\lambda_l(\mathbf{x}_i,y_i,\boldsymbol{\theta})\right\}\right|$$
$$+\sup_{\|\mathbf{t}\|\le C}\left|\frac{1}{\sqrt{n}}\sum_{i=1}^{n}\sum_{j=1}^{p} t_j(\partial/\partial\theta_j)\lambda_l(\mathbf{x}_i,y_i,\boldsymbol{\theta})\right.$$
$$+\frac{1}{n}\sum_{i=1}^{n}\sum_{j=1}^{p} t_j\left\{\frac{-\gamma_1}{\sigma^2(\mathbf{x}_i,\boldsymbol{\theta})}(\partial^2/\partial\theta_l\partial\theta_j)\sigma(\mathbf{x}_i,\boldsymbol{\theta})+\frac{\gamma_2}{\sigma^2(\mathbf{x}_i,\boldsymbol{\theta})}f_{\theta_l}(\mathbf{x}_i,\boldsymbol{\theta})f_{\theta_j}(\mathbf{x}_i,\boldsymbol{\theta})\right.$$
$$\left.\left.+\frac{\gamma_3+2\gamma_1\sigma(\mathbf{x}_i,\boldsymbol{\theta})}{\sigma^4(\mathbf{x}_i,\boldsymbol{\theta})}\sigma_{\theta_l}(\mathbf{x}_i,\boldsymbol{\theta})\sigma_{\theta_j}(\mathbf{x}_i,\boldsymbol{\theta})\right\}\right|.$$

Therefore, from Lemmas 9.1 and 9.2 it is concluded that:

$$\sup_{\|\mathbf{t}\|\le C}\left|\frac{1}{\sqrt{n}}\sum_{i=1}^{n}\left\{\lambda_l\left(\mathbf{x}_i,y_i,\boldsymbol{\theta}+\frac{\mathbf{t}}{\sqrt{n}}\right)-\lambda_l(\mathbf{x}_i,y_i,\boldsymbol{\theta})\right\}\right.$$
$$+\frac{1}{n}\sum_{i=1}^{n}\sum_{j=1}^{p} t_j\left\{\frac{-\gamma_1}{\sigma^2(\mathbf{x}_i,\boldsymbol{\theta})}(\partial^2/\partial\theta_l\partial\theta_j)\sigma(\mathbf{x}_i,\boldsymbol{\theta})+\frac{\gamma_2}{\sigma^2(\mathbf{x}_i,\boldsymbol{\theta})}f_{\theta_l}(\mathbf{x}_i,\boldsymbol{\theta})f_{\theta_j}(\mathbf{x}_i,\boldsymbol{\theta})\right.$$
$$\left.\left.+\frac{\gamma_3+2\gamma_1\sigma(\mathbf{x}_i,\boldsymbol{\theta})}{\sigma^4(\mathbf{x}_i,\boldsymbol{\theta})}\sigma_{\theta_l}(\mathbf{x}_i,\boldsymbol{\theta})\sigma_{\theta_j}(\mathbf{x}_i,\boldsymbol{\theta})\right\}\right|=o_p(1).\quad\square$$

The existence of a solution to Eq. (9.3) which is a $\sqrt{n}$-consistent estimator of $\boldsymbol{\theta}$ is now proved. Also in this theorem the asymptotic representation of the solution to Eq. (9.3) is provided.

Theorem 9.2. *Let the conditions* **[A1]**–**[A3]**, **[B1]**–**[B3]**, *and* **[C1]**–**[C2]** *hold. Then there exists a sequence* $\hat{\boldsymbol{\theta}}_n$ *of solutions of Eq. (9.3) such that:*

$$\hat{\boldsymbol{\theta}}_n=\boldsymbol{\theta}+\frac{1}{n}\left(\frac{1}{n}\boldsymbol{\Gamma}_{1n}(\boldsymbol{\theta})\right)^{-1}\sum_{i=1}^{n}\lambda(\mathbf{x}_i,y_i,\boldsymbol{\theta})+o_p(n^{-\frac{1}{2}}),\tag{9.10}$$

equivalently

$$\sqrt{n}(\hat{\boldsymbol{\theta}}_n-\boldsymbol{\theta})=\left(\frac{1}{n}\boldsymbol{\Gamma}_{1n}(\boldsymbol{\theta})\right)^{-1}\frac{1}{\sqrt{n}}\sum_{i=1}^{n}\lambda(\mathbf{x}_i,y_i,\boldsymbol{\theta})+o_p(1).\tag{9.11}$$

Thus,

$$\sqrt{n}\|\hat{\boldsymbol{\theta}}_n-\boldsymbol{\theta}\|=O_p(1).\tag{9.12}$$

Proof. From Theorem 9.1 it is known that the following system of equations:

$$\sum_{i=1}^{n} \lambda_l(\mathbf{x}_i, y_i, \boldsymbol{\theta} + n^{-\frac{1}{2}}\mathbf{t}) = 0$$

has a root $\mathbf{t}_n$ that lies in $\|\mathbf{t}\| \leq C$ with probability exceeding $1-\epsilon$ for $n \geq n_0$. Then $\hat{\boldsymbol{\theta}}_n = \boldsymbol{\theta} + n^{-\frac{1}{2}}\mathbf{t}_n$ is a solution of Eq. (9.3) satisfying:

$$P(\sqrt{n}\|\hat{\boldsymbol{\theta}}_n - \boldsymbol{\theta}\| \leq C) \geq 1-\epsilon \quad \text{for } n \geq n_0.$$

Inserting $\sqrt{n}(\hat{\boldsymbol{\theta}}_n - \boldsymbol{\theta})$ into $\mathbf{t}$ in Eq. (9.9), the expression in Eq. (9.10) or Eq. (9.11) is obtained. □

Theorem 9.3. *Let the conditions* **[A1]**, **[A2]***(i)-(ii),* **[B1]**, **[B2]** *(i)–(iv) hold. Then*

$$\frac{1}{\sqrt{n}}\sum_{i=1}^{n}\{\lambda(\mathbf{x}_i, y_i, \boldsymbol{\theta}) - \mu(\mathbf{x}_i, \boldsymbol{\theta})\} \longrightarrow N_p\big(0, \boldsymbol{\Gamma}_2(\boldsymbol{\theta})\big) \quad \text{as } n \to \infty, \tag{9.13}$$

where $\mu(\mathbf{x}_i, \boldsymbol{\theta}) = \gamma_1 \boldsymbol{\sigma}_{\boldsymbol{\theta}}(\mathbf{x}_i, \boldsymbol{\theta})/\sigma(\mathbf{x}_i, \boldsymbol{\theta})$.

Proof. Let

$$Z_n^* = \boldsymbol{\eta}^t \frac{1}{\sqrt{n}}\sum_{i=1}^{n}\{\lambda(\mathbf{x}_i, y_i, \boldsymbol{\theta}) - \mu(\mathbf{x}_i, \boldsymbol{\theta})\}, \quad \boldsymbol{\eta} \in \Re^p,$$

then

$$\begin{aligned} Z_n^* &= \frac{1}{\sqrt{n}}\sum_{i=1}^{n}\left[\psi(\epsilon_i)\left\{\frac{\boldsymbol{\eta}^t\mathbf{f}_{\boldsymbol{\theta}}(\mathbf{x}_i, \boldsymbol{\theta})}{\sigma(\mathbf{x}_i, \boldsymbol{\theta})} + \epsilon_i \frac{\boldsymbol{\eta}^t\boldsymbol{\sigma}_{\boldsymbol{\theta}}(\mathbf{x}_i, \boldsymbol{\theta})}{\sigma(\mathbf{x}_i, \boldsymbol{\theta})}\right\} - \mu(\mathbf{x}_i, \boldsymbol{\theta})\right] \\ &= \sum_{i=1}^{n}\frac{1}{\sqrt{n}}\left\{\frac{\boldsymbol{\eta}^t\mathbf{f}_{\boldsymbol{\theta}}(\mathbf{x}_i, \boldsymbol{\theta})}{\sigma(\mathbf{x}_i, \boldsymbol{\theta})}\psi(\epsilon_i) + \frac{\boldsymbol{\eta}^t\boldsymbol{\sigma}_{\boldsymbol{\theta}}(\mathbf{x}_i, \boldsymbol{\theta})}{\sigma(\mathbf{x}_i, \boldsymbol{\theta})}\left(\psi(\epsilon_i)\epsilon_i - \gamma_1\right)\right\}, \\ &= \sum_{i=1}^{n}\frac{1}{\sqrt{n}}(c_{i1}Z_{i1} + c_{i2}Z_{i2}), \end{aligned}$$

where

$$\epsilon_i = \frac{y_i - f(\mathbf{x}_i, \boldsymbol{\theta})}{\sigma(\mathbf{x}_i, \boldsymbol{\theta})}, \quad c_{i1} = \frac{\boldsymbol{\eta}^t\mathbf{f}_{\boldsymbol{\theta}}(\mathbf{x}_i, \boldsymbol{\theta})}{\sigma(\mathbf{x}_i, \boldsymbol{\theta})}, \quad c_{i2} = \frac{\boldsymbol{\eta}^t\boldsymbol{\sigma}_{\boldsymbol{\theta}}(\mathbf{x}_i, \boldsymbol{\theta})}{\sigma(\mathbf{x}_i, \boldsymbol{\theta})},$$

and

$$Z_{i1} = \psi(\epsilon_i), \quad Z_{i2} = \psi(\epsilon_i)\epsilon_i - \gamma_1.$$

Then,

$$EZ_{i1} = 0, \quad EZ_{i2} = 0,$$

and

$$EZ_{i1}^2 = \sigma_{\psi 1}^2 u(\mathbf{x}_i), \quad EZ_{i2}^2 = \sigma_{\psi 2}^2 v(\mathbf{x}_i), \quad EZ_{i1}Z_{i2} = 0.$$

Therefore, if we let

$$Z_n^* = \sum_{i=1}^{n} c_{ni} Z_{ni},$$

where

$$c_{ni} = \frac{1}{\sqrt{n}} \left\{ \boldsymbol{\eta}^t \boldsymbol{\Gamma}_3(\mathbf{x}_i, \boldsymbol{\theta}) \boldsymbol{\eta} \right\}^{\frac{1}{2}},$$

$$Z_{ni} = (c_{i1} Z_{i1} + c_{i2} Z_{i2}) / \left\{ \boldsymbol{\eta}^t \boldsymbol{\Gamma}_3(\mathbf{x}_i, \boldsymbol{\theta}) \boldsymbol{\eta} \right\}^{\frac{1}{2}},$$

and

$$\boldsymbol{\Gamma}_3(\mathbf{x}_i, \boldsymbol{\theta}) = \frac{\sigma_{\psi 1}^2 u(\mathbf{x}_i)}{\sigma^2(\mathbf{x}_i, \boldsymbol{\theta})} \mathbf{f}_{\boldsymbol{\theta}}(\mathbf{x}_i, \boldsymbol{\theta}) \mathbf{f}_{\boldsymbol{\theta}}^t(\mathbf{x}_i, \boldsymbol{\theta}) + \frac{\sigma_{\psi 2}^2 v(\mathbf{x}_i)}{\sigma^2(\mathbf{x}_i, \boldsymbol{\theta})} \boldsymbol{\sigma}_{\boldsymbol{\theta}}(\mathbf{x}_i, \boldsymbol{\theta}) \boldsymbol{\sigma}_{\boldsymbol{\theta}}^t(\mathbf{x}_i, \boldsymbol{\theta}),$$

then by using the Hájek-Šidak Central Limit Theorem, it is shown that Z_n^* converges in law to a normal distribution as $n \to \infty$. In order to use this theorem the following condition needs to be verified:

$$\max_i c_{ni}^2 / \sum_{i=1}^{n} c_{ni}^2 \longrightarrow 0, \quad \text{as} \quad n \to \infty.$$

The expression can be rewritten as

$$\sup_{\boldsymbol{\eta} \in \Re^p} \left[\max_i \boldsymbol{\eta}^t \boldsymbol{\Gamma}_3(\mathbf{x}_i, \boldsymbol{\theta}) \boldsymbol{\eta} \ / \ \boldsymbol{\eta}^t \boldsymbol{\Gamma}_{2n}(\boldsymbol{\theta}) \boldsymbol{\eta} \right] \longrightarrow 0, \quad \text{as} \quad n \to \infty.$$

Note that

$$\begin{aligned} \sup_{\boldsymbol{\eta} \in \Re^p} \left[\boldsymbol{\eta}^t \boldsymbol{\Gamma}_3(\mathbf{x}_i, \boldsymbol{\theta}) \boldsymbol{\eta} \ / \ \boldsymbol{\eta}^t \boldsymbol{\Gamma}_{2n}(\boldsymbol{\theta}) \boldsymbol{\eta} \right] &= ch_1(\boldsymbol{\Gamma}_3(\mathbf{x}_i, \boldsymbol{\theta}), \boldsymbol{\Gamma}_{2n}(\boldsymbol{\theta})) \\ &= ch_1(\boldsymbol{\Gamma}_3(\mathbf{x}_i, \boldsymbol{\theta}) \boldsymbol{\Gamma}_{2n}^{-1}(\boldsymbol{\theta})), \end{aligned}$$

where $ch_1(A)$ indicates the largest eigen value of the matrix A. Also, note that

$$\begin{aligned} ch_1(\boldsymbol{\Gamma}_3(\mathbf{x}_i, \boldsymbol{\theta}) \boldsymbol{\Gamma}_{2n}^{-1}(\boldsymbol{\theta})) = {} & \frac{\sigma_{\psi 1}^2 u(\mathbf{x}_i)}{\sigma^2(\mathbf{x}_i, \boldsymbol{\theta})} \mathbf{f}_{\boldsymbol{\theta}}^t(\mathbf{x}_i, \boldsymbol{\theta}) \mathbf{\Gamma_{2n}}^{-1}(\boldsymbol{\theta}) \mathbf{f}_{\boldsymbol{\theta}}(\mathbf{x}_i, \boldsymbol{\theta}) \\ & + \frac{\sigma_{\psi 2}^2 v(\mathbf{x}_i)}{\sigma^2(\mathbf{x}_i, \boldsymbol{\theta})} \boldsymbol{\sigma}_{\boldsymbol{\theta}}^t(\mathbf{x}_i, \boldsymbol{\theta}) \mathbf{\Gamma_{2n}}^{-1}(\boldsymbol{\theta}) \boldsymbol{\sigma}_{\boldsymbol{\theta}}(\mathbf{x}_i, \boldsymbol{\theta}). \end{aligned}$$

From the regularity conditions **[B2]** (iii)–(iv), the following is obtained:

$$Z_n^*/\Big(\sum_{i=1}^n c_{ni}^2\Big)^{\frac{1}{2}} \longrightarrow N(0,1) \ \text{ as } \ n \to \infty.$$

Appealing to Cramer-Wold Theorem and condition **[B2]** (ii) the expression in Eq. (9.13) is proved. □

Corollary 9.1. *Let the conditions* **[A1]**–**[A3]**, **[B1]**–**[B3]** *hold. Then*

$$\sqrt{n}(\hat{\boldsymbol{\theta}}_n - \boldsymbol{\theta} - \nu_n(\boldsymbol{\theta})) \longrightarrow N_p\big(\mathbf{0}, \boldsymbol{\Gamma}_1^{-1}(\boldsymbol{\theta})\boldsymbol{\Gamma}_2(\boldsymbol{\theta})\boldsymbol{\Gamma}_1^{-1}(\boldsymbol{\theta})\big) \quad \text{as } n \to \infty, \tag{9.14}$$

where

$$\nu_n(\boldsymbol{\theta}) = \left(\frac{1}{n}\boldsymbol{\Gamma}_{1n}(\boldsymbol{\theta})\right)^{-1} \frac{1}{n}\sum_{i=1}^n \mu(\mathbf{x}_i, \boldsymbol{\theta}).$$

Proof. From Theorem 9.2

$$\sqrt{n}(\hat{\boldsymbol{\theta}}_n - \boldsymbol{\theta}) = \Big(\frac{1}{n}\boldsymbol{\Gamma}_{1n}(\boldsymbol{\theta})\Big)^{-1} \frac{1}{\sqrt{n}}\sum_{i=1}^n \lambda(\mathbf{x}_i, y_i, \boldsymbol{\theta}) + o_p(1).$$

Then from Theorem 9.3 and Slutsky's Theorem the expression in Eq. (9.14) is obtained. □

Corollary 9.2. *Let the conditions* **[A1]**–**[A3]**, **[B1]**–**[B3]** *hold. Then*

$$\hat{\boldsymbol{\Gamma}}^{-\frac{1}{2}}\sqrt{n}(\hat{\boldsymbol{\theta}}_n - \boldsymbol{\theta} - \nu_n(\boldsymbol{\theta})) \longrightarrow N_p(\mathbf{0}, \mathbf{I}_p) \quad \text{as } n \to \infty, \tag{9.15}$$

where

$$\hat{\boldsymbol{\Gamma}} = \Big(\frac{1}{n}\boldsymbol{\Gamma}_{1n}(\hat{\boldsymbol{\theta}}_n)\Big)^{-1}\Big(\frac{1}{n}\boldsymbol{\Gamma}_{2n}(\hat{\boldsymbol{\theta}}_n)\Big)\Big(\frac{1}{n}\boldsymbol{\Gamma}_{1n}(\hat{\boldsymbol{\theta}}_n)\Big)^{-1}. \tag{9.16}$$

Proof. Using Eq. (9.14) and Slutsky's Theorem the expression in Eq. (9.15) is obtained. □

Corollary 9.3. *Let the conditions* **[A1]**–**[A3]**, **[B1]**–**[B3]** *hold. Then*

$$n(\hat{\boldsymbol{\theta}}_n - \boldsymbol{\theta} - \nu_n(\boldsymbol{\theta}))^t\hat{\boldsymbol{\Gamma}}^{-1}(\hat{\boldsymbol{\theta}}_n - \boldsymbol{\theta} - \nu_n(\boldsymbol{\theta})) \longrightarrow \chi_p^2 \quad \text{as } n \to \infty. \tag{9.17}$$

Proof. Using Eq. (9.15) and Cochran's Theorem, the expression in Eq. (9.17) is proved. □

9.4. Illustration

In this section the proposed estimator is applied to a real data from a study conducted by the US National Toxicology Program (NTP). In the study, three species of rodents (rats, mice, and guinea pigs) were exposed to water containing Hexavalent Chromium (CrVI) as sodium dichromate dihydrate and the researchers measured the total chromium concentration (y) in blood, kidneys, and femurs of the animals. The proposed methodology is illustrated using the guinea pig blood data. There were 7 dose groups (x) ranging from 0 to 300 and 4 observations at each dose level except for $x = 30$ (3 observations). Thus, total sample size is 27.

Hill model (Hill, 1910) was used to fit the data set, which is commonly used to investigate the concentration-response relationship:

$$y_{ij} = f(x_i, \boldsymbol{\theta}) + \sigma_i \epsilon_{ij} = \theta_0 + \frac{\theta_1 x_i^{\theta_2}}{x_i^{\theta_2} + \theta_3^{\theta_2}} + \sigma_i \epsilon_{ij},$$

where θ_0 is the intercept, θ_1 is the difference between the maximum and minimum response of the chemical, θ_2 is the slope, and θ_3 is the concentration producing 50% of the maximum response (ED_{50}). We modeled the standard deviation for the ith dose group using $\sigma_i = \sigma(x_i, \boldsymbol{\theta}) = \sigma_0 f(x_i, \boldsymbol{\theta})$. Parameters and their standard errors were estimated using the proposed estimator.

Results are summarized in Table 9.1 and the data and the fitted curve using the estimator are plotted in Fig. 9.1, which suggests that the data are heteroscedastic and the assumption of the error variance structure is reasonable. From Fig. 9.1 it appears that the proposed estimator fits the data well and the parameter estimates (and their estimated standard errors) provided in Table 9.1 seem reasonable.

Table 9.1. Estimate and Standard Error for parameters for chromium guinea pig blood data using the proposed estimator.

Parameter	Estimate	S.E.
θ_0	0.148	0.001
θ_1	2.732	0.200
θ_2	1.336	0.016
θ_3	94.97	5.546

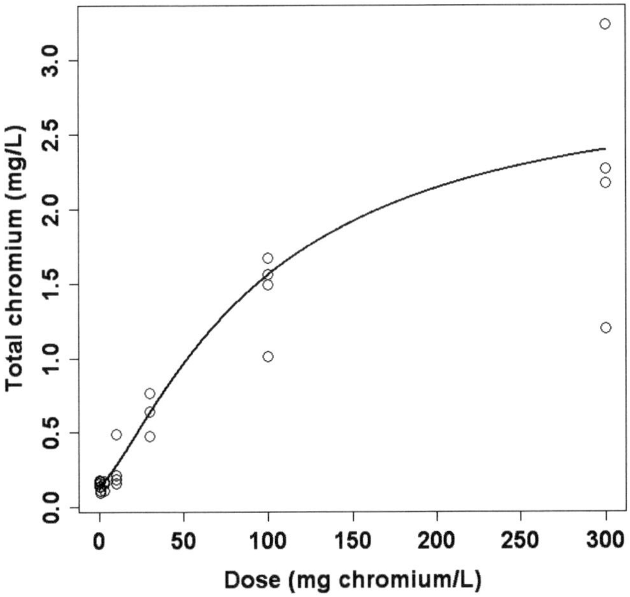

Fig. 9.1. Chromium concentration in blood for guinea pig using the proposed estimator.

9.5. Concluding Remarks and Ongoing Research

In this paper an M-estimator has been proposed for heteroscedastic nonlinear regression models. The methodology proposed here would allow researchers to use estimation procedures that are robust to potential influential or outlying observations and account for heteroscedasticity. The error variance is assumed to be a known function of unknown parameters of the nonlinear model, a reasonable assumption in many toxicological studies. For instance when modeling tumor incidence, one may use a binomial model where the variance is a function of the mean response. Under the some regularity conditions, the asymptotic normality of the estimator was proved. Although the asymptotic theory developed in this paper resembles the theory derived in Sanhueza and Sen (2001), there are differences in the basic setup. The problem of simultaneously estimating the variance function and the mean function was considered in this study. Since the mean as well as the variance of the response are nonlinear functions of an unknown parameter, the parameter estimates are asymptotically biased. However, an explicit form of the bias term is provided in this paper, which can be estimated using a resampling procedure.

The M-estimation methodology presented in this paper can be modified in various ways. For instance, in some applications a researcher may hypothesize that the error variance is a function of $\boldsymbol{\theta}$ and an additional parameter $\boldsymbol{\tau}$. The proposed methodology can be easily adapted to account for such additional parameters.

In some application, in particular the high throughput screening assays in toxicology, a research may not know *a priori* if the error variance is homoscedastic or heteroscedastic. In such cases, it will be useful to develop methods that perform a preliminary test of homoscedasticity and based on the outcome of the test the researcher may use the appropriate estimation procedure for $\boldsymbol{\theta}$. Such problems were discussed in Lim (2009).

Acknowledgments

This research was supported, in part, by the Intramural Research Program of the NIH, National Institute of Environmental Health Sciences [Z01 ES101744-04]. We thank the editor and the referee for several important comments and suggestions which helped improve the presentation of the manuscript.

Appendix. Proof of Lemmas

A.1. Proof of Lemma 9.1

For $j, l = 1, \ldots, p$, by differentiating,

$$\begin{aligned}
&(\partial/\partial\theta_j)\lambda_l(\mathbf{x}_i, y_i, \boldsymbol{\theta}) \\
&= \psi(\epsilon_i(\boldsymbol{\theta}))k(\mathbf{x}_i, \boldsymbol{\theta})\left\{(\partial^2/\partial\theta_l\partial\theta_j)f(\mathbf{x}_i, \boldsymbol{\theta}) - k(\mathbf{x}_i, \boldsymbol{\theta})f_{\theta_l}(\mathbf{x}_i, \boldsymbol{\theta})\sigma_{\theta_j}(\mathbf{x}_i, \boldsymbol{\theta})\right. \\
&\qquad\qquad \left. - k(\mathbf{x}_i, \boldsymbol{\theta})f_{\theta_j}(\mathbf{x}_i, \boldsymbol{\theta})\sigma_{\theta_l}(\mathbf{x}_i, \boldsymbol{\theta})\right\} \\
&+ \psi(\epsilon_i(\boldsymbol{\theta}))\epsilon_i(\boldsymbol{\theta})k^2(\mathbf{x}_i, \boldsymbol{\theta})\{(\partial^2/\partial\theta_l\partial\theta_j)\sigma(\mathbf{x}_i, \boldsymbol{\theta}) - 2k(\mathbf{x}_i, \boldsymbol{\theta})\sigma_{\theta_l}(\mathbf{x}_i, \boldsymbol{\theta})\sigma_{\theta_j}(\mathbf{x}_i, \boldsymbol{\theta})\} \\
&- \psi'(\epsilon_i(\boldsymbol{\theta}))k^2(\mathbf{x}_i, \boldsymbol{\theta})f_{\theta_l}(\mathbf{x}_i, \boldsymbol{\theta})f_{\theta_j}(\mathbf{x}_i, \boldsymbol{\theta}) \\
&+ \psi'(\epsilon_i(\boldsymbol{\theta}))\epsilon_i(\boldsymbol{\theta})k^2(\mathbf{x}_i, \boldsymbol{\theta})\left\{f_{\theta_l}(\mathbf{x}_i, \boldsymbol{\theta})\sigma_{\theta_j}(\mathbf{x}_i, \boldsymbol{\theta}) + f_{\theta_j}(\mathbf{x}_i, \boldsymbol{\theta})\sigma_{\theta_l}(\mathbf{x}_i, \boldsymbol{\theta})\right\} \\
&+ \psi'(\epsilon_i(\boldsymbol{\theta}))\epsilon_i^2(\boldsymbol{\theta})k^4(\mathbf{x}_i, \boldsymbol{\theta})\sigma_{\theta_l}(\mathbf{x}_i, \boldsymbol{\theta})\sigma_{\theta_j}(\mathbf{x}_i, \boldsymbol{\theta}),
\end{aligned} \tag{A.1}$$

where $\epsilon_i(\boldsymbol{\theta}) = \{y_i - f(\mathbf{x}_i, \boldsymbol{\theta})\}/\sigma(\mathbf{x}_i, \boldsymbol{\theta})$ and $k(\mathbf{x}_i, \boldsymbol{\theta}) = 1/\sigma(\mathbf{x}_i, \boldsymbol{\theta})$. Then,

$$\sup_{\|\mathbf{t}\|\le C}\left|\frac{1}{n}\sum_{i=1}^{n}\sum_{j=1}^{p}t_j\left\{(\partial/\partial\theta_j)\lambda_l\left(\mathbf{x}_i,y_i,\boldsymbol{\theta}+\frac{s\mathbf{t}}{\sqrt{n}}\right)-(\partial/\partial\theta_j)\lambda_l(\mathbf{x}_i,y_i,\boldsymbol{\theta})\right\}\right|$$
$$\le\frac{1}{n}C\sum_{i=1}^{n}\sum_{j=1}^{p}\sup_{\|\mathbf{t}\|\le C}\left|(\partial/\partial\theta_j)\lambda_l\left(\mathbf{x}_i,y_i,\boldsymbol{\theta}+\frac{s\mathbf{t}}{\sqrt{n}}\right)-(\partial/\partial\theta_j)\lambda_l(\mathbf{x}_i,y_i,\boldsymbol{\theta})\right|,$$

and using standard algebra the following inequality is derived:

$$\sup_{\|\mathbf{t}\|\le C}\left|(\partial/\partial\theta_j)\lambda_l\left(\mathbf{x}_i,y_i,\boldsymbol{\theta}+\frac{s\mathbf{t}}{\sqrt{n}}\right)-(\partial/\partial\theta_j)\lambda_l(\mathbf{x}_i,y_i,\boldsymbol{\theta})\right|$$
$$\le\sup_{\|\mathbf{t}\|\le C}\left\{\left|\psi\left(\epsilon_i\left(\boldsymbol{\theta}+\frac{s\mathbf{t}}{\sqrt{n}}\right)\right)-\psi(\epsilon_i(\boldsymbol{\theta}))\right|\left|k\left(\mathbf{x}_i,\boldsymbol{\theta}+\frac{s\mathbf{t}}{\sqrt{n}}\right)\right.\right.$$
$$\left.\left.\times(\partial^2/\partial\theta_l\partial\theta_j)f\left(\mathbf{x}_i,\boldsymbol{\theta}+\frac{s\mathbf{t}}{\sqrt{n}}\right)\right|\right\}$$
$$+\sup_{\|\mathbf{t}\|\le C}\left\{\left|k\left(\mathbf{x}_i,\boldsymbol{\theta}+\frac{s\mathbf{t}}{\sqrt{n}}\right)-k(\mathbf{x}_i,\boldsymbol{\theta})\right|\left|(\partial^2/\partial\theta_l\partial\theta_j)f\left(\mathbf{x}_i,\boldsymbol{\theta}+\frac{s\mathbf{t}}{\sqrt{n}}\right)\right|\right.$$
$$\left.\times|\psi(\epsilon_i(\boldsymbol{\theta}))|\right\}$$
$$+\sup_{\|\mathbf{t}\|\le C}\left\{\left|(\partial^2/\partial\theta_l\partial\theta_j)f\left(\mathbf{x}_i,\boldsymbol{\theta}+\frac{s\mathbf{t}}{\sqrt{n}}\right)-(\partial^2/\partial\theta_l\partial\theta_j)f(\mathbf{x}_i,\boldsymbol{\theta})\right|\right.$$
$$\left.\times|k(\mathbf{x}_i,\boldsymbol{\theta})|\,|\psi(\epsilon_i(\boldsymbol{\theta}))|\right\}$$
$$+\sup_{\|\mathbf{t}\|\le C}\left\{\left|\psi\left(\epsilon_i\left(\boldsymbol{\theta}+\frac{s\mathbf{t}}{\sqrt{n}}\right)\right)-\psi(\epsilon_i(\boldsymbol{\theta}))\right|\right.$$
$$\left.\times\left|k^2\left(\mathbf{x}_i,\boldsymbol{\theta}+\frac{s\mathbf{t}}{\sqrt{n}}\right)f_{\theta_l}\left(\mathbf{x}_i,\boldsymbol{\theta}+\frac{s\mathbf{t}}{\sqrt{n}}\right)\sigma_{\theta_j}\left(\mathbf{x}_i,\boldsymbol{\theta}+\frac{s\mathbf{t}}{\sqrt{n}}\right)\right|\right\}$$
$$+\sup_{\|\mathbf{t}\|\le C}\left\{\left|k^2\left(\mathbf{x}_i,\boldsymbol{\theta}+\frac{s\mathbf{t}}{\sqrt{n}}\right)-k^2(\mathbf{x}_i,\boldsymbol{\theta})\right|\right.$$
$$\left.\times\left|f_{\theta_l}\left(\mathbf{x}_i,\boldsymbol{\theta}+\frac{s\mathbf{t}}{\sqrt{n}}\right)\sigma_{\theta_j}\left(\mathbf{x}_i,\boldsymbol{\theta}+\frac{s\mathbf{t}}{\sqrt{n}}\right)\right|\,|\psi(\epsilon_i(\boldsymbol{\theta}))|\right\}$$
$$+\sup_{\|\mathbf{t}\|\le C}\left\{\left|f_{\theta_l}\left(\mathbf{x}_i,\boldsymbol{\theta}+\frac{s\mathbf{t}}{\sqrt{n}}\right)-f_{\theta_l}(\mathbf{x}_i,\boldsymbol{\theta})\right|\left|\sigma_{\theta_j}\left(\mathbf{x}_i,\boldsymbol{\theta}+\frac{s\mathbf{t}}{\sqrt{n}}\right)k^2(\mathbf{x}_i,\boldsymbol{\theta})\right|\right.$$
$$\left.\times|\psi(\epsilon_i(\boldsymbol{\theta}))|\right\}$$
$$+\sup_{\|\mathbf{t}\|\le C}\left\{\left|\sigma_{\theta_l}\left(\mathbf{x}_i,\boldsymbol{\theta}+\frac{s\mathbf{t}}{\sqrt{n}}\right)-\sigma_{\theta_l}(\mathbf{x}_i,\boldsymbol{\theta})\right||k^2(\mathbf{x}_i,\boldsymbol{\theta})f_{\theta_l}(\mathbf{x}_i,\boldsymbol{\theta})||\psi(\epsilon_i(\boldsymbol{\theta}))|\right\}$$
$$+\sup_{\|\mathbf{t}\|\le C}\left\{\left|\psi\left(\epsilon_i\left(\boldsymbol{\theta}+\frac{s\mathbf{t}}{\sqrt{n}}\right)\right)-\psi(\epsilon_i(\boldsymbol{\theta}))\right|\right.$$

$$
\begin{aligned}
&\qquad\qquad \times\Big|k^2\Big(\mathbf{x}_i,\boldsymbol{\theta}+\frac{s\mathbf{t}}{\sqrt{n}}\Big)f_{\theta_j}\Big(\mathbf{x}_i,\boldsymbol{\theta}+\frac{s\mathbf{t}}{\sqrt{n}}\Big)\sigma_{\theta_l}\Big(\mathbf{x}_i,\boldsymbol{\theta}+\frac{s\mathbf{t}}{\sqrt{n}}\Big)\Big|\Big\} \\
&+\sup_{\|\mathbf{t}\|\le C}\Big\{\Big|k^2\Big(\mathbf{x}_i,\boldsymbol{\theta}+\frac{s\mathbf{t}}{\sqrt{n}}\Big)-k^2(\mathbf{x}_i,\boldsymbol{\theta})\Big|\Big|f_{\theta_j}\Big(\mathbf{x}_i,\boldsymbol{\theta}+\frac{s\mathbf{t}}{\sqrt{n}}\Big)\sigma_{\theta_l}\Big(\mathbf{x}_i,\boldsymbol{\theta}+\frac{s\mathbf{t}}{\sqrt{n}}\Big)\Big| \\
&\qquad\qquad \times|\psi(\epsilon_i(\boldsymbol{\theta}))|\Big\} \\
&+\sup_{\|\mathbf{t}\|\le C}\Big\{\Big|f_{\theta_j}\Big(\mathbf{x}_i,\boldsymbol{\theta}+\frac{s\mathbf{t}}{\sqrt{n}}\Big)-f_{\theta_j}(\mathbf{x}_i,\boldsymbol{\theta})\Big|\Big|\sigma_{\theta_l}\Big(\mathbf{x}_i,\boldsymbol{\theta}+\frac{s\mathbf{t}}{\sqrt{n}}\Big)k^2(\mathbf{x}_i,\boldsymbol{\theta})\Big| \\
&\qquad\qquad \times|\psi(\epsilon_i(\boldsymbol{\theta}))|\Big\} \\
&+\sup_{\|\mathbf{t}\|\le C}\Big\{\Big|\sigma_{\theta_l}\Big(\mathbf{x}_i,\boldsymbol{\theta}+\frac{s\mathbf{t}}{\sqrt{n}}\Big)-\sigma_{\theta_l}(\mathbf{x}_i,\boldsymbol{\theta})\Big||k^2(\mathbf{x}_i,\boldsymbol{\theta})f_{\theta_j}(\mathbf{x}_i,\boldsymbol{\theta})||\psi(\epsilon_i(\boldsymbol{\theta}))|\Big\} \\
&+\sup_{\|\mathbf{t}\|\le C}\Big\{\Big|\psi\Big(\epsilon_i\Big(\boldsymbol{\theta}+\frac{s\mathbf{t}}{\sqrt{n}}\Big)\Big)\epsilon_i\Big(\boldsymbol{\theta}+\frac{s\mathbf{t}}{\sqrt{n}}\Big)-\psi(\epsilon_i(\boldsymbol{\theta}))\epsilon_i(\boldsymbol{\theta})\Big| \\
&\qquad\qquad \times\Big|k^2\Big(\mathbf{x}_i,\boldsymbol{\theta}+\frac{s\mathbf{t}}{\sqrt{n}}\Big)(\partial^2/\partial\theta_l\partial\theta_j)\sigma\Big(\mathbf{x}_i,\boldsymbol{\theta}+\frac{s\mathbf{t}}{\sqrt{n}}\Big)\Big|\Big\} \\
&+\sup_{\|\mathbf{t}\|\le C}\Big\{\Big|k^2\Big(\mathbf{x}_i,\boldsymbol{\theta}+\frac{s\mathbf{t}}{\sqrt{n}}\Big)-k^2(\mathbf{x}_i,\boldsymbol{\theta})\Big|\Big|(\partial^2/\partial\theta_l\partial\theta_j)\sigma\Big(\mathbf{x}_i,\boldsymbol{\theta}+\frac{s\mathbf{t}}{\sqrt{n}}\Big)\Big| \\
&\qquad\qquad \times|\psi(\epsilon_i(\boldsymbol{\theta}))\epsilon_i(\boldsymbol{\theta})|\Big\} \\
&+\sup_{\|\mathbf{t}\|\le C}\Big\{\Big|(\partial^2/\partial\theta_l\partial\theta_j)\sigma\Big(\mathbf{x}_i,\boldsymbol{\theta}+\frac{s\mathbf{t}}{\sqrt{n}}\Big)-(\partial^2/\partial\theta_l\partial\theta_j)\sigma(\mathbf{x}_i,\boldsymbol{\theta})\Big| \\
&\qquad\qquad \times|k^2(\mathbf{x}_i,\boldsymbol{\theta})||\psi(\epsilon_i(\boldsymbol{\theta}))\epsilon_i(\boldsymbol{\theta})|\Big\} \\
&+2\sup_{\|\mathbf{t}\|\le C}\Big\{\Big|\psi\Big(\epsilon_i\Big(\boldsymbol{\theta}+\frac{s\mathbf{t}}{\sqrt{n}}\Big)\Big)\epsilon_i\Big(\boldsymbol{\theta}+\frac{s\mathbf{t}}{\sqrt{n}}\Big)-\psi(\epsilon_i(\boldsymbol{\theta}))\epsilon_i(\boldsymbol{\theta})\Big| \\
&\qquad\qquad \times\Big|k^3\Big(\mathbf{x}_i,\boldsymbol{\theta}+\frac{s\mathbf{t}}{\sqrt{n}}\Big)\sigma_{\theta_l}\Big(\mathbf{x}_i,\boldsymbol{\theta}+\frac{s\mathbf{t}}{\sqrt{n}}\Big)\sigma_{\theta_j}\Big(\mathbf{x}_i,\boldsymbol{\theta}+\frac{s\mathbf{t}}{\sqrt{n}}\Big)\Big|\Big\} \\
&+2\sup_{\|\mathbf{t}\|\le C}\Big\{\Big|k^3\Big(\mathbf{x}_i,\theta+\frac{s\mathbf{t}}{\sqrt{n}}\Big)-k^3(\mathbf{x}_i,\boldsymbol{\theta})\Big| \\
&\qquad\qquad \times\Big|\sigma_{\theta_l}\Big(\mathbf{x}_i,\boldsymbol{\theta}+\frac{s\mathbf{t}}{\sqrt{n}}\Big)\sigma_{\theta_j}\Big(\mathbf{x}_i,\boldsymbol{\theta}+\frac{s\mathbf{t}}{\sqrt{n}}\Big)\Big||\psi(\epsilon_i(\boldsymbol{\theta}))\epsilon_i(\boldsymbol{\theta})|\Big\}
\end{aligned}
$$

$$+2\sup_{\|\mathbf{t}\|\le C}\left\{\left|\sigma_{\theta_l}\left(\mathbf{x}_i,\boldsymbol{\theta}+\frac{s\mathbf{t}}{\sqrt{n}}\right)\sigma_{\theta_j}\left(\mathbf{x}_i,\boldsymbol{\theta}+\frac{s\mathbf{t}}{\sqrt{n}}\right)-\sigma_{\theta_l}(\mathbf{x}_i,\boldsymbol{\theta})\sigma_{\theta_j}(\mathbf{x}_i,\boldsymbol{\theta})\right|\right.$$

$$\left.\times|k^3(\mathbf{x}_i,\boldsymbol{\theta})||\psi(\epsilon_i(\boldsymbol{\theta}))\epsilon_i(\boldsymbol{\theta})|\right\}$$

$$+\sup_{\|\mathbf{t}\|\le C}\left\{\left|\psi'\left(\epsilon_i\left(\boldsymbol{\theta}+\frac{s\mathbf{t}}{\sqrt{n}}\right)\right)-\psi'(\epsilon_i(\boldsymbol{\theta}))\right|\right.$$

$$\left.\times\left|k^2\left(\mathbf{x}_i,\boldsymbol{\theta}+\frac{s\mathbf{t}}{\sqrt{n}}\right)f_{\theta_l}\left(\mathbf{x}_i,\boldsymbol{\theta}+\frac{s\mathbf{t}}{\sqrt{n}}\right)f_{\theta_j}\left(\mathbf{x}_i,\boldsymbol{\theta}+\frac{s\mathbf{t}}{\sqrt{n}}\right)\right|\right\}$$

$$+\sup_{\|\mathbf{t}\|\le C}\left\{\left|k^2\left(\mathbf{x}_i,\boldsymbol{\theta}+\frac{s\mathbf{t}}{\sqrt{n}}\right)-k^2(\mathbf{x}_i,\boldsymbol{\theta})\right|\right.$$

$$\left.\times\left|f_{\theta_l}\left(\mathbf{x}_i,\boldsymbol{\theta}+\frac{s\mathbf{t}}{\sqrt{n}}\right)f_{\theta_j}\left(\mathbf{x}_i,\boldsymbol{\theta}+\frac{s\mathbf{t}}{\sqrt{n}}\right)\right||\psi'(\epsilon_i(\boldsymbol{\theta}))|\right\}$$

$$+\sup_{\|\mathbf{t}\|\le C}\left\{\left|f_{\theta_l}\left(\mathbf{x}_i,\boldsymbol{\theta}+\frac{s\mathbf{t}}{\sqrt{n}}\right)f_{\theta_j}\left(\mathbf{x}_i,\boldsymbol{\theta}+\frac{s\mathbf{t}}{\sqrt{n}}\right)-f_{\theta_l}(\mathbf{x}_i,\boldsymbol{\theta})f_{\theta_j}(\mathbf{x}_i,\boldsymbol{\theta})\right|\right.$$

$$\left.\times|k^2(\mathbf{x}_i,\boldsymbol{\theta})||\psi'(\epsilon_i(\boldsymbol{\theta}))|\right\}$$

$$+\sup_{\|\mathbf{t}\|\le C}\left\{\left|\psi'\left(\epsilon_i\left(\boldsymbol{\theta}+\frac{s\mathbf{t}}{\sqrt{n}}\right)\right)\epsilon_i\left(\boldsymbol{\theta}+\frac{s\mathbf{t}}{\sqrt{n}}\right)-\psi'(\epsilon_i(\boldsymbol{\theta}))\epsilon_i(\boldsymbol{\theta})\right|\right.$$

$$\left.\times\left|k^3\left(\mathbf{x}_i,\boldsymbol{\theta}+\frac{s\mathbf{t}}{\sqrt{n}}\right)f_{\theta_l}\left(\mathbf{x}_i,\boldsymbol{\theta}+\frac{s\mathbf{t}}{\sqrt{n}}\right)\sigma_{\theta_j}\left(\mathbf{x}_i,\boldsymbol{\theta}+\frac{s\mathbf{t}}{\sqrt{n}}\right)\right|\right\}$$

$$+\sup_{\|\mathbf{t}\|\le C}\left\{\left|k^3\left(\mathbf{x}_i,\boldsymbol{\theta}+\frac{s\mathbf{t}}{\sqrt{n}}\right)-k^3(\mathbf{x}_i,\boldsymbol{\theta})\right|\right.$$

$$\left.\times\left|f_{\theta_l}\left(\mathbf{x}_i,\boldsymbol{\theta}+\frac{s\mathbf{t}}{\sqrt{n}}\right)\sigma_{\theta_j}\left(\mathbf{x}_i,\boldsymbol{\theta}+\frac{s\mathbf{t}}{\sqrt{n}}\right)\right||\psi'(\epsilon_i(\boldsymbol{\theta}))\epsilon_i(\boldsymbol{\theta})|\right\}$$

$$+\sup_{\|\mathbf{t}\|\le C}\left\{\left|f_{\theta_l}\left(\mathbf{x}_i,\boldsymbol{\theta}+\frac{s\mathbf{t}}{\sqrt{n}}\right)-f_{\theta_l}(\mathbf{x}_i,\boldsymbol{\theta})\right|\left|\sigma_{\theta_j}\left(\mathbf{x}_i,\boldsymbol{\theta}+\frac{s\mathbf{t}}{\sqrt{n}}\right)k^3(\mathbf{x}_i,\boldsymbol{\theta})\right|\right.$$

$$\left.\times|\psi'(\epsilon_i(\boldsymbol{\theta}))\epsilon_i(\boldsymbol{\theta})|\right\}$$

$$+\sup_{\|\mathbf{t}\|\le C}\left\{\left|\sigma_{\theta_j}\left(\mathbf{x}_i,\boldsymbol{\theta}+\frac{s\mathbf{t}}{\sqrt{n}}\right)-\sigma_{\theta_j}(\mathbf{x}_i,\boldsymbol{\theta})\right||k^3(\mathbf{x}_i,\boldsymbol{\theta})f_{\theta_l}(\mathbf{x}_i,\boldsymbol{\theta})|\right.$$

$$\left.\times|\psi'(\epsilon_i(\boldsymbol{\theta}))\epsilon_i(\boldsymbol{\theta})|\right\}$$

$$+\sup_{\|\mathbf{t}\|\le C}\left\{\left|\psi'\left(\epsilon_i\left(\boldsymbol{\theta}+\frac{s\mathbf{t}}{\sqrt{n}}\right)\right)\epsilon_i\left(\boldsymbol{\theta}+\frac{s\mathbf{t}}{\sqrt{n}}\right)-\psi'(\epsilon_i(\boldsymbol{\theta}))\epsilon_i(\boldsymbol{\theta})\right|\right.$$

$$\times\left|k^3\Big(\mathbf{x}_i,\boldsymbol{\theta}+\frac{s\mathbf{t}}{\sqrt{n}}\Big)f_{\theta_j}\Big(\mathbf{x}_i,\boldsymbol{\theta}+\frac{s\mathbf{t}}{\sqrt{n}}\Big)\sigma_{\theta_l}\Big(\mathbf{x}_i,\boldsymbol{\theta}+\frac{s\mathbf{t}}{\sqrt{n}}\Big)\right|\Bigg\}$$

$$+\sup_{\|\mathbf{t}\|\le C}\Bigg\{\left|k^3\Big(\mathbf{x}_i,\boldsymbol{\theta}+\frac{s\mathbf{t}}{\sqrt{n}}\Big)-k^3(\mathbf{x}_i,\boldsymbol{\theta})\right|$$

$$\times\left|f_{\theta_j}\Big(\mathbf{x}_i,\boldsymbol{\theta}+\frac{s\mathbf{t}}{\sqrt{n}}\Big)\sigma_{\theta_l}\Big(\mathbf{x}_i,\boldsymbol{\theta}+\frac{s\mathbf{t}}{\sqrt{n}}\Big)\right||\psi'(\epsilon_i(\boldsymbol{\theta}))\epsilon_i(\boldsymbol{\theta})|\Bigg\}$$

$$+\sup_{\|\mathbf{t}\|\le C}\Bigg\{\left|f_{\theta_j}\Big(\mathbf{x}_i,\boldsymbol{\theta}+\frac{s\mathbf{t}}{\sqrt{n}}\Big)-f_{\theta_j}(\mathbf{x}_i,\boldsymbol{\theta})\right|\left|\sigma_{\theta_l}\Big(\mathbf{x}_i,\boldsymbol{\theta}+\frac{s\mathbf{t}}{\sqrt{n}}\Big)k^3(\mathbf{x}_i,\boldsymbol{\theta})\right|$$

$$\times|\psi'(\epsilon_i(\boldsymbol{\theta}))\epsilon_i(\boldsymbol{\theta})|\Bigg\}$$

$$+\sup_{\|\mathbf{t}\|\le C}\Bigg\{\left|\sigma_{\theta_l}\Big(\mathbf{x}_i,\boldsymbol{\theta}+\frac{s\mathbf{t}}{\sqrt{n}}\Big)-\sigma_{\theta_l}(\mathbf{x}_i,\boldsymbol{\theta})\right||k^3(\mathbf{x}_i,\boldsymbol{\theta})f_{\theta_j}(\mathbf{x}_i,\boldsymbol{\theta})|$$

$$\times|\psi'(\epsilon_i(\boldsymbol{\theta}))\epsilon_i(\boldsymbol{\theta})|\Bigg\}$$

$$+\sup_{\|\mathbf{t}\|\le C}\Bigg\{\left|\psi'\Big(\epsilon_i\Big(\boldsymbol{\theta}+\frac{s\mathbf{t}}{\sqrt{n}}\Big)\Big)\epsilon_i^2\Big(\boldsymbol{\theta}+\frac{s\mathbf{t}}{\sqrt{n}}\Big)-\psi'(\epsilon_i(\boldsymbol{\theta}))\epsilon_i^2(\boldsymbol{\theta})\right|$$

$$\times\left|k^4\Big(\mathbf{x}_i,\boldsymbol{\theta}+\frac{s\mathbf{t}}{\sqrt{n}}\Big)\sigma_{\theta_l}\Big(\mathbf{x}_i,\boldsymbol{\theta}+\frac{s\mathbf{t}}{\sqrt{n}}\Big)\sigma_{\theta_j}\Big(\mathbf{x}_i,\boldsymbol{\theta}+\frac{s\mathbf{t}}{\sqrt{n}}\Big)\right|\Bigg\}$$

$$+\sup_{\|\mathbf{t}\|\le C}\Bigg\{\left|k^4\Big(\mathbf{x}_i,\boldsymbol{\theta}+\frac{s\mathbf{t}}{\sqrt{n}}\Big)-k^4(\mathbf{x}_i,\boldsymbol{\theta})\right|$$

$$\times\left|\sigma_{\theta_l}\Big(\mathbf{x}_i,\boldsymbol{\theta}+\frac{s\mathbf{t}}{\sqrt{n}}\Big)\sigma_{\theta_j}\Big(\mathbf{x}_i,\boldsymbol{\theta}+\frac{s\mathbf{t}}{\sqrt{n}}\Big)\right||\psi'(\epsilon_i(\boldsymbol{\theta}))\epsilon_i^2(\boldsymbol{\theta})|\Bigg\}$$

$$+\sup_{\|\mathbf{t}\|\le C}\Bigg\{\left|\sigma_{\theta_l}\Big(\mathbf{x}_i,\boldsymbol{\theta}+\frac{s\mathbf{t}}{\sqrt{n}}\Big)\sigma_{\theta_j}\Big(\mathbf{x}_i,\boldsymbol{\theta}+\frac{s\mathbf{t}}{\sqrt{n}}\Big)-\sigma_{\theta_l}(\mathbf{x}_i,\boldsymbol{\theta})\sigma_{\theta_j}(\mathbf{x}_i,\boldsymbol{\theta})\right|$$

$$\times|k^4(\mathbf{x}_i,\boldsymbol{\theta})||\psi'(\epsilon_i(\boldsymbol{\theta}))\epsilon_i^2(\boldsymbol{\theta})|\Bigg\}.$$

Then, by taking the expectation at both sides and conditions **[A3]** (i)-(v), **[B3]** (i)-(ii), and **[C2]** (i)-(ii),

$$E\left\{\sup_{\|\mathbf{t}\|\le C}\left|(\partial/\partial\theta_j)\lambda_l\Big(\mathbf{x}_i,y_i,\boldsymbol{\theta}+\frac{s\mathbf{t}}{\sqrt{n}}\Big)-(\partial/\partial\theta_j)\lambda_l(\mathbf{x}_i,y_i,\boldsymbol{\theta})\right|\right\}\longrightarrow 0,\quad\forall i,$$

and

$$E\left[\sup_{\|\mathbf{t}\|\le C}\left|\frac{1}{n}\sum_{i=1}^{n}\sum_{j=1}^{p}t_j\Big\{(\partial/\partial\theta_j)\lambda_l\Big(\mathbf{x}_i,y_i,\boldsymbol{\theta}+\frac{s\mathbf{t}}{\sqrt{n}}\Big)-(\partial/\partial\theta_j)\lambda_l(\mathbf{x}_i,y_i,\boldsymbol{\theta})\Big\}\right|\right]\longrightarrow 0.$$

Also,

$$\begin{aligned}
&\mathrm{Var}\left[\sup_{\|\mathbf{t}\|\le C}\left|\frac{1}{n}\sum_{i=1}^{n}\sum_{j=1}^{p}t_j\Big\{(\partial/\partial\theta_j)\lambda_l\Big(\mathbf{x}_i,y_i,\boldsymbol{\theta}+\frac{s\mathbf{t}}{\sqrt{n}}\Big)-(\partial/\partial\theta_j)\lambda_l(\mathbf{x}_i,y_i,\boldsymbol{\theta})\Big\}\right|\right]\\
&\le\frac{C^2}{n^2}\sum_{i=1}^{n}\mathrm{Var}\left\{\sum_{j=1}^{p}\sup_{\|\mathbf{t}\|\le C}\left|(\partial/\partial\theta_j)\lambda_l\Big(\mathbf{x}_i,y_i,\boldsymbol{\theta}+\frac{s\mathbf{t}}{\sqrt{n}}\Big)-(\partial/\partial\theta_j)\lambda_l(\mathbf{x}_i,y_i,\boldsymbol{\theta})\right|\right\}\\
&\le C^2K/n\longrightarrow 0.
\end{aligned}$$

Therefore, the result in Eq. (9.6) is obtained.

A.2. Proof of Lemma 9.2

From Eq. (A.1),

$$\begin{aligned}
&\sup_{\|\mathbf{t}\|\le C}\left|\frac{1}{n}\sum_{i=1}^{n}\sum_{j=1}^{p}t_j(\partial/\partial\theta_j)\lambda_l(\mathbf{x}_i,y_i,\boldsymbol{\theta})\right.\\
&\quad+\frac{1}{n}\sum_{i=1}^{n}\sum_{j=1}^{p}t_j\left\{\frac{-\gamma_1}{\sigma^2(\mathbf{x}_i,\boldsymbol{\theta})}(\partial^2/\partial\theta_l\partial\theta_j)\sigma(\mathbf{x}_i,\boldsymbol{\theta})+\frac{\gamma_2}{\sigma^2(\mathbf{x}_i,\boldsymbol{\theta})}f_{\theta_l}(\mathbf{x}_i,\boldsymbol{\theta})f_{\theta_j}(\mathbf{x}_i,\boldsymbol{\theta})\right.\\
&\qquad\qquad\left.\left.+\frac{\gamma_3+2\gamma_1\sigma(\mathbf{x}_i,\boldsymbol{\theta})}{\sigma^4(\mathbf{x}_i,\boldsymbol{\theta})}\sigma_{\theta_l}(\mathbf{x}_i,\boldsymbol{\theta})\sigma_{\theta_j}(\mathbf{x}_i,\boldsymbol{\theta})\right\}\right|\\
&=\sup_{\|\mathbf{t}\|\le C}\left|\frac{1}{n}\sum_{i=1}^{n}\sum_{j=1}^{p}t_j\psi(\epsilon_i(\boldsymbol{\theta}))k(\mathbf{x}_i,\boldsymbol{\theta})\left\{(\partial^2/\partial\theta_l\partial\theta_j)f(\mathbf{x}_i,\boldsymbol{\theta})\right.\right.\\
&\qquad\left.-k(\mathbf{x}_i,\boldsymbol{\theta})f_{\theta_l}(\mathbf{x}_i,\boldsymbol{\theta})\sigma_{\theta_j}(\mathbf{x}_i,\boldsymbol{\theta})-k(\mathbf{x}_i,\boldsymbol{\theta})f_{\theta_j}(\mathbf{x}_i,\boldsymbol{\theta})\sigma_{\theta_l}(\mathbf{x}_i,\boldsymbol{\theta})\right\}\\
&\quad+\frac{1}{n}\sum_{i=1}^{n}\sum_{j=1}^{p}t_j\left\{\psi(\epsilon_i(\boldsymbol{\theta}))\epsilon_i(\boldsymbol{\theta})-\gamma_1\right\}k^2(\mathbf{x}_i,\boldsymbol{\theta})\left\{(\partial^2/\partial\theta_l\partial\theta_j)\sigma(\mathbf{x}_i,\boldsymbol{\theta})\right.\\
&\qquad\left.-2k(\mathbf{x}_i,\boldsymbol{\theta})\sigma_{\theta_l}(\mathbf{x}_i,\boldsymbol{\theta})\sigma_{\theta_j}(\mathbf{x}_i,\boldsymbol{\theta})\right\}\\
&\quad-\frac{1}{n}\sum_{i=1}^{n}\sum_{j=1}^{p}t_j\left\{\psi'(\epsilon_i(\boldsymbol{\theta}))-\gamma_2\right\}k^2(\mathbf{x}_i,\boldsymbol{\theta})f_{\theta_l}(\mathbf{x}_i,\boldsymbol{\theta})f_{\theta_j}(\mathbf{x}_i,\boldsymbol{\theta})\\
&\quad+\frac{1}{n}\sum_{i=1}^{n}\sum_{j=1}^{p}t_j\psi'(\epsilon_i(\boldsymbol{\theta}))\epsilon_i(\boldsymbol{\theta})k^2(\mathbf{x}_i,\boldsymbol{\theta})\left\{f_{\theta_l}(\mathbf{x}_i,\boldsymbol{\theta})\sigma_{\theta_j}(\mathbf{x}_i,\boldsymbol{\theta})+f_{\theta_j}(\mathbf{x}_i,\boldsymbol{\theta})\sigma_{\theta_l}(\mathbf{x}_i,\boldsymbol{\theta})\right\}\\
&\quad\left.+\frac{1}{n}\sum_{i=1}^{n}\sum_{j=1}^{p}t_j\left\{\psi'(\epsilon_i(\boldsymbol{\theta}))\epsilon_i^2(\boldsymbol{\theta})-\gamma_3\right\}k^4(\mathbf{x}_i,\boldsymbol{\theta})\sigma_{\theta_l}(\mathbf{x}_i,\boldsymbol{\theta})\sigma_{\theta_j}(\mathbf{x}_i,\boldsymbol{\theta})\right|\\
&\le C\sum_{j=1}^{p}\left|\frac{1}{n}\sum_{i=1}^{n}\psi(\epsilon_i(\boldsymbol{\theta}))k(\mathbf{x}_i,\boldsymbol{\theta})\left\{(\partial^2/\partial\theta_l\partial\theta_j)f(\mathbf{x}_i,\boldsymbol{\theta})-k(\mathbf{x}_i,\boldsymbol{\theta})f_{\theta_l}(\mathbf{x}_i,\boldsymbol{\theta})\sigma_{\theta_j}(\mathbf{x}_i,\boldsymbol{\theta})\right.\right.\\
&\qquad\left.\left.-k(\mathbf{x}_i,\boldsymbol{\theta})f_{\theta_j}(\mathbf{x}_i,\boldsymbol{\theta})\sigma_{\theta_l}(\mathbf{x}_i,\boldsymbol{\theta})\right\}\right|
\end{aligned}$$

$$+C\sum_{j=1}^{p}\left|\frac{1}{n}\sum_{i=1}^{n}\{\psi(\epsilon_i(\boldsymbol{\theta}))\epsilon_i(\boldsymbol{\theta})-\gamma_1\}k^2(\mathbf{x}_i,\boldsymbol{\theta})\left\{(\partial^2/\partial\theta_l\partial\theta_j)\sigma(\mathbf{x}_i,\boldsymbol{\theta})\right.\right.$$
$$\left.\left.-2k(\mathbf{x}_i,\boldsymbol{\theta})\sigma_{\theta_l}(\mathbf{x}_i,\boldsymbol{\theta})\sigma_{\theta_j}(\mathbf{x}_i,\boldsymbol{\theta})\right\}\right|$$

$$+C\sum_{j=1}^{p}\left|\frac{1}{n}\sum_{i=1}^{n}\left\{\psi'(\epsilon_i(\boldsymbol{\theta}))-\gamma_2\right\}k^2(\mathbf{x}_i,\boldsymbol{\theta})f_{\theta_l}(\mathbf{x}_i,\boldsymbol{\theta})f_{\theta_j}(\mathbf{x}_i,\boldsymbol{\theta})\right|$$
$$+C\sum_{j=1}^{p}\left|\frac{1}{n}\sum_{i=1}^{n}\psi'(\epsilon_i(\boldsymbol{\theta}))\epsilon_i(\boldsymbol{\theta})k^2(\mathbf{x}_i,\boldsymbol{\theta})\left\{f_{\theta_l}(\mathbf{x}_i,\boldsymbol{\theta})\sigma_{\theta_j}(\mathbf{x}_i,\boldsymbol{\theta})\right.\right.$$
$$\left.\left.+f_{\theta_j}(\mathbf{x}_i,\boldsymbol{\theta})\sigma_{\theta_l}(\mathbf{x}_i,\boldsymbol{\theta})\right\}\right|$$

$$+C\sum_{j=1}^{p}\left|\frac{1}{n}\sum_{i=1}^{n}\left\{\psi'(\epsilon_i(\boldsymbol{\theta}))\epsilon_i^2(\boldsymbol{\theta})-\gamma_3\right\}k^4(\mathbf{x}_i,\boldsymbol{\theta})\sigma_{\theta_l}(\mathbf{x}_i,\boldsymbol{\theta}),\sigma_{\theta_j}(\mathbf{x}_i,\boldsymbol{\theta})\right|$$

which by using the Markov WLLN and conditions **[A2]** (i)-(ii) yields:

$$\frac{1}{n}\sum_{i=1}^{n}\psi(\epsilon_i(\boldsymbol{\theta}))k(\mathbf{x}_i,\boldsymbol{\theta})\left\{(\partial^2/\partial\theta_l\partial\theta_j)f(\mathbf{x}_i,\boldsymbol{\theta})-k(\mathbf{x}_i,\boldsymbol{\theta})f_{\theta_l}(\mathbf{x}_i,\boldsymbol{\theta})\sigma_{\theta_j}(\mathbf{x}_i,\boldsymbol{\theta})\right.$$
$$\left.-k(\mathbf{x}_i,\boldsymbol{\theta})f_{\theta_j}(\mathbf{x}_i,\boldsymbol{\theta})\sigma_{\theta_l}(\mathbf{x}_i,\boldsymbol{\theta})\right\}=o_p(1),$$
$$\frac{1}{n}\sum_{i=1}^{n}\{\psi(\epsilon_i(\boldsymbol{\theta}))\epsilon_i(\boldsymbol{\theta})-\gamma_1\}k^2(\mathbf{x}_i,\boldsymbol{\theta})\left\{(\partial^2/\partial\theta_l\partial\theta_j)\sigma(\mathbf{x}_i,\boldsymbol{\theta})\right.$$
$$\left.-2k(\mathbf{x}_i,\boldsymbol{\theta})\sigma_{\theta_l}(\mathbf{x}_i,\boldsymbol{\theta})\sigma_{\theta_j}(\mathbf{x}_i,\boldsymbol{\theta})\right\}=o_p(1),$$

$$\frac{1}{n}\sum_{i=1}^{n}\left\{\psi'(\epsilon_i(\boldsymbol{\theta}))-\gamma_2\right\}k^2(\mathbf{x}_i,\boldsymbol{\theta})f_{\theta_l}(\mathbf{x}_i,\boldsymbol{\theta})f_{\theta_j}(\mathbf{x}_i,\boldsymbol{\theta})=o_p(1),$$

$$\frac{1}{n}\sum_{i=1}^{n}\psi'(\epsilon_i(\boldsymbol{\theta}))\epsilon_i(\boldsymbol{\theta})k^2(\mathbf{x}_i,\boldsymbol{\theta})\left\{f_{\theta_l}(\mathbf{x}_i,\boldsymbol{\theta})\sigma_{\theta_j}(\mathbf{x}_i,\boldsymbol{\theta})+f_{\theta_j}(\mathbf{x}_i,\boldsymbol{\theta})\sigma_{\theta_l}(\mathbf{x}_i,\boldsymbol{\theta})\right\}=o_p(1),$$

and

$$\frac{1}{n}\sum_{i=1}^{n}\left\{\psi'(\epsilon_i(\boldsymbol{\theta}))\epsilon_i^2(\boldsymbol{\theta})-\gamma_3\right\}k^4(\mathbf{x}_i,\boldsymbol{\theta})\sigma_{\theta_l}(\mathbf{x}_i,\boldsymbol{\theta})\sigma_{\theta_j}(\mathbf{x}_i,\boldsymbol{\theta})=o_p(1).$$

Thus, the result in Eq. (9.8) is obtained.

References

1. Avalos, M., Mak, C., Randall, P. K., Trzeciakowski, J. P., Abell, C., Kwan, S.-W. and Wilcox, R. E. (2001). Nonlinear analysis of partial dopamine agonist effects on cAMP in C6 glioma cells. *Journal of Pharmacological and Toxicological Methods* **45**, 17–37.

2. Barata, C., Baird, D. J., Nogueira, A. J. A., Soares, A. M. V. M. and Riva, M. C. (2006). Toxicity of binary mixtures of metals and pyrethroid insecticides to Daphnia magna Straus. Implications for multi-substance risks assessment. *Aquatic Toxicology* **78**, 1–14.
3. Bose, A. and Chatterjee, S. (2001). Generalised bootstrap in non-regular M-estimation problems. *Statistics & Probability Letters* **55**, 319–328.
4. El Bantli, F. (2004). M-estimation in linear models under nonstandard conditions. *Journal of Statistical Planning and Inference* **121**, 231–248.
5. El Bantli, F. and Hallin M. (2001). Asymptotic behaviour of M-estimators in AR(p) models under nonstandard conditions. *The Canadian Journal of Statistics* **29**(1), 155–168.
6. Hampel, F. R., Rousseeuw, P. J., Ronchetti, E., and Stabel, W. (1986). *Robust Statistics – The Approach Based on Influence Functions.* New York: John Wiley & Sons.
7. Hill, A. V. (1910). The possible Effects of the Aggregation of the Molecules of Haemoglobin on its Dissociation Curves. *Journal of Physiology* **40**(Suppl), iv–vii.
8. Huber, P. J. (1973). Robust regression: asymptotics, conjectures and Monte-Carlo. *Ann. Statist.* **1**, 799–821.
9. Huber, P. J. (1981). *Robust Statistics.* New York: John Wiley & Sons.
10. Gaylor, D. W. and Aylward, L. L. (2004). An evaluation of benchmark dose methodology for non-cancer continuous-data health effects in animals due to exposures to dioxin (TCDD). *Regulatory Toxicology and Pharmacology* **40**, 9–17.
11. Jureĉková, J. and Sen, P. K. (1996). *Robust Statistical Procedures, Asymptotics and Interrelations.* New York: John Wiley & Sons.
12. Kent, J. T. and Tyler, D. E. (2001). Regularity and uniqueness for constrained M-estimates and redescending M-estimates. *Ann. Statist.* **29**(1), 252–265.
13. Klein, R. and Yohai, V. J. (1981). Asymptotic behavior of iterative M-estimators for the linear model. *Comm. Statist.* **A10**, 2373–2388.
14. Lim, C. (2009). *Statistical Theory and Robust Methodology for Nonlinear models with Application to Toxicology.* Ph.D. dissertation, University of North Carolina at Chapel Hill.
15. Maronna, R. A. (1976). Robust M-estimators of multivariate location and scatter. *Ann. Statist.* **4**, 163–169.
16. Maronna, R. A., Martin, D. R. and Yohai, V. J. (2006). *Robust Statistics: Theory and Methods.* New York: John Wiley & Sons.
17. Maronna, R. A. and Yohai, V. J. (2008). Robust low-rank approximation of data matrices with elementwise contamination. *Technometrics.* **50**(3), 295–304.
18. Morris, J. B., Symanowicz, P. and Sarangapani, R. (2002). Regional distribution and kinetics of vinyl acetate hydrolysis in the oral cavity of the rat and mouse. *Toxicology Letters* **126**, 31–39.
19. National Toxicology Program. (2007). NTP Toxicity Studies of Sodium Dichromate Dihydrate (CAS No. 7789-12-0) Administered in Drinking

Water to Male and Female F344/N Rats and B6C3F1 Mice and Male BALB/c and am3-C57BL/6 Mice. *Toxicity Report Series* **72**, 1–G4, U.S. Department of Health and Human Services, Public Health Service, National Institutes of Health, RTP, NC.
20. Pounds, J. G., Haider, J., Chen, D. G. and Mumtaz, M. (2004). Interactive toxicity of simple chemical mixtures of cadmium, mercury, methylmercury and trimethyltin: model-dependent responses. *Environmental Toxicology and Pharmacology* **18**, 101–113.
21. Relles, D. (1968). *Robust regression by modified least squares.* PhD dissertation, Yale University.
22. Sanhueza, A. I. (2000). *Robust M-procedures in Nonlinear Regression Models.* PhD dissertation, University of North Carolina at Chapel Hill.
23. Sanhueza, A. I. and Sen, P. K. (2001). M-methods in generalized nonlinear models. In *Probability and Statistical Models with Applications.* Eds. Ch. A. Charalambides, M. V. Koutras and N. Balakrishnan, pp. 359–375. New York: Chapman & Hall/CRC.
24. Sanhueza, A. I. and Sen, P. K. (2004). Robust M-procedures in univariate nonlinear regression models. *Brazilian Journal of Probability and Statistics* **18**, 183–200.
25. Sanhueza, A. I., Sen, P. K. and Leiva, V. (2009). A robust procedure in nonlinear for repeated measurements. *Communications in Statistics – Theory and Methods* **38**, 138–155.
26. Seber, G. A. F. and Wild, C. H. (1989). *Nonlinear Regression.* New York: Wiley.
27. Singer, J. M. and Sen, P. K. (1985). M-methods in multivariate linear models. *J. Multivar. Anal.* **17**, 168–184.
28. Tyler, D. E. (2002). High breakdown point multivariate M-estimation. *Estadística.* **54**, 213–247.
29. Velarde, G., Ait-aissa, S., Gillet, C., Rogerieux, F., Lambre, C., Vindimian, E. and Porcher, J. M. (1999). Use of transepithelial electrical resistance in the study of pentachlorophenol toxicity. *Toxicology in Vitro* **13**, 723–727.
30. Yohai, V. J. and Maronna, R. A. (1979). Asymptotic behaviour of M-estimators for the linear model. *Ann. Statist.* **7**, 258–268.

Chapter 10

The Inverse Censoring Weighted Approach for Estimation of Survival Functions from Left and Right Censored Data

Sundarraman Subramanian* and Peixin Zhang†

Center for Applied Mathematics and Statistics
Department of Mathematical Sciences
New Jersey Institute of Technology
Newark, New Jersey USA
**sundars@njit.edu*
†*pz7@njit.edu*

We propose an inverse censoring weighted estimator of a survival function when the data are doubly censored but the left censoring variable is always observed. The proposed estimator reduces to the standard inverse censoring weighted estimator for right censored data, where there is no left censoring, and in that sense may be viewed as an extension of the latter estimator to the special double censoring scenario considered in this paper. However, the equivalence exhibited by the Kaplan–Meier and inverse censoring weighted estimators in the case of right censoring does not apply any more for the scenario studied here. Specifically, a Kaplan–Meier approach based on a modified risk set and the proposed inverse censoring approach lead to different estimators. Furthermore, when both censoring variables are always observed, as in the case of an AIDS clinical trial data set analyzed in this paper, an alternative inverse censoring weighted estimator can be computed using the additional available censoring information. We present the results of a numerical comparison study between the Kaplan–Meier type and the two inverse censoring weighted estimators.

10.1. Introduction

Double censoring arises in survival studies when an outcome variable T, typically a failure time, can only be accurately observed or measured within a certain range (L, U). While left censoring results whenever T falls below L — so that only L is observed — right censoring happens when the failure has not occurred before U, in which case only U is observed. More specifically, the conventional double censoring framework supposes the observation of the pair (X, δ), where $X = \max\{\min(T, U), L\}$, and δ denotes a double censoring indicator: $\delta = 1^{I(T<L)}2^{I(L \le T \le U)}3^{I(T>U)}$. Note that only one of T, L and U is observed for each of a sample of n items. We assume that L and U are random variables with marginal distribution functions $Q(t)$ and $W(t)$ respectively, and that $L \le U$ with probability 1. The dependence between L and U is otherwise left unspecified. For grouped data, Ref. [1] constructed a nonparametric maximum likelihood estimator (NPMLE) of $S(t)$, the survival function of T, which solves a self consistency equation. Ref. [2] considered the problem for ungrouped data and proposed the NPMLE of $S(t)$. The self consistency equation does not produce a unique solution, however [3, 4]. The large sample properties of the NPMLE are well studied [2, 3, 5–7]. For related work concerning smooth survival function estimation, see Ref. [8]. It has been noted, however, that one-sample estimation, two-sample testing, and regression methods from doubly censored data are quite complicated because the censoring variables are observable only when T is censored [9]. In this article, we consider estimation of $S(t)$ under the relaxed condition that one or both of the censoring variables are always observed. Ref. [10] considered the setting in which L is always observed and introduced a Kaplan–Meier type estimator of $S(t)$ based on a reduced sample that does not take the left censored values into account while forming the risk set essential for its computation. This estimator was employed as the initial value in the iteration process needed to obtain the NPMLE for the general case. Ref. [11] also considered this scenario when investigating nonparametric two-sample approximate tests for discrete interval censored data.

Ref. [9] proposed semiparametric transformation models for analyzing doubly censored data, which required that both L and U are always observed. They presented analysis of data from a randomized pediatric AIDS clinical trial, in which the main outcome variable, plasma HIV-1 RNA level, is considered unreliable below 400 copies/ml and above 750,000 copies/ml. Taking T to be the change in the log10 RNA level from a baseline value to the value after 24 weeks, T is observed between $L = l_0 - 5.88$ and $U = l_0 - 2.6$, where l_k denotes the log10 RNA value at week k. Out of the 195 subjects included in their analysis, there were about 4% left and 42% right censored cases. A feature of the data is that both L and U are always observed. As discussed above, with conventional doubly censored data L and U are not always observed, however.

The objective of our investigation is two fold, both relating to the performance of an inverse censoring weighted (ICW) estimator of $F(t) = 1 - S(t)$ that we propose in the context of doubly censored data with L always observed. More specifically, we are concerned with its performance relative to the above-mentioned Kaplan–Meier type estimator, and also with an ICW counterpart that utilizes the available additional information whenever *both* L and U are always observed. For standard right censored data, it is well known that the ICW estimator of $F(t)$ scales the jump size of the empirical estimator of the subdistribution function corresponding to the uncensored failures using estimated inverse censoring weights. In particular, if $G(t)$ denoted the censoring distribution and $1 - \hat{G}(t)$ its Kaplan–Meier estimator, the ICW estimator assigns a mass of $1/(1 - \hat{G}(X_i))$ at each uncensored failure X_i. This approach works because of the independence of failure and censoring, also called random censoring. In a regression context, the ICW approach has been exploited for simple linear regression [12] and for median regression [13, 14], and obviates curse of dimensionality under the additional condition of covariate free censoring, permitting use of the global Kaplan–Meier estimator of the censoring distribution. More recently, for standard right censored data, Ref. [15] derived the equivalence between the Kaplan–Meier and ICW estimators of $S(t)$.

When estimating $S(t)$ from doubly censored data where L is always observed, the rationale of the ICW approach is the same as well, depending on an estimated censoring weight $\hat{G}(t)$, where $G(t)$ will now denote the censoring probability $P(L < t \leq U)$. Analogous to right censoring, the estimator for this situation, which we shall call the ICW Type I estimator, assigns a mass of $1/(1-\hat{G}(X_i))$ at each uncensored failure X_i. The censoring probability arises through a unique solution of a certain Volterra integral equation, and can be estimated using empirical estimators and the product integral, see Ref. [14]. Our first objective, therefore, is investigation of the equivalence between the Kaplan–Meier type and ICW estimators for doubly censored data when only L is always observed. We show that, although our proposed estimator reduces to its right censored ICW version when there is no left censoring, equivalence between the two estimators does not hold. Asymptotic equivalence of the two estimators is an open problem. Numerical results, however, indicate that the Kaplan–Meier type estimator may have the superior moderate sample performance, especially for moderate left and right censoring rates.

When L and U are both always observed, an estimator of $G(t)$ may be computed as the difference between the empirical estimators of the marginal distributions of L and U respectively. Using this estimate $\hat{G}(t)$, an ICW estimator for this situation can be computed and we call it ICW Type II estimator. Based on our simulation studies, this ICW Type II estimator is inferior to the ICW Type I estimator, which leads to the surprising result that always knowing the right censoring time does not contribute to increased efficiency. We also present plots of the three estimators and a semiparametric estimator for the AIDS clinical trial data described above.

The article is organized as follows. In Sec. 10.2, we introduce the estimators and derive the associated large sample properties. In Sec. 10.3, we present the AIDS clinical trial data analysis. In Sec. 10.4, we present our simulation results. We end with a brief concluding section. Detailed calculations are given in the Appendix.

10.2. Survival Function Estimators

In this section we first review the double censoring specific Kaplan–Meier type estimator introduced by Ref. [10], and then propose our ICW estimator.

10.2.1. *The KM type estimator*

We denote this estimator by $\hat{S}_{KM}(t)$. Let $H_2(t)$ and $y_1(t)$ denote the limits of the quantities

$$\hat{H}_2(t) = \frac{1}{n}\sum_{i=1}^{n} I(X_i \le t, \delta_i = 2), \qquad Y_1(t) = \frac{1}{n}\sum_{i=1}^{n} I(L_i < t \le T_i \wedge U_i),$$

respectively. Let τ^* be such that $y_1(t) > 0$ for $t \in (0, \tau^*]$. The estimator $\hat{S}_{KM}(t)$ is obtained from the product integral of a Nelson–Aalen type estimator of the cumulative hazard $\Lambda(t)$, see Ref. [10]:

$$\hat{\Lambda}_{\mathrm{NA}}(t) = \int_0^t \frac{d\hat{H}_2(u)}{Y_1(u)}. \tag{10.1}$$

Suppose also that there exists a $\tau > 0$ such that $\int_0^\tau dW(u)/G^2(u) < \infty$, see Ref. [16]. Using standard methods (e.g., [17]), we can show that $\hat{\Lambda}_{\mathrm{NA}}(t) - \Lambda(t)$ is asymptotically linear with influence function, $\alpha(t)$, taking the form

$$\alpha(t) = \int_0^t \frac{dI(X \le s, \delta = 2)}{y_1(s)} - \int_0^t \frac{I(L < s \le T \wedge U)}{y_1(s)} d\Lambda(s).$$

Defining $C(s) = \int_0^s dH_2(u)/y_1^2(u) = \int_0^s d\Lambda(u)/y_1(u)$, it can be shown that the process $n^{1/2}\left(\hat{\Lambda}_{\mathrm{NA}}(s) - \Lambda(s)\right)$ converges weakly in $D[0, \tau^*]$ to a

Gaussian process with zero mean and covariance given, for $s \le t$, by $E(\alpha(s)\alpha(t)) = C(s)$, see Refs. [17–20], among others. The asymptotic variance of $n^{1/2}\left(\hat{S}_{\mathrm{KM}}(t) - S(t)\right)$ is then given by the quantity $S^2(t)C(t)$.

10.2.2. *The ICW estimator*

Analogous to the case of right censored data, the ICW Type I estimator is obtained by integrating over $[0, t]$ the ratio of two estimators. The numerator of this ratio is $d\hat{H}_2(s)$, and the denominator is an estimator of the censoring probability $G(s) = P(L < s \le U)$. Let $\hat{H}_3(t) = n^{-1}\sum_{i=1}^{n} I(X_i \le t, \delta_i = 3)$ denote the empirical estimator of the subdistribution function $H_3(t) = P(X \le t, \delta = 3)$. Let $\hat{Q}(t)$ denote the empirical estimator of $Q(t)$, and $\hat{G}(t)$ an estimator of $G(t)$. Note that G has the representation [14]

$$G(t) = \int_0^t \left(\prod_{u \in (s,t]} (1 + dA(u)) \right) dQ(s), \tag{10.2}$$

where π denotes the product integral and $A(t)$ is given by

$$A(t) \equiv -\int_0^t \frac{1}{y_1(s)} dH_3(s) = -\int_0^t \frac{dW(s)}{G(s)}. \tag{10.3}$$

We obtain $\hat{G}(t)$ by plugging in $\hat{Q}(t)$ and the estimator $\hat{A}(t) = -\int_0^t d\hat{H}_3(s)/Y_1(s)$ for $A(t)$ into Eq. (10.2). The ICW Type I estimator of $S(t)$, therefore, takes the form

$$\hat{S}_{\mathrm{ICW}}(t) = 1 - \int_0^t \frac{d\hat{H}_2(s)}{\hat{G}(s)} \equiv 1 - \hat{F}_{\mathrm{ICW}}(t). \tag{10.4}$$

When there is no left censoring, $L = -\infty$, and $\hat{G}(t)$ reduces to the standard Kaplan–Meier estimator of the censoring survival probability (right censored case). In turn, Eq. (10.4) is just the ICW estimator of the censoring survival distribution. In this way, we are able to extend the ICW estimator for right censored data to doubly censored data with L always observed.

The first issue concerns the equivalence of $\hat{S}_{\mathrm{KM}}$ and $\hat{S}_{\mathrm{ICW}}$. For right censored data, the Kaplan–Meier estimators of the failure time and censoring time survival distributions do not jump together, so it is relatively straightforward that their product is the empirical survival function of the minimum, which in turn is the basis for proving the equivalence of the two estimators. For the setting considered here, however, it is more difficult.

Clearly, the estimators $\hat{S}_{\mathrm{KM}}$ and $\hat{S}_{\mathrm{ICW}}$ jump only at uncensored points ($\delta = 2$). Consider an uncensored observation X_j (assuming no ties). The size of the jump of $\hat{S}_{\mathrm{KM}}(t)$ at X_j equals

$$\begin{aligned}\hat{S}_{\mathrm{KM}}(X_j-) - \hat{S}_{\mathrm{KM}}(X_j) &= \hat{S}_{\mathrm{KM}}(X_j-) - \hat{S}_{\mathrm{KM}}(X_j-)\left(1 - \frac{1}{Y_1(X_j)}\right)\\ &= \frac{\hat{S}_{\mathrm{KM}}(X_j-)}{Y_1(X_j)}.\end{aligned}$$

On the other hand, the size of the jump of $\hat{S}_{\mathrm{ICW}}(t)$ at X_j equals $1/\hat{G}(X_j)$. The jump sizes are equal only if $\hat{S}_{\mathrm{KM}}(X_j-)\hat{G}(X_j) = Y_1(X_j)$. In the following counter example we show that this is not the case, which serves to dispel any notion that the estimators $\hat{S}_{\mathrm{KM}}$ and $\hat{S}_{\mathrm{ICW}}$ may be the exact same.

Table 10.1. A left and right censored data set.

X	10	14^-	21^-	24	27^+	31^+	33	35^+	39^+	40
L	4	14	21	8	5	7	15	12	18	20

Note: Censoring indicated by "+" (right) and "−" (left).

From Table 10.1, clearly, $\hat{S}_{\mathrm{KM}}(10-) = 1$ since there are no jumps before the first observation. We have that

$$\begin{aligned}\hat{G}(10) &= \frac{1}{10}\left[\prod_{u\in(4,10]}\left(1 - \frac{d\hat{H}_3(u)}{Y_1(u)}\right) + \prod_{u\in(5,10]}\left(1 - \frac{d\hat{H}_3(u)}{Y_1(u)}\right)\right.\\ &\quad \left. + \prod_{u\in(7,10]}\left(1 - \frac{d\hat{H}_3(u)}{Y_1(u)}\right) + \prod_{u\in(8,10]}\left(1 - \frac{d\hat{H}_3(u)}{Y_1(u)}\right)\right]\\ &= \frac{1}{10}(1+1+1+1) \;=\; \frac{4}{10}.\end{aligned}$$

Since $Y_1(10) = 4/10$, we have that $\hat{S}_{\mathrm{KM}}(10-)\hat{G}(10) = Y_1(10)$. Next consider the observation 24, for which $\delta = 2$. We have $\hat{S}_{\mathrm{KM}}(24-) = \hat{S}_{\mathrm{KM}}(10) = 1(1 - 1/4) = 3/4$. Furthermore, $\hat{G}(24) = 1$ and $Y_1(24) = 7/10$. Therefore, $\hat{S}_{\mathrm{KM}}(24-)\hat{G}(24) \neq Y_1(24)$. The two estimators are not equivalent.

For proving the asymptotic normality of the ICW estimator, we define

$$B(t) = \prod_{(0,t]}(1+dA(s)), \quad R(t) = \int_0^t dQ(s)/B(s), \quad \rho_t(s) = \int_s^t dF(u)/R(u).$$

We also define the influence function

$$\beta(t) = \int_0^t \frac{dI(X \le s, \delta = 2)}{G(s)} + \int_0^t \frac{R(s)}{y_1(s)} \rho_t(s) dI(X \le s, \delta = 3) + \\ + \int_0^t \frac{R(s)}{y_1(s)} \rho_t(s) I(L < s \le T \wedge U) dA(s) \\ - \int_0^t \frac{\rho_t(s)}{B(s)} dI(L \le s). \tag{10.5}$$

Note that $E(\beta(t)) = 0$. Write the second moment of $\beta(t)$ as $V_{\mathrm{ICW}}(t)$. Denote the conditional distributions of L given $U = u$, and U given $L = v$, by $Q_{L|U}(v|u)$ and $W_{U|L}(u|v)$ respectively. Assuming continuity of the conditional distributions, it will be shown that

$$\begin{aligned} V_{\mathrm{ICW}}(t) = & \int_0^t \frac{dF(s)}{G(s)} + \int_0^t \rho_t^2(s) \left(\frac{dQ(s)}{B^2(s)} + \frac{dW(s)}{S(s)B^2(s)} \right) \\ & + 2 \int_0^t \left[\left\{ \int_0^s \frac{\rho_t(u)}{B(u)} \frac{P(L < u, U \ge s)}{S(u)} dA(u) \right\} \right. \\ & \qquad \left. \times \left\{ \frac{dF(s)}{G(s)} - \frac{\rho_t(s)}{B(s)} \frac{dW(s)}{G(s)} \right\} \right] \\ & + 2 \int_0^t \left[\left\{ \int_0^s \frac{\rho_t(u)}{B(u)} (1 - W_{U|L}(s|u)) dQ(u) \right\} \right. \\ & \qquad \left. \times \left\{ \frac{\rho_t(s)}{B(s)} \frac{dW(s)}{G(s)} - \frac{dF(s)}{G(s)} \right\} \right] \\ & + 2 \int_0^t \left[\left\{ \int_0^s \frac{\rho_t(u)}{B(u)} \left(\frac{Q_{L|U}(u|s)}{S(u)} dA(u) - dQ_{L|U}(u|s) \right) \right\} \right. \\ & \qquad \left. \times \frac{\rho_t(s)}{B(s)} \right] dW(s). \end{aligned} \tag{10.6}$$

Theorem 10.1. *Suppose that τ and τ^* are as defined in Sec. 10.2.1. Then, $\hat{F}_{\mathrm{ICW}}(t) - F(t)$ is asymptotically linear with influence function $\beta(t)$ given by Eq. (10.5).*

Proof. Some of the details needed for the proof here can be found in Ref. [14]. From Eq. (10.2), $G(t) = B(t)R(t)$. Furthermore, the process $\hat{A}(t) - A(t)$ has the following asymptotic representation, where the remainder term is $o_p(n^{-1/2})$ uniformly for $t \in (0, \tau^*]$:

$$\hat{A}(t) - A(t) = -\frac{1}{n}\sum_{i=1}^{n}\int_0^t \left[\frac{1}{y_1(s)}\{dI(X_i \le s, \delta_i = 3) + I(L_i < s \le T_i \wedge U_i)dA(s)\}\right] + o_p(n^{-1/2}).$$

From the convergence rate of $\hat{G}(t)$ to $G(t)$ (see, Ref. [14]), it follows that

$$\hat{F}_{\mathrm{ICW}}(t) - F(t) = \frac{1}{n}\sum_{i=1}^{n}\int_0^t \frac{d\{I(X_i \le s, \delta_i = 2) - H_2(s)\}}{G(s)} - \int_0^t \frac{\hat{G}(s) - G(s)}{G(s)} dF(s) + o_p(n^{-1/2}). \quad (10.7)$$

We can write the second term of (10.7) as $I_1(t) + I_2(t) + o_p(n^{-1/2})$, where

$$I_1(t) = -\int_0^t \frac{B(s)}{G(s)}\left[\int_0^s R(u)d\left\{\hat{A}(u) - A(u)\right\}\right] dF(s),$$

$$I_2(t) = -\int_0^t \frac{B(s)}{G(s)}\left[\int_0^s \frac{1}{B(u)} d\left\{\hat{Q}(u) - Q(u)\right\}\right] dF(s).$$

Now interchange the order of integration and use the representation for $\hat{A}(t) - A(t)$ to obtain

$$\begin{aligned} I_1(t) &= -\int_0^t \frac{1}{R(s)}\left[\int_0^s R(u)d\left\{\hat{A}(u) - A(u)\right\}\right] dF(s) \\ &= -\int_0^t \rho_t(s)R(s)d\left\{\hat{A}(s) - A(s)\right\} \\ &= \frac{1}{n}\sum_{i=1}^{n}\int_0^t \frac{\rho_t(s)R(s)}{y_1(s)}\left\{dI(X_i \le s, \delta_i = 3) + I(L_i < s \le T_i \wedge U_i)dA(s)\right\} + o_p(n^{-1/2}). \end{aligned}$$

Likewise, interchange the order of integration to obtain

$$I_2(t) = -\int_0^t \rho_t(s)d\hat{Q}(s)/B(s) + F(t)$$

Plugging in these asymptotic representations for $I_1(t)$ and $I_2(t)$ into (10.7), it follows that $\hat{F}_{\mathrm{ICW}}(t) - F(t)$ is asymptotically linear with influence function $\beta(t)$ given by Eq. (10.5). □

Writing the four components of $\beta(t)$ in Eq. (10.5) as $\beta_i(t), i = 1, 2, 3, 4$, the following expressions hold; see the Appendix for detailed calculations.

$$E(\beta_1^2(t)) = \int_0^t \frac{1}{G(s)} dF(s), \qquad E(\beta_1(t)\beta_2(t)) = 0,$$

$$E(\beta_2^2(t)) = \int_0^t \frac{\rho_t^2(s)}{B^2(s)} \frac{dW(s)}{S(s)}, \qquad E(\beta_4^2(t)) = \int_0^t \frac{\rho_t^2(s)}{B^2(s)} dQ(s),$$

$$E(\beta_3^2(t)) = -2 \int_0^t \left\{ \int_0^s \frac{\rho_t(u)}{B(u)} \frac{P(L < u, U \geq s)}{S(u)} dA(u) \right\} \frac{\rho_t(s)}{B(s)} \frac{dW(s)}{G(s)},$$

$$E(\beta_1(t)\beta_3(t)) = \int_0^t \left\{ \int_0^s \frac{\rho_t(u)}{B(u)} \frac{P(L < u, U \geq s)}{S(u)} dA(u) \right\} \frac{dF(s)}{G(s)},$$

$$E(\beta_1(t)\beta_4(t)) = -\int_0^t \left\{ \int_0^s \frac{\rho_t(u)}{B(u)} (1 - W_{U|L}(s|u)) dQ(u) \right\} \frac{dF(s)}{G(s)},$$

$$E(\beta_3(t)\beta_4(t)) = -\int_0^t \left\{ \int_0^s \frac{\rho_t(u)}{B(u)} (1 - W_{U|L}(s|u)) dQ(u) \right\} \frac{\rho_t(s)}{B(s)} dA(s),$$

$$E(\beta_2(t)\beta_3(t)) = \int_0^t \left\{ \int_0^s \frac{\rho_t(u)}{B(u)} \frac{Q_{L|U}(u|s)}{S(u)} dA(u) \right\} \frac{\rho_t(s)}{B(s)} dW(s),$$

$$E(\beta_2(t)\beta_4(t)) = -\int_0^t \left\{ \int_0^s \frac{\rho_t(u)}{B(u)} dQ_{L|U}(u|s) \right\} \frac{\rho_t(s)}{B(s)} dW(s).$$

Collecting all the terms we obtain $V_{\mathrm{ICW}}(t)$ given by Eq. (10.6).

10.3. An Illustration

We illustrate the proposed estimators using data from an AIDS clinical trial [9]. The response T is the change in $\log_{10}$ RNA level from a baseline value to the value after 24 weeks. The response $T = l_0 - l_{24}$ is left censored by $L = l_0 - 5.88$ and right censored by $U = l_0 - 2.6$, where l_k denotes the $\log_{10}$ RNA value at week k. Out of the 196 subjects chosen for analysis, there are about 4% left and 42% right censored cases.

For the data, since L and U are always observed, we are able to estimate the censoring probability $G(t) = P(L < t \leq U)$ in two ways, namely, using Eq. (10.2), and by $\hat{G}(t) = \hat{Q}(t) - \hat{W}(t)$. As mentioned before, the resulting estimators are designated as ICW Type I and ICW Type II. Along with the Kaplan–Meier (KM) type estimator we also plotted a semiparametric estimator [21]. For this estimator, we employed a logistic model $p(x) = exp(\beta_1 + \beta_2 x^2)/(1 + exp(\beta_1 + \beta_2 x^2))$ for the conditional probability $p(x) = P(\delta = 2|X = x)$. The linear term was ignored as it was found, using R, not to be significant. The R software based analysis returned $\hat{\beta}_1 = 1.5091$

and $\hat{\beta}_2 = -1.1913$. To obtain the semiparametric estimator, the empirical estimator of $H(x) = P(X \leq x)$, denoted by $\hat{H}(x)$, and the estimated conditional probability, denoted by $\hat{p}(x)$, were plugged into the equation

$$\hat{S}_{\mathrm{D}}(t) = \prod_{0 \leq s \leq t} \left\{1 - d\hat{\Lambda}_{\mathrm{D}}(s)\right\} = \prod_{0 \leq s \leq t} \left\{1 - \hat{p}(s)d\hat{H}(s)/Y_1(s)\right\}.$$

The survival function estimators are shown in Fig. 10.1. The two estimators of the censoring probability G, used for estimating the ICW Type I and Type II estimators are shown in Fig. 10.2. It is interesting to note that G [and, by extension, $\hat{S}_{\mathrm{ICW}}(t)$] is estimated equally well with only L always observed. In particular, *always* observing U does not seem to improve inference, see also the numerical results presented in the next section.

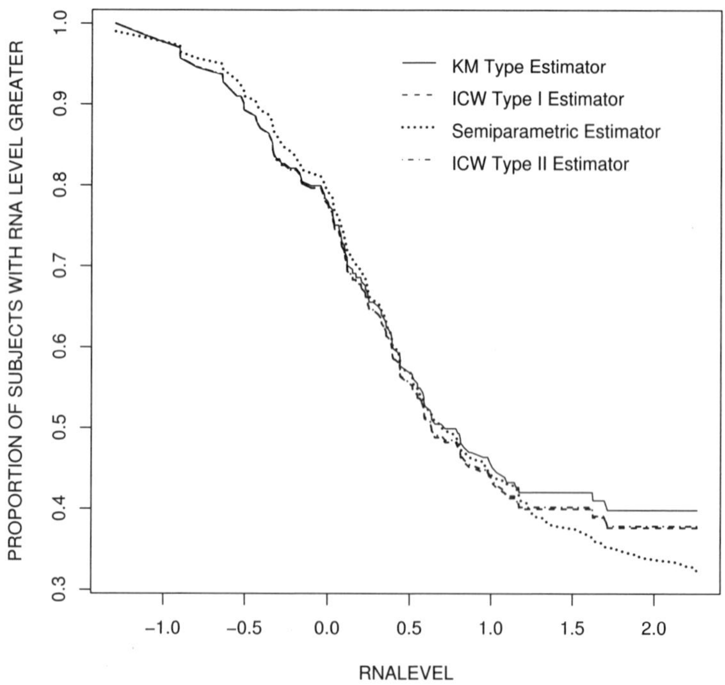

Fig. 10.1. Survival function estimators for AIDS Clinical Trial Data.

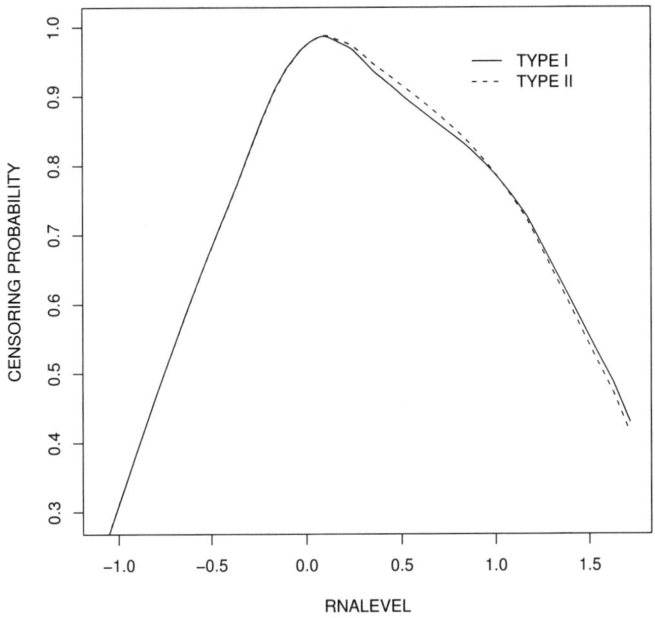

Fig. 10.2. Comparison of two estimators of censoring probability.

10.4. Numerical Results

For our simulation study, the failure time was normal with mean 5 and unit variance. The left censoring L was exponential with parameter θ. The right censoring U was $L + K$, where K is distributed independently as exponential with parameter θ_1. The parameters θ and θ_1 were chosen to get several different left and right censoring rates, denoted here by LCR and RCR respectively. The mean and standard deviation of the mean integrated squared errors (MISEs) of the estimators are presented in Table 10.2 and Table 10.3 below. The MISEs were calculated over the interval $[1, 6]$ and were based on 10,000 Monte Carlo replications at sample sizes $n = 100$ and $n = 500$. The KM type estimator performs best among the three estimators, while the ICW Type II is the least preferable, implying that knowledge of U does not help in improving the ICW Type I estimator.

Table 10.2. Mean and standard deviation (SD) of mean integrated squared errors of the KM type and ICW estimators. Sample size 100.

		KM Type		ICW Type I		ICW Type II	
LCR	RCR	Mean	SD	Mean	SD	Mean	SD
10%	10%	0.0112	0.0053	0.0120	0.0058	0.0121	0.0058
	20%	0.0125	0.0059	0.0134	0.0065	0.0137	0.0066
	30%	0.0141	0.0067	0.0152	0.0074	0.0156	0.0075
	40%	0.0167	0.0079	0.0182	0.0088	0.0189	0.0091
20%	10%	0.0134	0.0064	0.0150	0.0074	0.0151	0.0074
	20%	0.0152	0.0073	0.0171	0.0084	0.0175	0.0086
	30%	0.0174	0.0082	0.0198	0.0097	0.0202	0.0099
	40%	0.0215	0.0102	0.0250	0.0123	0.0258	0.0127
30%	10%	0.0160	0.0078	0.0182	0.0092	0.0183	0.0092
	20%	0.0185	0.0089	0.0214	0.0107	0.0219	0.0109
	30%	0.0221	0.0105	0.0262	0.0130	0.0272	0.0136
	40%	0.0281	0.0136	0.0342	0.0174	0.0357	0.0181
40%	10%	0.0200	0.0096	0.0232	0.0116	0.0234	0.0117
	30%	0.0301	0.0143	0.0367	0.0184	0.0385	0.0193

Table 10.3. Mean and standard deviation (SD) of mean integrated squared errors of the KM type and ICW estimators. Sample size 500.

		KM Type		ICW Type I		ICW Type II	
LCR	RCR	Mean	SD	Mean	SD	Mean	SD
10%	10%	0.0023	0.0011	0.0025	0.0012	0.0025	0.0012
	20%	0.0025	0.0012	0.0027	0.0013	0.0028	0.0013
	30%	0.0029	0.0013	0.0031	0.0015	0.0032	0.0015
	40%	0.0033	0.0015	0.0036	0.0017	0.0038	0.0018
20%	10%	0.0027	0.0013	0.0030	0.0014	0.0030	0.0014
	20%	0.0030	0.0014	0.0034	0.0016	0.0034	0.0016
	30%	0.0035	0.0016	0.0039	0.0019	0.0041	0.0019
	40%	0.0042	0.0020	0.0049	0.0023	0.0051	0.0024
30%	10%	0.0033	0.0015	0.0037	0.0018	0.0037	0.0018
	20%	0.0037	0.0018	0.0043	0.0021	0.0044	0.0021
	30%	0.0044	0.0021	0.0052	0.0025	0.0054	0.0026
	40%	0.0056	0.0026	0.0069	0.0034	0.0072	0.0035
40%	10%	0.0040	0.0019	0.0046	0.0022	0.0047	0.0023
	30%	0.0059	0.0028	0.0071	0.0036	0.0074	0.0037

10.5. Concluding Discussion

First, a distinction must be made between the double censoring scenario investigated in this paper and data that arise from doubly interval censored data [22]. In the latter case, the start and terminal events are either right or interval censored, see also Refs. [23, 24]. Second, the relaxed condition that one or both censoring variables are always observed facilitates implementable inference unlike in the case of conventional doubly censored data where the limiting covariance function of the NPMLE is "too complicated to even calculate numerically" [20]. This has also been pointed out by Ref. [9], whose proposed procedure is one instance where inference is feasible. However, in cases where only L is always observed, their procedure will not be applicable, and modified estimating functions would need to consider additional assumptions concerning the censoring variables to avoid curse of dimensionality. One approach would be to assume that the censoring is free of the covariate [13, 14, 25, 26], which allows the estimating function proposed by Ref. [9] to be modified using the ICW Type I estimator proposed here. The resulting ICW estimating function would share the spirit of its counterparts in the above mentioned papers; see also Refs. [27–29] for related work concerning the ICW approach, which also require a covariate free censoring distribution. This would be a worthwhile direction for future research.

Acknowledgments

The first author's research was partly supported by a National Institute of Health grant CA 103845. The authors thank Dr. Tianxi Cai for providing the AIDS clinical trial data set analyzed in this paper. The authors also thank a reviewer for constructive comments.

Appendix.

A.1. The ICW Type I Estimator

Here we present the details of the moment calculations of the four components of $\beta(t)$, which we denoted by $\beta_i(t), i = 1, \ldots, 4$. Recall that

$$B(t) = \prod_{(0,t]}(1+dA(s)), \quad R(t) = \int_0^t dQ(s)/B(s), \quad \rho_t(s) = \int_s^t dF(u)/R(u).$$

It is easy to see that $E(\beta_1^2(t)) = \int_0^t dF(s)/G(s)$ and that $E(\beta_1(t)\beta_2(t)) = 0$. Also, since $G(t) = R(t)B(t)$, it is straightforward to show that

$$E(\beta_2^2(t)) = \int_0^t \frac{\rho_t^2(s)}{B^2(s)} \frac{dW(s)}{S(s)}, \qquad E(\beta_4^2(t)) = \int_0^t \frac{\rho_t^2(s)}{B^2(s)} dQ(s).$$

After a routine interchange of the order of integration, we have that

$$\begin{aligned} E(\beta_1(t)\beta_3(t)) &= \int_0^t \frac{R(s)}{y_1(s)} \rho_t(s) E\left\{ \frac{I(L < s \le X \le t, \delta = 2)}{G(X)} \right\} dA(s) \\ &= \int_0^t \frac{R(s)}{y_1(s)} \rho_t(s) \left\{ \int_s^t P(L < s, U \ge u) \frac{dF(u)}{G(u)} \right\} dA(s), \\ &= \int_0^t \left\{ \int_0^s \frac{\rho_t(u)}{B(u)} \frac{P(L < u, U \ge s)}{S(u)} dA(u) \right\} \frac{dF(s)}{G(s)}. \end{aligned}$$

Also, it is easy to see that

$$E(\beta_1(t)\beta_4(t)) = -\int_0^t \left\{ \int_0^s \frac{\rho_t(u)}{B(u)} P(U \ge s | L = u) dQ(u) \right\} \frac{dF(s)}{G(s)}.$$

Again, the following cross moment expressions rely on interchanging the order of integrations:

$$\begin{aligned} E(\beta_2(t)\beta_3(t)) &= \int_0^t \frac{R(s)}{y_1(s)} \rho_t(s) \bigg\{ \int_s^t \frac{R(u)}{y_1(u)} \rho_t(u) S(u) \\ &\qquad \times P(L < s | U = u) dW(u) \bigg\} dA(s) \\ &= \int_0^t \left\{ \int_0^s \frac{\rho_t(u)}{B(u)} \frac{P(L < u | U = s)}{S(u)} dA(u) \right\} \frac{\rho_t(s)}{B(s)} dW(s), \\ E(\beta_2(t)\beta_4(t)) &= -\int_0^t \left\{ \int_0^s \frac{\rho_t(u)}{B(u)} P(L = u | U = s) du \right\} \frac{\rho_t(s)}{B(s)} dW(s), \\ E(\beta_3(t)\beta_4(t)) &= -\int_0^t \left\{ \int_0^s \frac{\rho_t(u)}{B(u)} P(U \ge s | L = u) dQ(u) \right\} \frac{\rho_t(s)}{B(s)} dA(s). \end{aligned}$$

The next moment calculation is laborious but routine: Note that

$$\begin{aligned} E(\beta_3^2(t)) = E\bigg\{ &\int_0^t \frac{R(s)}{y_1(s)} \rho_t(s) I(L < s \le T \wedge U) dA(s) \\ &\times \int_0^t \frac{R(s')}{y_1(s')} \rho_t(s') I(L < s' \le T \wedge U) dA(s') \bigg\}. \end{aligned}$$

We can split the range of the second integral as s' from 0 to s and as s' from s to t. When $0 < s' \le s$, we have

$$E(I(L < s \le T \wedge U) I(L < s' \le T \wedge U)) = P(L < s', U \ge s) S(s).$$

When $s < s' \le t$, we have

$$E(I(L < s \le T \wedge U)I(L < s' \le T \wedge U)) = P(L < s, U \ge s')S(s').$$

Then

$$E(\beta_3^2(t)) = \int_0^t \left\{ \int_0^s \frac{\rho_t(u)}{B(u)} \frac{P(L < u, U \ge s)}{S(u)} dA(u) \right\} \frac{\rho_t(s)}{B(s)} dA(s) + \int_0^t \left\{ \int_s^t \frac{\rho_t(u)}{B(u)} P(L < s, U \ge u) dA(u) \right\} \frac{\rho_t(s)}{B(s)} \frac{dA(s)}{S(s)}.$$

The second integral after changing the order of integration, can be seen to be exactly equal to the first. Since $dA(s) = -dW(s)/G(s)$, we can see that

$$E(\beta_3^2(t) = 2 \int_0^t \left\{ \int_0^s \frac{\rho_t(u)}{B(u)} \frac{P(L < u, U \ge s)}{S(u)} dA(u) \right\} \frac{\rho_t(s)}{B(s)} dA(s)$$
$$= -2 \int_0^t \left\{ \int_0^s \frac{\rho_t(u)}{B(u)} \frac{P(L < u, U \ge s)}{S(u)} dA(u) \right\} \frac{\rho_t(s)}{B(s)} \frac{dW(s)}{G(s)}.$$

References

[1] B. W. Turnbull, Nonparametric estimation of a survivorship function with doubly censored data, *J. Ameri. Statist. Assoc.* **69**, 169–173, (1974).

[2] W. Tsai and J. Crowley, A large sample study of generalized maximum likelihood estimators from incomplete data via self-consistency, *Ann. Statist.* **13**, 1317–1334, (1985).

[3] M. G. Gu and C. H. Zhang, Asymptotic properties of self-consistent estimators based on doubly censored data, *Ann. Statist.* **21**, 611–624, (1993).

[4] P. A. Mykland and J. J. Ren, Algorithms for computing self-consistent and maximum likelihood estimators with doubly censored data, *Ann. Statist.* **24**, 1740–1764, (1996).

[5] M. N. Chang and G. L. Yang, Strong consistency of a nonparametric estimator of the survival function with doubly censored data, *Ann. Statist.* **15**, 1536–1547, (1987).

[6] M. N. Chang, Weak convergence of a self-consistent estimator of the survival function with doubly censored data, *Ann. Statist.* **18**, 391–404, (1990).

[7] M. van der Laan, M. and R. D. Gill, Efficiency of the NPMLE in nonparametric missing data models, *Math. Meth. Statist.* **8**, 251–276, (1999).

[8] A. Biswas and R. Sundaram, Kernel survival function estimation based on doubly censored data, *Comm. Statist. Theory Meth.* **35**, 1293–1307, (2006).

[9] T. Cai and S. Cheng, Semiparametric regression analysis for doubly censored data, *Biometrika* **91**, 277–290, (2004).

[10] S. O. Samuelsen, Asymptotic theory for non-parametric estimators from doubly censored data, *Scand. J. Statist.* **16**, 1–21, (1989).

[11] G. R. Petroni and R. A. Wolfe, A two-sample test for stochastic ordering with interval-censored data, *Biometrics* **50**, 77–87, (1988).
[12] H. Koul, V. Susarla, and J. Van Ryzin, Regression analysis of randomly right censored data, *Ann. Statist.* **9**, 1276–1288, (1981).
[13] Z. Ying, S. Jung, and L. J. Wei, Survival analysis with median regression models, *J. Ameri. Statist. Assoc.* **90**, 178–184, (1995).
[14] S. Subramanian, Median regression analysis from data with left and right censored observations, *Statist. Meth.* **4**, 121-131, (2007).
[15] G. A. Satten and S. Datta, The Kaplan-Meier estimator as an inverse-probability-of-censoring weighted average, *Ameri. Statist.* **55**, 207–210, (2001).
[16] N. Keiding and R. D. Gill, Random truncation models and Markov processes, *Ann. Statist.* **18**, 582-602, (1990).
[17] P. Major and L. Rejtő, Strong embedding of the estimator of the distribution function under random censorship, *Ann. Statist.* **16**, 1113–1132, (1988).
[18] N. Breslow and J. Crowley, A large sample study of the life table and product-limit estimates under random censorship, *Ann. Statist.* **2**, 437–453, (1974).
[19] R. D. Gill and S. Johansen, A survey of product-integration with a view toward application in survival analysis, *Ann. Statist.* **18**, 1501–1555, (1990).
[20] P. K. Andersen, O. Borgan, R. D. Gill, and N. Keiding, *Statistical models based on counting processes*, (Springer-Verlag, 1993).
[21] G. Dikta, On semiparametric random censorship models, *J. Statist. Plann. Inference* **66**, 253–279, (1998).
[22] J. Sun, Empirical estimation of a distribution function with truncated and doubly interval-censored data and its application to AIDS studies, *Biometrics* **51**, 1096–1104, (1995).
[23] M. Y. Kim, V. G. De Gruttola, and S. W. Lagakos, Analyzing doubly censored data with covariates with application to AIDS, *Biometrics* **49**, 13–22, (1993).
[24] R. B. Geskus, Methods for estimating the AIDS incubation time distribution when data of sero-conversion is censored, *Statist. Med.* **20**, 795812, (2001).
[25] G. Yin, D. Zeng, and H. Li, Power-transformed linear quantile regression with censored data, *J. Ameri. Statist. Assoc.* **103**, 1214–1224, (2008).
[26] S. Subramanian and G. Dikta, Inverse censoring weighted median regression, *Statist. Meth.* **6**, 594–603, (2009).
[27] S. C. Cheng, L. J. Wei, and Z. Ying, Analysis of transformation models with censored data, *Biometrika* **83**, 835–846, (1995).
[28] S. C. Cheng, L. J. Wei, and Z. Ying, Prediction of survival probabilities with semiparametric transformation models, *J. Ameri. Statist. Assoc.* **92**, 227–235, (1997).
[29] J. P. Fine, L. J. Wei, and Z. Ying, On the linear transformation model for censored data, *Biometrika* **85**, 980–986, (1998).

Chapter 11

Analysis and Design of Competing Risks Data in Clinical Research

Haesook T. Kim

Department of Biostatistics and Computational Biology
Dana-Farber Cancer Institute, Boston, MA 02115, USA

kim.haesook@jimmy.harvard.edu

As competing risks occur commonly in medical research, increased attention has been paid to competing risks data in recent years. In this article, we review fundamentals of cumulative incidence function, Gray test, and Fine and Gray model and illustrate competing risks data analysis using clinical datasets of hematopoietic stem cell transplantation. In addition, we present limitations of Fine and Gray model, model selection in competing risks regression analysis, power calculation, and computing tools.

11.1. Introduction

Competing risks arise when individuals can experience any one of J distinct event types and the occurrence of one type of event prevents the occurrence of other types of events. Competing risks data are inherent to cancer clinical trials in which failure can be classified by its types, and the information on each type of failure is as important as the overall survival probability. For instance, patients who undergo allogeneic hematopoietic stem cell transplantation (HSCT) die from either recurrence of disease (*relapse*) or complications related to transplantation (transplant-related mortality, treatment-related mortality or TRM) if the transplant does not cure the underlying disease. Disease recurrence is an important event of interest as is TRM. If disease recurrence is the event of interest and if an individual dies from TRM, this competing risk removes the individual from being at risk for disease recurrence. Therefore, applying methods of standard survival analysis to an event of interest when a competing risk is present would lead to biased results since standard survival analysis does not take types of failure into account.

The goal of this article is to give an overview of competing risks data analysis and power calculation for competing risks data. Throughout the presentation, allogeneic HSCT studies are used to illustrate competing risks data analysis although competing risks occur commonly in other cancer clinical trials, such as breast cancer, cervical cancer, melanoma, leukemia, or lymphoma trials.

11.2. Estimation and Comparison of Cumulative Incidence Curves

Suppose there are k distinct types of failure. The hazard of failing from cause i ($i = 1, ..., k$) is

$$\lambda_i(t) = \lim_{u \to 0} \frac{\text{Prob}(t \leq T < t + u, I = i | T \geq t)}{u}. \tag{11.1}$$

The cumulative incidence function for failure i is[11,16]

$$\text{F}_i(t) = \text{Prob}(T \leq t, I = i) = \int_0^t f_i(u)du = \int_0^t \lambda_i(u)S(u)du, \tag{11.2}$$

where $S(t)$ is the overall survival probability. (11.2) is also known as *sub-distribution function.* The estimate of (11.2) is

$$\hat{F}_i(t) = \sum_{j:t_j<t} \frac{d_{ij}}{n_j} \hat{S}(t_{j-1}), \tag{11.3}$$

where d_{ij} is the number of failures of cause i at time t_j, n_j is the number of subjects at risk just prior to t_j, and $\hat{S}(t)$ is the Kaplan-Meier estimate of $S(t)$. If there are k failures, $\sum_{i=1}^k \hat{\text{F}}_i(t) + \hat{S}(t) = 1$.

Much of early theoretical development in competing risks data focused on a set of latent failure times and cause-specific hazard, $\lambda_i(t)$, that can be formulated as the marginal distribution of latent failure times.[4–6,9,16] However, the marginal distribution is unidentifiable unless independence is assumed between competing risks.[29] Cumulative incidence function, on the other hand, is often more appealing and circumvents the un-identifiability issue. Since there exits no one-to-one relationship between the cause-specific hazard (11.1) and the cumulative incidence function (11.2), Gray[13] proposed a class of K-group tests (11.4) for comparison of the cumulative incidence functions of competing risks.

$$\int_0^{\tau} w(t)\left(\frac{d\hat{F}_{i1}(t)}{1-\hat{F}_{i1}(t-)}-\frac{d\hat{F}_{i2}(t)}{1-\hat{F}_{i2}(t-)}\right), \tag{11.4}$$

where $\hat{F}_{il}(t)$ is an estimate of F_{il} for failure i in group l ($l = 1, 2$ for the purpose of simplicity) as defined in (11.2), and $w(t)$ is a weight function. (11.4) is basically comparing weighted averages of the subdistribution hazards $\gamma_i(t)$, where

$$\begin{aligned}\gamma_i(t) &= \lim_{u\to 0}\frac{1}{u}\Pr\{t \le T < t+u, I = i | T \ge t \cup (T \le t \cap I \ne i)\}\\ &= \frac{f_i(t)}{1-F_i(t)}.\end{aligned} \tag{11.5}$$

$\gamma_i(t)$ (suppressing the second subscript for simplicity) is the hazard of failing from failure i in the presence of competing risks, given that a subject has survived or has already failed from a competing cause. This hazard, known as *subdistribution hazard*, corresponds to the cumulative incidence function (11.2) and is the key concept of Gray test and the Fine and Gray model.[12]

Example 1: Myeloablative vs. Non-myeloablative allogeneic hematopoietic stem cell transplantation (HSCT) for patients older than 50 years of age

The therapeutic benefit and potential cure achieved by allogeneic HSCT is mediated by the graft-vs-tumor effect. In the beginning, healthy cells from an allogeneic donor was considered a rescue therapy for replacing ablated host(patient)-derived hematopoiesis. Thus, the conventional HSCT (called 'myeloablative transplant' (MT)) has used high doses of chemotherapy and/or total body irradiation for both eradication of underlying malignancies and suppression of the host's immune system to allow donor engraftment. However, the conventional MT has an impact on the increased tissue damage and intensified inflammatory state in the host, leading to significant transplant-related morbidity and mortality. Reduced intensity conditioning regimen (RIC, or called 'non-myeloablative transplant' (NT)), on the other hand, decreases the TRM through the reduction in regimen-related toxicity and graft-versus-host disease(GVHD), and would rely on the adoptively transferred donor immune cells for the eradication of host malignancies. To investigate the impact of conditioning regimen on the two important outcomes in allogeneic HSCT, namely relapse and TRM, we use the data set presented in Alyea et al.[2]

One hundred and fifty two patients older than 50 years who underwent MT or NT between 1997 and 2002 at the Dana-Farber Cancer Institute were included in the analysis. Figure 11.1 shows the cumulative incidence of relapse and TRM among 81 patients who received MT using the standard Kaplan-Meier (KM) method that ignores competing risks. The 3-year cumulative incidence rate was 50% for relapse and 58% for TRM, and thus the combined rate of 108%. Figure 11.2 shows the cumulative incidence of relapse and TRM using the competing risks method (CR) (11.2). The 3-year cumulative incidence rate was 30% for relapse and 50% for TRM, so the combined rate of 80%, which corresponds to one minus the event-free survival at 3 year post transplant (i.e., $1 - \hat{S}(3)$). It is noticed that even though these two events are mutually exclusive, and some patients were still alive and relapse-free at 3 years post transplant, the combined rate of relapse and TRM is more than 100% if we use the naive KM method, indicating the amount of bias introduced by the KM method. Table 11.1 summarizes the cumulative incidences of relapse and TRM and compared

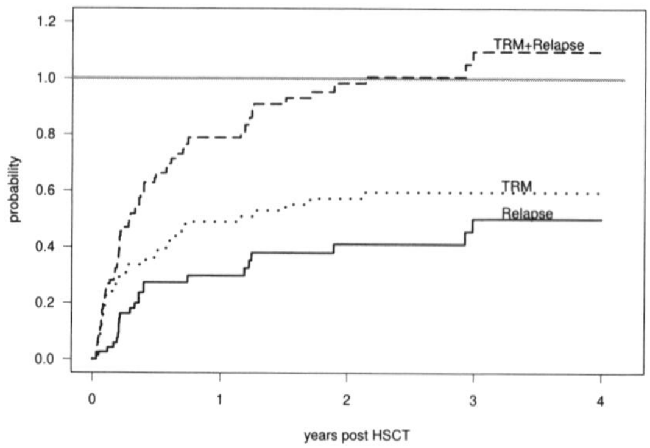

Fig. 11.1. Cumulative incidence of relapse (solid line), TRM (dotted line), and TRM and relapse combined (dashed line) using the Kaplan-Meier method. Gray horizontal line marks the maximum probability of 1.

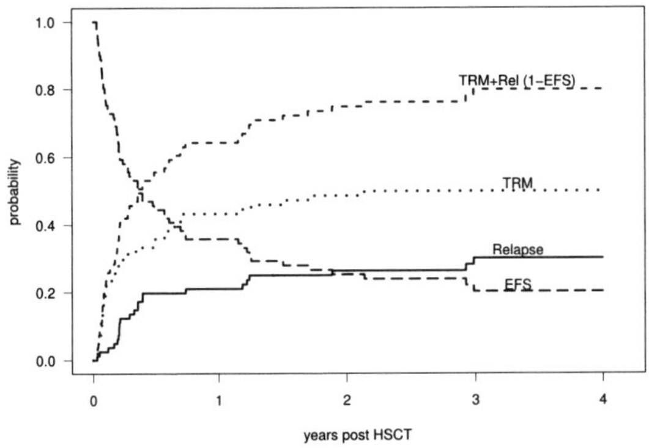

Fig. 11.2. Cumulative incidence of relapse (solid line), TRM (dotted line) and TRM and relapse combined (dashed line) using the competing risks method. The long dashed line represents the Kaplan-Meier curve of event-free survival.

them between NT and MT using the log-rank test and Gray test. The results suggest that the KM method overestimated the cumulative incidences of these events and thus the difference between MT and NT in cumulative incidence of relapse is not significant ($p = 0.35$ using the log-rank test). The results also suggest that even though the cumulative incidence of combined events (relapse and TRM) is similar between two types of transplantation (80% vs. 78%), NT is associated with a high risk of relapse but low risk of TRM, indicating that the immunologic effects of these two types of transplant on relapse and TRM are very different (Fig. 11.3).

Table 11.1. Summary of 3-year cumulative incidence of relapse and TRM.

	KM method			CR method		
	NT	MT	p-value (log-rank)	NT	MT	p-value (Gray test)
3-yr CI of relapse	61%	50%	0.35	46%	30%	0.05
3-yr CI of TRM	38%	58%	0.008	32%	50%	0.01
3-yr CI of relapse and TRM combined	99%	108%		78%	80%	

NT: non-myeloablative transplantation. MT: myeloablative transplantation.

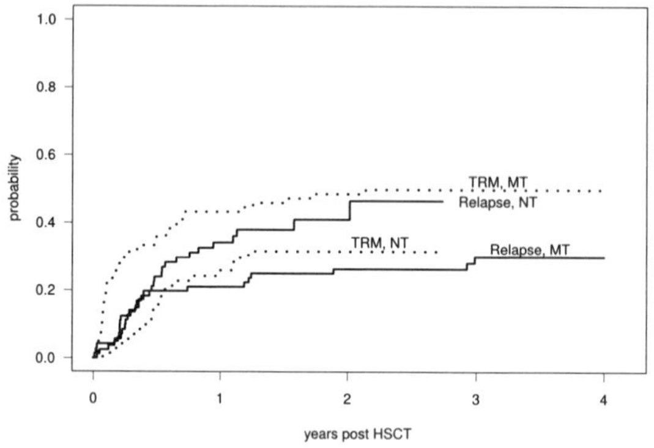

Fig. 11.3. Cumulative incidence of relapse and TRM for MT and NT using the CR method.

11.3. Competing Risks Regression Analysis

In the presence of competing risks, a standard Cox proportional hazards model[8] is not appropriate for assessment of covariates because the cause-specific Cox model treats competing risks as censored observations, and the cause-specific hazard function does not have a direct interpretation in terms of survival probability. Since the simple relationship between a single endpoint and a single cause-specific hazard does not hold in the presence of competing risks, Fine and Gray[12] and Klein and Andersen[20] proposed a direct regression modeling of the effect of covariates on the cumulative incidence function for competing risks data. The Fine and Gray method is based on proportional hazards model, whereas the Klein and Andersen method is based on the psuedo values from a jackknife statistic from the cumulative incidence curve. We here present the Fine and Gray model and refer readers to their paper[20] for the Klein and Andersen method.

11.3.1. *Fine and Gray model*

The Fine and Gray model is a Cox proportional hazards like model for the subdistribution hazard (11.5). The model uses the partial likelihood principle and weighted estimated equations to obtain consistent estimators of the covariate effects. As in Cox model, the Fine and Gray model for failure 1, (11.6), is a relative risk model that decomposes into the baseline hazard and regression effect of covariates.

$$\begin{aligned}\gamma_1(t;X) &= \lim_{u\to 0}\frac{1}{u}\Pr\{t \le T < t+u, I=1|T \ge t \cup (T \le t \cap I \ne 1), X\}\\ &= \frac{f_1(t;X)}{1-F_1(t;X)} = \gamma_0(t)e^{X'\beta},\end{aligned} \tag{11.6}$$

where $\gamma_0(t)$ is an unspecified baseline hazard, X is the vector of covariates, and β is the vector of coefficients. The risk set is

$$R_j = \{r : (min(C_j, T_j) \ge T_r) \cup (T_j \le T_r \cap I \ne 1 \cap C_j \ge T_r)\}. \tag{11.7}$$

The first part of the risk set represents subjects who have not failed from any cause, and the second part represents subjects who have failed from a competing risk and censoring failure 1 has not occurred. In other words, subjects who failed from competing causes remain in the risk set indefinitely.

This model handles time by covariate interactions, but not time-dependent covariates (i.e., $\beta X(t)$). As time-dependent covariates are common in competing risks data, this is a limitation of the model.

Example 2: HSCT for patients with T cell lymphoma

Allogeneic hematopoietic stem cell transplantation is an accepted therapy for B cell lymphoma but the indications for allogeneic HSCT and the outcomes of the procedure are less clear in T cell lymphoma. To investigate clinical outcomes after HSCT for T cell lymphoma, we identified 52 patients who underwent transplant between April 4, 1997 and February 17, 2009 at the Dana-Farber/Brigham and Women's Cancer Center. Table 11.2 shows the results of the standard Cox model for the combined events of TRM and relapse and the Fine and Gray model for TRM and relapse as competing risks.

The competing risks model in Table 11.2 reveals that the effects of NT and MT on relapse and TRM point to the opposite direction even though the combined effect in Cox model is similar between NT and MT. In deed, NT is significantly associated with an increased risk of relapse ($\hat{\beta}$=1.64, HR=5.13, p = 0.01), whereas MT is associated with an increased risk of TRM ($\hat{\beta}$=1.19, HR=3.28, p = 0.05). This result is consistent with the result shown in Example 1. It again suggests different immunologic effects of NT and MT on relapse and TRM.

In addition, competing risks regression analysis is also useful in identifying prognostic factors, other than the factor of interest, for each type of failure. For example, in Table 11.2, good prognosis is associated with the

Table 11.2. Summary of competing risks regression model and Cox model.

	Relapse: CRR			TRM: CRR			Relapse & TRM: Cox		
	$\hat{\beta}$	HR	p-value	$\hat{\beta}$	HR	p-value	$\hat{\beta}$	HR	p-value
Age ($\geq$50 vs <50)	-0.38	0.68	0.52	0.71	2.02	0.33	0.51	1.66	0.26
Patient-donor sex mismatch (MF vs Other)	0.92	2.51	0.08	0.14	1.15	0.81	0.86	2.36	0.02
NT vs. MT	1.64	5.13	0.01	-1.19	0.31	0.05	0.37	1.45	0.43
Good vs. Poor prognosis	-1.66	0.19	0.005	-1.18	0.31	0.07	-1.34	0.26	0.002
Year ($\geq$2002 vs. <2002)	-1.98	0.14	0.014	-1.65	0.19	0.009	-1.83	0.16	0.002

HR: hazard ratio. Cox: Cox proportional hazards model for Relapse and TRM combined as a single event. CRR: Fine and Gray's competing risks regression model. NT: Non-myeloablative transplant. MT: Myeloablative transplant. MF: Male patient with female donor. Year: year of transplant.

Table 11.3. Summary of BIC and AIC.

model	n	L	Df.fit	BIC	BIC.diff	AICc
null	52	-77.61	0	155.22	0	155.22
1 (full)	52	-71.28	5	162.32	7.10	153.87
2	52	-71.52	4	158.84	3.61	151.88
3	52	-72.82	3	157.5	2.28	152.15
4	52	-74.99	2	157.88	2.66	154.23
5	52	-76.59	1	157.13	1.91	155.26

Df.fit: the number of parameters. BIC.diff or AIC.diff: difference with the respect to the minimum value observed from the set of candidate models.

event-free survival in Cox model ($\hat{\beta}$=-1.34, HR=0.26, $p = 0.002$), but it is unknown from this model whether this is for relapse or TRM prevention. The competing risks model reveals that good prognosis is associated with an decreased risk of relapse, but not with TRM.

Over-fitting can be an issue in regression analysis. Analogous to other regression analysis, a backward or forward model selection can be done in Fine and Gray model. In the above example, we consider a 'backward' approach for relapse using Akaike information criteria (AIC)[1] and Baysian information criteria (BIC).[28] We first fit the full model, then remove one variable each time on the basis of significance in the full model. Table 11.3 shows results of BIC and AIC scores for 5 models. The first 5 columns were generated by the add-on R program **modsel.crr**(see Sec. 11.5 Computing Tools), and corrected AIC (AICc) was calculated based on the formula, $-2 * L + 2 * p * [1 + \frac{p+1}{n-p-1}]$,[14] where L is the maximized log likelihood value and p is the number of parameters. Applying the general rule-of-thumb[15,18] BIC suggests model 5, which includes prognosis only, whereas AICc suggests model 2, which excludes age only.

11.4. Power Calculation

One important aspect of designing a clinical trial comparing two treatments is determining an adequate sample size so that the study will be properly powered to test the main hypothesis of a treatment difference.

Suppose that a randomized clinical trial is planned comparing two treatment difference for failure 1 in the presence of competing risks. Under the proportional hazards assumption, the number of events necessary to detect a specific subdistribution hazard ratio (HR_{sub}) with Type I and II error rates of α and β is

$$n_1 = \frac{(Z_{1-\alpha/2} + Z_{1-\beta})^2}{\sigma^2 \log(\mathrm{HR_{sub}})}, \tag{11.8}$$

where $z_{1-\gamma}$ is the usual r-th quantile of the standard normal distribution, and σ^2 is the variance of the covariate of interest. In the case of two treatment arms, $\sigma^2 = p(1-p)$, where p is the proportion of patients in the experimental arm. Then the total sample size required is

$$N = \frac{n_1}{P_1}, \tag{11.9}$$

where P_1 is the probability of occurrence of failure 1 at a specific time point.[23,24] Analogous to the sample size formula in the absense of competing risks, P_1 can be calculated as

$$P_1 = 1 - \frac{1}{a}\int_f^{a+f} (1 - F_1(u)du)$$

where a is the accrual time, and f is the additional follow-up time after the completion of the accrual. Since HR_{sub} of treatment group 1 to treatment group 2 for faiulre 1 is

$$\frac{\log(1 - F_{11}(t))}{\log(1 - F_{12}(t))},$$

$F_{11} = 1 - (1 - F_{12})^{1/\mathrm{HR}_{sub}}$. Typically the cumulative incidence of failure 1 for the control arm at a specified time point is known from a previous study, and HR_{sub} is a hypothesized value at the time of study design.

Mimicking the HSCT study presented in Example 1, Table 11.4 below shows power for various scenarios of cumulative incidence in the presence and absence of competing risks. This power calculation is based on a two-sided significance level of 0.05 for a sample size 400 assuming $f = 2$. As indicated in this table, power can be very different between presence and absence (i.e., CIF_c=0%) of a competing risk as well as the magnitude of a competing risk.

11.5. Computing Tools

Computing tools are available for competing risks data analysis. The R package **cmprsk**, developed by Robert Gray, can be down-loaded from the website (http://www.r-project.org). **Cmprsk** has two components: **cuminc** and **crr**. **cuminc** is to estimate cumulative incidence functions and to test equality across groups using the Gray test. **crr** is to perform

Table 11.4. Power in the presence and absence of a competing risk (N=400).

Group 1		Group 2		power	
CIF_e	CIF_c	CIF_e	CIF_c	a=2	a=3
30%	50%	45%	30%	50%	52%
50%	30%	30%	45%	96%	97%
30%	0%	45%	0%	88%	91%
50%	0%	30%	0%	99%	99%

a: accrual time in years. CIF_e: cumulative incidence rate of an event of interest. CIF_c: cumulative incidence rate of a competing event.

regression modeling of subdistribution functions in competing risks data as described in Fine and Gray.[12] **crr** fits time-varying non-proportional covariates, $\beta(t)X$, via time by covariate interaction, but does not handle time-dependent covariates, $\beta X(t)$, nor allow stratification. **crr** outputs a matrix of Schoenfeld residuals, and plots of these residuals against failure times can be used for checking the proportional hazards assumption.

In addition, two adds-on functions to **crr** are available on http://www.stat.unipg.it/ luca/R/. **CumIncidence**[26] is a R program to calculate the confidence intervals of cumulative incidence functions (11.2). **modsel.crr**[27] is a model selection tool among candidate competing risks models. For power calculation, **power**, a R program, is available on http://www.uhnres.utoronto.ca/labs/hill/People_Pintilie.htm.

11.6. Conclusion

Much of theoretical development in competing risks data was established a few decades ago, focusing on latent model and cause-specific hazard.[4–6,11,29] However, recognition and application of competing risks data in medical research have been only recent through the development of subdistribution hazard[13] and regression models.[12,20] The widespread use of Gray test and Fine and Gray model is in part due to the readily available R package, **cmprsk**. Competing risks regression analysis has been, in deed, very useful for determining clinical utility of treatment. Limitations of the Fine and Gray model, however, are that it does not handle time-dependent covariates nor allow stratified models. To address this issue, Cortese and Andersen[7] recently proposed a few approaches to deal with time-dependent covariates. One approach was a multi-state model on cause-specific hazard, not on cumulative incidence function. The other approach was use of the landmark analysis which also comes with shortcomings.[3] As time dependent covarites are common in competing risks data, further research is needed to develop models for time dependent covarites on the cumulative incidence.

Pintilie[24] modified the existing sample size formula and extended to the competing risks setting. If a phase III clinical trial is planned accounting for competing events, the interim monitoring plan for the event of interest as well as for competing risks also needs to be planned. To this end, further development of sequential monitoring is needed in the future.

Acknowledgments

We thank Dr. David Harrington for his helpful comments during the preparation of this manuscript, and gratefully acknowledge the anonymous referee's suggestions and comments that were valuable in preparing a final revised version. The research was suported by CA142106-06A2 from the National Cancer Institute and AI029530 from the National Institute of Allergy and Infectious Diseases.

References

1. Akaike, Hirotugu (1974). A new look at the statistical model identification. *IEEE Transactions on Automatic Control* **19** (6): 716-723.
2. Alyea, E. P., Kim, H. T., Ho, V., Cutler, C., Gribben, J., DeAngelo, D. J. , Lee, S. J., Windawi, S., Ritz, J., Stone, R. M., Antin, J. H., Soiffer, R. J. (2005) Comparative outcome of nonmyeloablative and myeloablative allogeneic hematopoietic cell transplantation for patients older than 50 years of age. *Blood* **105**:1810-1814.
3. Anderson JR, Cain KC, Gelber RD. (1983). Analysis of survival by tumor response and other comparisons of time-to-event by outcome variables. *J. Clin Oncol.* **1**(11):710-9.
4. Basu, A. P., Ghosh, J. K. (1978). Identifiability of the multinormal distribution under competing risks model (with J.K. Ghosh). *J. Multivariate Analysis*, **8**:413-429.
5. Basu, A. P., Ghosh, J. K. (1980). Identifiability of distributions under competing risks and complementary risks model. *Communications in Statistics, A: Theory and Methods*, **9**:1515-1525.
6. Basu, A. P., Klein, J. P. (1982). Some recent results in competing risks theory. *Survival Analysis*, 216–229 Crowley, John (ed.) and Johnson, Richard A. (ed.) Institute of Mathematical Statistics (Hayward).
7. Cortese, G., Andersen, P.K. (2009). Competing Risks and Time-Dependent Covariates. *Biometrical Journal*, **51**:138-158.

8. Cox, D.R., Oakes, D. (1984). *Analysis of Survival Data.* Chapman and Hall p.91-110.
9. Crowder, M. (1994). identifiability crises in competing risks. *Int. Statist. Rev.* **62**: 379-391.
10. Crowder, M. (2001). *Classical Competing Risks.* Chapman & Hall/CRC.
11. David, H. A., Moeschberger, M.L. (1978). *The Theory of Competing Risks.* London: Griffin.
12. Fine, J. P., Gray, R.J.(1999). A proportional hazards model for the subdistribution of a competing risk. *J. Am. Stat. Assoc.* **94**:496-509.
13. Gray, R. J. A class of K-sample tests for comparing the cumulative incidence of a competing risk(1988). *Ann. Statist.* **16**:1140-1154.
14. Hurvich, C. M., Tsai, C (1995) Model selection for extended quasi-likelihood models in small samples. *Biometrics*, **51**:1077-1084.
15. Jeffreys H. (1961). *Theory of Probability*, 3rd ed., Oxford University Press.
16. Kalbfleisch, J. D., Prentice, R. L. (2002). *The Statistical Analysis of Failure Time Data.* John Wiley & Sons (New York; Chichester).
17. Kaplan, E. L., Meier P. (1958). Nonparametric estimation from incomplete observations. *J. Am. Stat. Assoc.* **53**:457-81.
18. Kass, R. E., Raftery, A. E. (1995). Bayes Factors. *J. Am. Stat. Assoc.*, **90**:773-795.
19. Kim, H.T. (2007). Cumulative incidence in a competing risks setting and competing risks regression analysis. *Clinical Cancer Research.* **13**(2):559-65.
20. Klein J.P., Andersen P.K. (2005). Regression modeling of competing risks data based on pseudovalues of the cumulative incidence function. *Biometrics.* **61**:223-229.
21. Latouche, A., Porcher, R. (2007). Sample size calculations in the presence of competing risks. *Stat. Med.* **26**(30):5370-80.
22. Maki, E. (2006). Power and sample size considerations in clinical trials with competing risk endpoints. *Pharm. Stat.* **5**(3):159-71.
23. Latouche, A., Porcher, R, Chevret, S. (2004). Sample size formula for proportional hazards modelling of competing risks. *Stat Med.* **23**(21):3263-74.
24. Pintilie, M. (2002). Dealing with competing risks: Testing covariates and calculating sample size. *Statistics in Medicine*, **21**:3317-3324.
25. Pintilie, M. (2006). *Competing Risks: A Practical Perspective.* Wiley, New York.
26. Scrucca L, Santucci A, Aversa F. (2007). Competing risk analysis using R: an easy guide for clinicians. *Bone Marrow Transplantation.* **40**(4):381-7. Epub 2007 Jun 11.
27. Scrucca L, Santucci A, Aversa F (2010). Regression Modeling of Competing Risk Using R: An In Depth Guide for Clinicians. *Bone Marrow Transplantation.* Jan 11. [Epub ahead of print].
28. Schwarz, Gideon (1978).Estimating the dimension of a model. *The Annals of Statistics*, **6**:461-464.
29. Tsiatis, A. (1975). A nonidentifiability aspect of the problem of competing risks. *Proc. Natl. Acad. Sci. U.S.A.* **72**(1):20-22.

Part III

Related Topics: Genomics/Bioinformatics, Medical Imaging and Diagnosis, Clinical Trials

Chapter 12

Comparative Genomic Analysis Using Information Theory

Sarosh N. Fatakia, Stefano Costanzi and Carson C. Chow*

Laboratory of Biological Modeling,
National Institute of Diabetes and Digestive and Kidney Diseases
National Institutes of Health
Bethesda, MD 20892, USA
**carsonc@mail.nih.gov*

Comparative genomic analyses of homologous protein sequences provide insight to molecular evolution of amino acid residues within species and across species. Here, we review our method of identifying statistically related amino acid positions from multiple sequence alignments of non-olfactory human G protein-coupled receptors. We apply the same algorithm to the homeodomain of homeobox proteins from different species. The method uncovers statistically related amino acid positions that may potentially be important for its structure and/or function.

12.1. Introduction

The past decade has seen an explosion in the number of complete genome sequences for mammalian and non-mammalian species. High-throughput genome sequencing commenced with the Human Genome Project.[1,2] The challenge now is to interpret and analyze this plethora of data. However, even before high throughput genome sequencing began, Doolittle proposed that sequence information from protein superfamilies could help unravel the molecular evolution of proteins.[3,4] Comparative genomics studies have used information theory as a tool to uncover common and distinct traits within species and across species. For example, common characteristics may help identify amino acid residues (AAs) essential for protein structure and/or function and distinct features may show how a protein family maintains its basic structure as the sequences diverge.[5] During protein evolution the attributes pertinent to function diverge slowly.[6] It was first reported by

Chothia and Lesk[7] that despite sequence diversity, structural conservation was salient for protein evolution. Subsequent studies [8–11] confirmed those results, and keeping pace with the ever increasing rate of additional sequence and structure information, high throughput studies have supported and enhanced our understanding of protein structure evolution.[12]

The sequence of base pairs in a gene determines the protein product the gene can encode. Genes within the same species, arising out of common ancestry due to gene duplication events are referred to as paralogs, e.g. the superfamily of human G protein-coupled receptors (GPCRs) are paralogs which have evolved from some common ancestorial gene.[13–16] The genes associated with the same function (e.g. OPSD gene that encodes rhodopsin, which is expressed in retinal rod cells) in different species are termed orthologs. Both paralogs and orthologs may be traced to a common ancestor and are referred to as homologous.

The GPCR superfamily is one of the most diverse and the largest superfamily of membrane proteins in humans. They are involved in a variety of essential physiological functions and are a vital target for pharmaceutical intervention. The different receptors have very diverse amino acid (AA) sequence compositions. The superfamily of GPCRs can be subdivided into numerous subfamilies in accordance with their sequence similarity. The common feature of the GPCR superfamily is the seven transmembrane (7TM) alpha-helical bundle. The subfamilies have preserved common ancestral structure and/or function (e.g. the 7TMs and their lengths embedded in the bilayer cell membrane), while incorporating specific and selective traits (e.g. ligand specificity).

It is hypothesized that statistically related positions are important for the protein structure and/or function. Investigations from W.M. Fitch and E. Markowitz[17] four decades ago, to recent studies by S. Chakrabarti, S. and A. R. Panchenko,[18] highlight that coevolving AA substitutions are salient. In a recent review Pazos and Valencia have discussed at length the phenomenon of coevolution and coadaption, pertinent to AAs.[19] In a recent comparative genomic analysis from class A, B and C GPCRs, we identified a cohort of functionally important residues that were not conserved across the different receptors.[20] Here, we also apply our method to the homeodomain protein, which is found within hox genes that specify the body plan and regulate development.[21] It is well-conserved across various lineages, from fruit flies to humans.[22]

To study the evolutionary traits of families of proteins, the protein sequences are aligned such that the AAs of common ancestry (i.e.

homologous) are brought into a vertical register in a multiple sequence alignment (MSA). The MSA positions that consist of identical AAs are conserved. A conserved AA position is likely to be important for the structure and/or function of the protein since it has resisted mutations throughout evolution. Experiments that mutate conserved residues at those positions can validate the relevance of theoretical prediction. Proteins may also have positions important for structure and/or function that are not conserved in the alignment. For example, a pair or a group of positions could evolve together, each mutating to compensate for mutations in others. Here, we review our previously published method[20] to uncover such a cohort of co-evolving positions. Our method utilizes concepts from information theory and graph theory.

12.2. Materials and Methods

12.2.1. *Multi-sequence alignments*

MSAs of the non-olfactory human GPCRs due to Surgand *et al.*[23,24] had 189 AA positions. The sets were:

(1) 7TMs from class A GPCRs (287 sequences),
(2) 7TMs from class B GPCRs (49 sequences),
(3) 7TMs from class C GPCRs (22 sequences).

An MSA of the 7TMs and a portion of the second extracellular loop (EL2) from positions EL2.49 to EL2.53 (AA positions denoted by Ballesteros-Weinstein indexing scheme of GPCRs[25]) of 249 class A GPCRs was obtained. The homeodomain MSA (with 237 sequences) was obtained for a comparative study from Gloor *et al.*[26,27]

12.2.2. *Shannon entropy and mutual information*

Consider a MSA with N_{total} AA positions (or columns) and S different sequences. For the jth position in the MSA, ($0 \leq j < N_{total}$), the Shannon entropy or informational entropy,[28] denoted by $H(j)$, is

$$H(j) = -\sum_{x=1}^{N_{AA}} p_j(x) \log_2 \, p_j(x), \tag{12.1}$$

where $p_j(x)$ is the measured probability of AA x occurring at position j. The logarithm base 2 is used. Individual AA probabilities $p_j(x)$ were obtained for the N_{AA} ($= 20$) naturally available AA using the corresponding AA frequency. Conserved positions of the MSA have 0 bits of entropy (no variability or disorder exists in the AA composition), while non-conserved positions will have higher entropy up to a maximum of $\log_2 N_{AA} = 4.32$ bits.

The joint entropy[28] for the pair of MSA positions (j, k) is defined as

$$H(j,k) = -\sum_{x=1}^{N_{AA}} \sum_{y=1}^{N_{AA}} p_{j,k}(x,y)\ log_2\ p_{j,k}(x,y), \tag{12.2}$$

where $p_{j,k}(x, y)$ is the joint probability of AAs x and y occurring at positions j and k respectively. Summations are over the N_{AA} (= 20) naturally occurring AAs. ($N_{AA} = 21$ when gaps "-" are included in the MSA). For an ordered (j, k) pair, where $j < k$, there are $\frac{1}{2}N_{total}(N_{total} - 1)$ unique MI pairs, because $0 \leq j < N_{total}$, $0 \leq k < N_{total}$, and $H(j, k) = H(k, j)$. If two AA positions in an MSA are statistically related then knowledge of the AA residue in one position will reduce the uncertainty of identity of the other AA residue. Thus, the joint entropy will be reduced. Mutual information (MI)[28,29] quantifies this reduction in uncertainty. The mutual information $MI(j, k)$ for the (j, k) position pair is

$$MI(j,k) = H(j) + H(k) - H(j,k). \tag{12.3}$$

Using Eqs. (12.1–12.3), $MI(j, k)$ may be expressed as

$$MI(j,k) = \sum_{x=1}^{N_{AA}} \sum_{y=1}^{N_{AA}} p_{j,k}(x,y) \log_2 \frac{p_{j,k}(x,y)}{p_j(x)p_k(y)}. \tag{12.4}$$

As before, $p_{j,k}(x, y)$ is the joint probability of AAs x and y occurring at positions j and k respectively. The range of values for $MI(j, k)$ is given by the inequality $0 \leq MI(j,k) \leq \log_2 N_{AA}$ and the greater the statistical relation between the positions (j, k) the greater is the $MI(j, k)$ value. Using a MSA of the class A 7TMs, for $0 \leq j < N_{total} = 189$, a total of $\binom{189}{2}$ = 17766 $MI(j, k)$ values (for $j < k$) were computed. To avoid evolutionary correlations among positions within the same TMs and nearest neighbor interactions among them, we used the MI computed for inter-TM pairs. The MI distributions from intra-TM (j, k) pairs versus the inter-TM (j, k) pairs are revisited in the Discussion section. All 15255 inter-TM MI values are shown in Fig. 12.1 with violet representing lower $MI(j, k)$ values and red representing higher values. A portion of Fig. 12.1, corresponding to positions from TM3 and TM7 is shown in Fig. 12.2. The results from class B and class C are shown in Fig. 12.3 and Fig. 12.4 respectively.

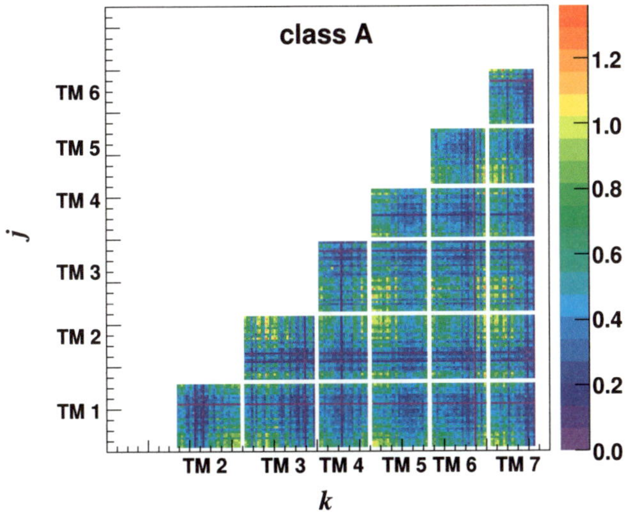

Fig. 12.1. Mutual Information values from class A 7TMs. The $MI(j,k)$ values from all 15255 inter-TM position pairs are represented in the rainbow color scheme with the high $MI(j,k)$ values in red.

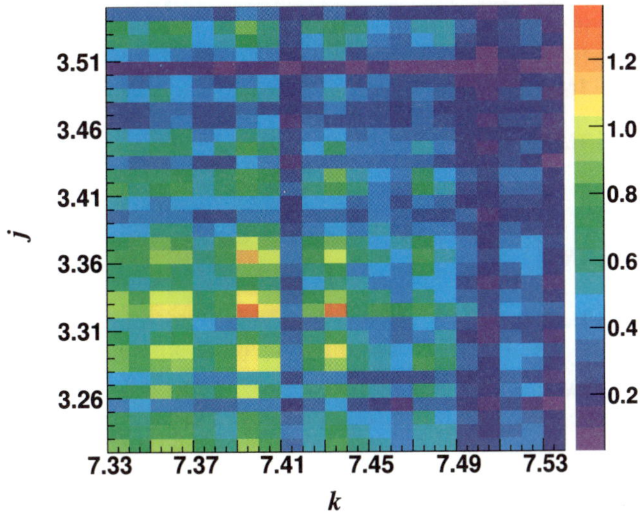

Fig. 12.2. Mutual Information values obtained from TM3 and TM7 for class A GPCRs. This is a zoomed image for j positions that correspond to TM3 and k positions that correspond to TM7 from the spectrum from Fig. 12.1. The Ballesteros-Weinstein scheme[25] to represent GPCR positions has been used.

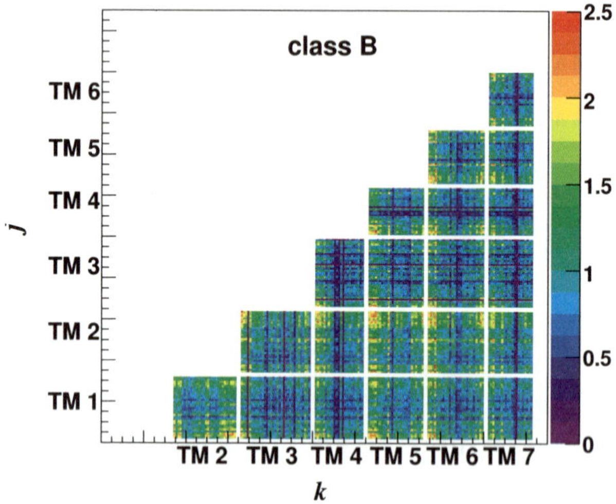

Fig. 12.3. Mutual Information values from class B 7TMs. The $MI(j,k)$ values from all 15255 inter-TM position pairs are represented in the rainbow color scheme with the high $MI(j,k)$ values in red.

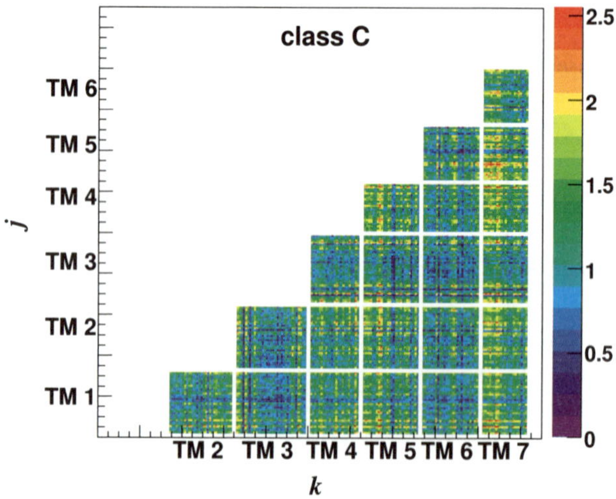

Fig. 12.4. Mutual Information values from class C 7TMs. The $MI(j,k)$ values from all 15255 inter-TM position pairs are represented in the rainbow color scheme with the high $MI(j,k)$ values in red.

Table 12.1. Total number of candidate positions from studies of class A GPCRs.

M edges	P_M	$P_D < 0.050$	$P_D < 0.010$	$P_D < 0.005$	$P_D < 0.001$
280	6.0×10^{-5}	8	5	5	4
374	1.9×10^{-4}	10	5	5	4
428	7.2×10^{-4}	10	7	6	4
526	2.6×10^{-3}	13	8	7	6
599	6.8×10^{-3}	18	10	9	6
674	1.5×10^{-2}	19	14	12	9
789	3.5×10^{-2}	21	17	13	11
812	4.6×10^{-2}	23	17	15	10

Note: Fewer candidate positions are identified for progressively stringent threshold cuts in P_D, as one follows the number of candidate positions identified from left to right in a row. For a given P_D, as P_M is relaxed (along a column), a greater number of candidate positions are obtained.

The probabilities for Eq. 12.4 were estimated from the frequencies of AAs at the respective positions. However, the use of a finite set of sequences will result in finite-size error for MI measurements. In theory, assuming statistical independence between positions (j, k) and given an infinite number of random sequences, $MI(j, k)$ is zero since $p_{j,k}(x, y) = p_j(x)p_k(y)$. For a small number of sequences S ($S \sim N_{AA}$), we estimated that MI may be nonzero and scales as $-\log S$.[20] Figure 12.5 shows results from MSAs of progressively increasing number of random sequences S, which have been simulated using the AA probability density derived from (i) the concatenated 7TMs, and (ii) equally probable occurrences of AA residues. In every simulation we kept $N_{total} = 189$, which was consistent with the GPCR 7TM dataset.

In order to obtain the statistical significance of the estimated MI values, we generated surrogate MSAs which preserved the entropic information along the MSA columns but destroyed the joint entropy (within the original dataset) by independently shuffling the AA columns at random.[31] As a null hypothesis, we ascribed the non zero MI values to arise due to finite-size effects within surrogate MSAs. The alternative hypothesis was that pairs of positions with high MI represented the non-random evolutionary correlations we sought. Figure 12.6 shows the probability density distributions (PDFs) for MI computed using class A MSA and from surrogate MSAs. It shows that the MI PDFs from the class A dataset and the surrogates can be highly skewed. In addition, Fig. 3 of Ref. 20 shows that spurious high MI from surrogates of class B and class C GPCRs, which are much smaller datasets, are comparable to the high MI obtained from the dataset itself (finite-size effects). Statistical significance of the MI was then deter-

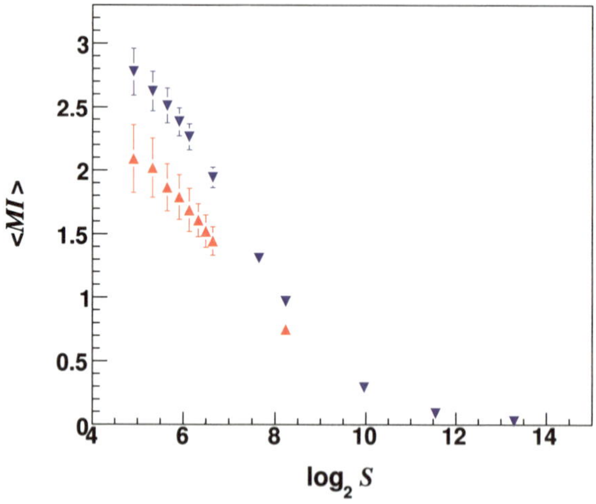

Fig. 12.5. Average MI as a function of the number of the logarithm of the number of input sequences (**S**) in simulated MSAs. The inverted triangles represent results from MSAs generated with AA sequences selected with equal probability while the upright triangles represent the results where the probability distribution function of AAs is derived from that of the 7TMs. For an MSA with 10^4 random sequences the average MI from all pairs of positions is negligibly small. For small S ($S \sim N_{AA}$), $\langle MI \rangle \sim -\log_2 S$.

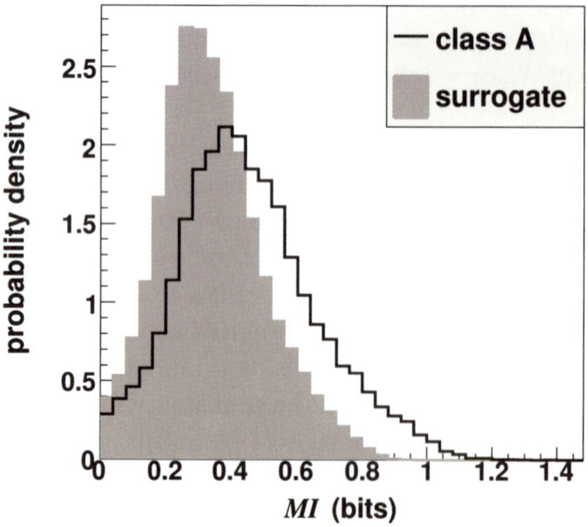

Fig. 12.6. Probability density distribution of MI obtained from class A MSA contrasted with surrogates. The average obtained from all surrogate MSAs is represented by the shaded distribution.

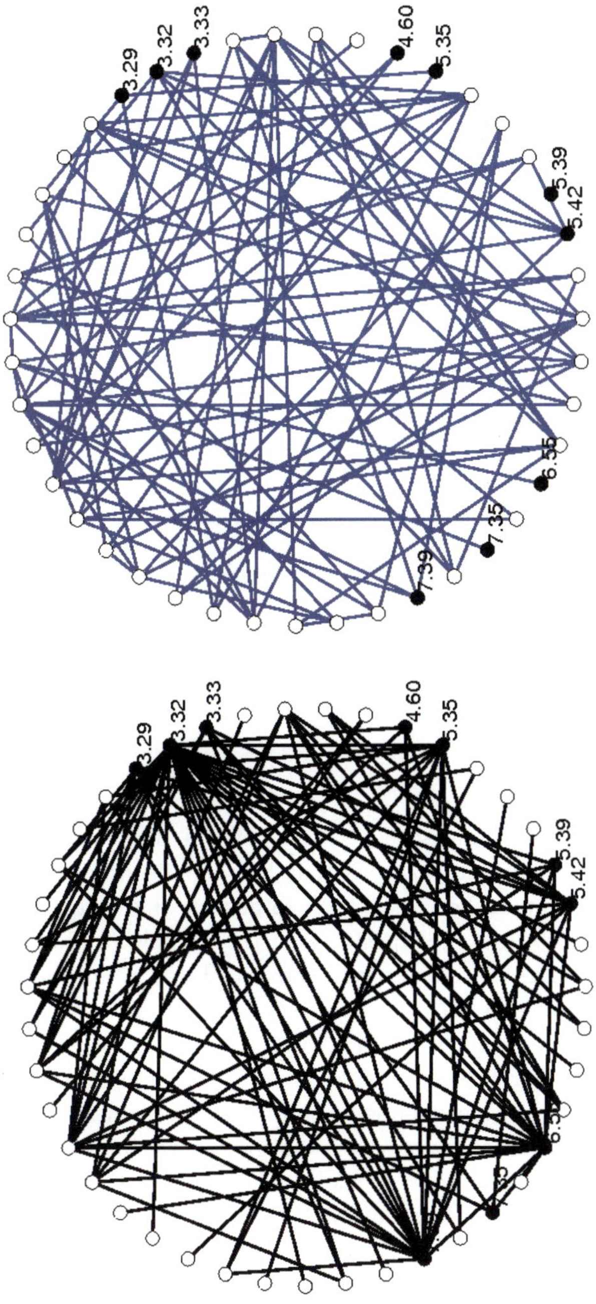

Fig. 12.7. MI graph representing the class A 7TM dataset with **M = 100** and **N = 43** (left panel) contrasted with randomly generated graph with identical M and N values (right panel).

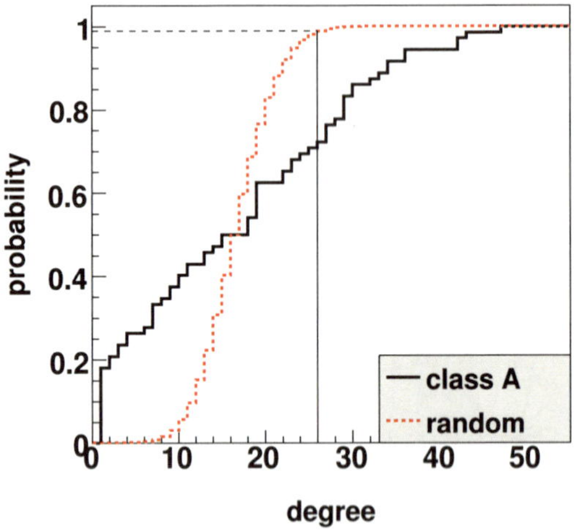

Fig. 12.8. Cumulative probability of vertices in $\mathbf{G(M,N)}$ with given degree. The solid curve is obtained from class A MI graph $G(M = 600, N = 72)$. The dotted curve represents results from the ensemble average of randomly computed $G(M = 600, N = 72)$s.

mined with respect to the probability of occurrence (i.e. P value) from the surrogate set.

12.2.3. *MI graphs*

Mutual information is a measure between pairs of positions. Our goal was to extract a set of *key positions* that had significantly high mutual information with many other positions. To achieve this end, we created MI graphs where the N_{total} AA positions of the MSA were designated as vertices and the edges of two vertices were connected if and only if the (j,k) pairs had significant $MI(j,k)$ (i.e. greater than a threshold MI set by the P value). The result was a MI graph with $N \leq N_{total}$ vertices and M edges. The graph will have missing edges because only the significant MI values for a given P value compared to the surrogate set were retained. The number of edges in the graph M is an increasing monotone function of the P value denoted by P_M. A range of P_M was used to construct a set of different MI graphs. As an example, the left panel in Fig. 12.7 is a MI graph constructed from class A GPCRs with $M = 100$ (corresponding to $P_M < 0.00006$) and $N = 43$. A random graph with $M = 100$ edges and $N = 43$ vertices is shown on the panel to the right. The number of

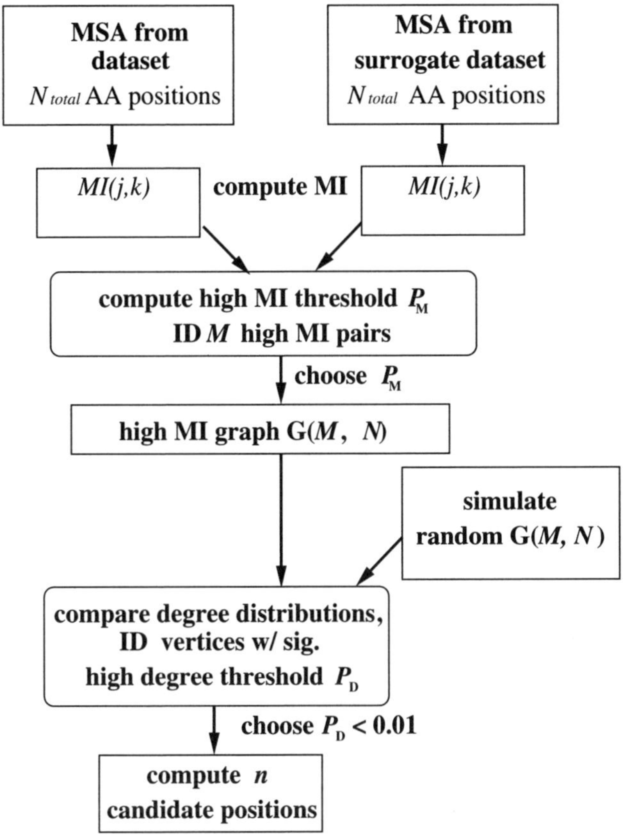

Fig. 12.9. Identification of candidate positions. Flowchart describing the steps leading to the identification of cadidate positions using MSAs from the dataset and surrogate sets.

edges incident to a vertex of the graph is referred to as the degree of the vertex. The distribution (or histogram) of degree values from all vertices of the graph is called the degree distribution. Hence, our goal corresponds to identifying vertices with significantly high degree.

To assess significance of high degree vertices, the degree distribution of vertices obtained from the dataset MI graph was compared with an ensemble average of random graphs with the same number of edges and vertices. Figure 12.8 shows the cumulative probability distribution for degree from MI graphs using class A dataset with $M = 600$. From the Fig. we see that in this example, it is very unlikely for a random graph to have more than

Table 12.2. Number of candidate positions from class A, B and C GPCRs.

M edges	class A	class B	class C
300	5	3	6
350	5	4	8
400	7	3	9
450	7	3	9
500	7	8	10
550	8	9	10
600	10	10	10
650	13	11	10
700	15	15	11
750	16	15	12
800	17	16	12

Note: Number of candidate positions identified by varying threshold P_M from $M = 400$ to $M = 800$ to compute MI graphs. The degree threshold used is $P_D < 0.010$. Key positions were subsequent obtained from candidate positions.

30 vertices. A P value P_D could be assigned to a statistically significant number of edges attached to a given vertex compared to the null hypothesis of a random graph. Each value of P_D resulted in a set of statistically significant vertices with high degree. Similarly, for a fixed P_D, a range of P_M values resulted in a different set of vertices which were statistically significant. We called these positions, which are associated with the two P values, candidate positions.[20] A flowchart summarizing the method of extraction of candidate positions is shown in Fig. 12.9. Table 12.1 represents the number of candidate positions obtained from class A. Table 12.2 represents the number of candidate positions obtained using $P_D < 0.010$ for a range of P_M values for all three classes.

12.2.4. *Key position identification*

Key positions were selected out of the candidate positions if they remained invariant to changes in P_M and P_D as well as to a leave-one-out procedure, where the analysis was performed with one sequence removed. As P_M was increased (i.e. M increased), the number of identified candidate positions also increased but the identity of these positions could change for different M. We examined graphs at increments of 50 starting at $M = 50$. For each M we identified a cohort of positions that were invariant to the leave-one-out analysis. An invariant cohort of positions was found over the ranges of $500 \leq M < 650$ with the largest cohort of ten positions found at $M =$

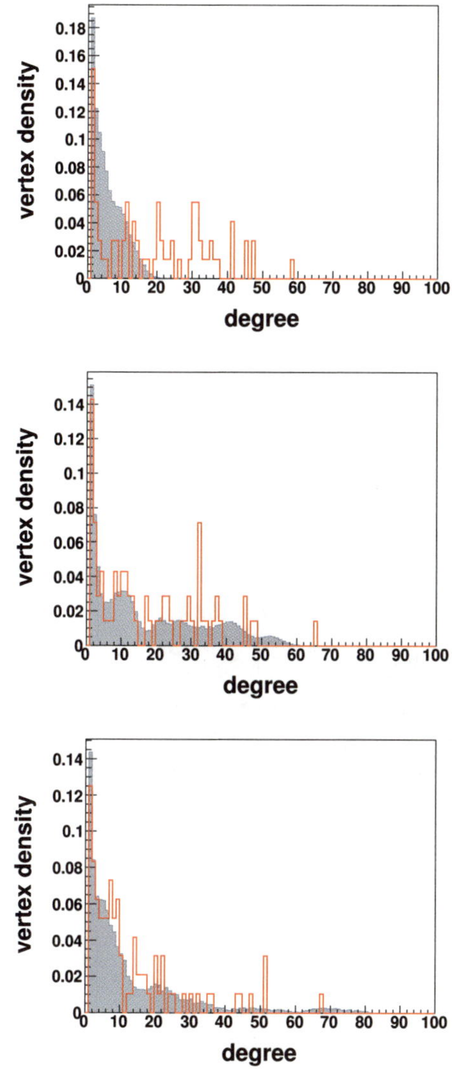

Fig. 12.10. Degree distributions from dataset and surrogates. (Top panel) Results of degree distributions from class A (with 7TMs and EL2) are shown, along with those from class B (middle panel) and class C (bottom panel). Although class C MSA has less than half the number of sequences of class B, the level of spurious signal due to high degree is less than than of class B.

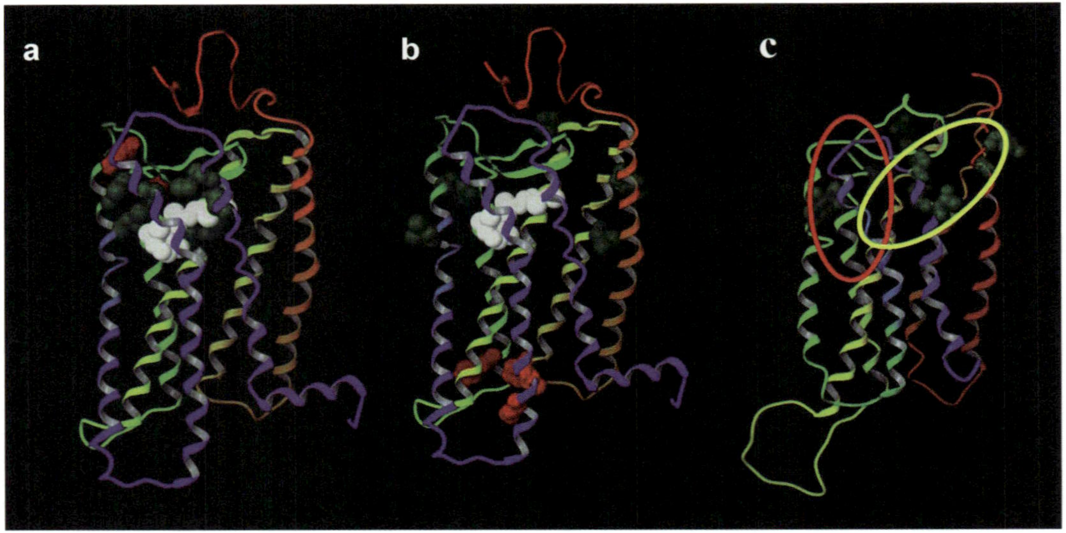

Fig. 12.11. Key positions identified from class A, class B and class C receptors. Panel a: class A key positions visualized in the crystal structures of bovine rhodopsin. All key positions are located in the exofacial 7TM binding cavity. Residues at 4.60 and 5.35 (in red) can be considered hinges for EL2. All the residues at the remaining key positions (in green) directly line the pockets for the co-crystallized ligands (retinal shown in white). Panel b: class B key positions visualized in the crystal structures of rhodopsin (a class A receptor). The key positions are localized in two different areas: five of them (in green) are located within the exofacial 7TM cavity, while the remaining four (in red) are located near the intracellular loops. Panel c: class C key positions visualized in a rhodopsin based molecular model of the calcium sensing receptor.[32] The key positions (in green) are located in correspondence with the two predicted adjacent sites for different classes of synthetic allosteric modulators. Ligands and residues at key positions are represented as Corey-Pauling-Koltun space filling models. The backbone of the receptor is schematically represented as a ribbon, depicted with the colors of the rainbow from the N-terminus to the C-terminus (TM1: red; TM2: orange; TM3: yellow; TM4: yellow-green; TM5: green; TM6: cyan; TM7: blue).

Table 12.3. Identified key positions from class A GPCRs.

	7TM	7TM+EL2	rhodopsin
TM2	-	2.67	His100
TM3	3.29	3.29	Gly114
	3.32	3.32	Ala117
	3.33	3.33	Thr118
TM4	4.60	4.60	Pro171
EL2	NA	EL2.49	Ser186
	NA	EL2.51	Gly188
TM5	5.35	5.35	Asn200
	5.39	5.39	Val204
	5.42	5.42	Met207
TM6	6.55	6.55	Ala272
TM7	7.35	7.35	Met288
	7.39	7.39	Ala292

Note: Ballesteros-Weinstein indices of key positions and residue number in the rhodopsin sequences. NA indicates position not included in the analysis.

600. These ten positions were selected to be our key positions for Class A. The algorithm was also applied to class B and C. The largest cohorts for class B and C receptors were coincidentally also found for $M = 600$ edges. Similar to class A, ten candidate positions were found for classes B and C, shown in Table 12.2. (Nine of those ten positions made the cut for key positions). Using an identical procedure for class A MSA with 7TMs and the 2nd extracellular loop (EL2), the largest cohort of candidate positions was found for $M = 700$. The degree distributions from classes A, B and C for the lowest P_M threshold for obtaining the largest number of key positions and their associated surrogate MSAs are shown in Fig. 12.10. The EL2 fragment from class A MSA was incorporated in this analysis but only 7TMs were used for classes B and C.

12.3. Results

Ten key positions were obtained from class A GPCR 7TMs. For the class A MSA which included EL2, 13 key positions were identified. In that set, the

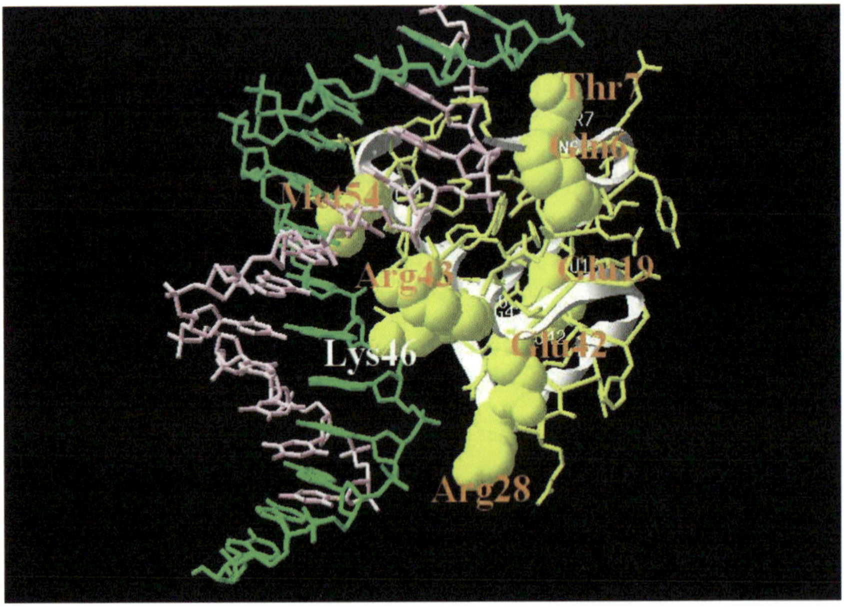

Fig. 12.12. Key positions shown in resolved structure of 9ANT: Antennapedia homeodomain-DNA complex (PDB ID: 9ANT.pdb). The residues identified as key positions are shown in yellow (for chain A). Labeled in white is Lys46 which has the highest degree in the MI graph. Also shown in yellow are key positions Gln6, Thr7, Glu19, Arg28, Glu42, Arg 43 and Met54. One of the identified key positions Gly4 is not represented in the pdb file.

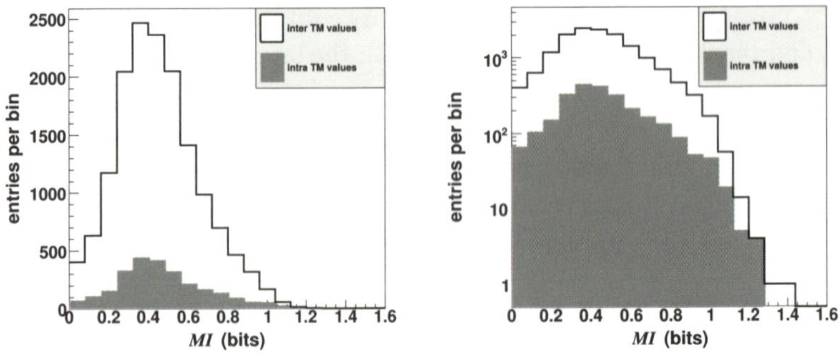

Fig. 12.13. Intra-TM MI distribution. The distribution of MI values from intra-TM pairs of positions compared with that obtained from inter-TM pairs.

Table 12.4. Identified key positions[20] and specificity determining residues[46] obtained from homeodomain homologs.

MSA position	9ANT.pdb file	9ANT.fasta.txt file	rank of key position	SDR	structural information	missense mutation
3	NA	Gly6	6		(loop)	
5	Gln6	Gln8	2	x	p (loop)	xx
6	Thr7	Thr9	3	x	m (loop)	xx
18	Glu19	Glu21	7		p (helix 1)	
23	Arg24	Arg26			(loop)	
27	Arg28	Arg30	4	x	p (helix 2)	
30	Arg31	Arg33		x	p (helix 2)	
62	Glu42	Glu44	5	x	(helix 3)	xx
63	Arg43	Arg45	9		p (helix 3)	
65	Ile45	Ile47		x	(helix 3)	
66	Lys46	Lys48	1	x	p (helix 3)	xx
74	Met54	Met56	8	x	M (helix 3)	xx

Note: Indices of key positions and AA information from the resolved Antennapedia homeodomain-DNA complex 9ANT PDB structure. Shown alongside are the identified specificity determining residues (SDR).[46,48] Positions common to both methods are shown in x.
The PDB file uses the standard numbering for homeodomains. The FASTA sequence corresponds to the peptide that was expressed, and hence the indices are off by two.[54] The xx indicate known missense mutation at that position and letters M, m and p respectively indicate major groove, minor groove and backbone contacts respectively. (Relevant information was obtained from Ref. 34 and 35). Information from the pdb file reveals three helices (helix 1, helix 2 and helix 3) that span AA positions 10 – 22, 28 – 38 and 42 – 58 respectively.

10 previously identified key positions and three additional key positions were obtained. Two key positions were adjacent to the conserved Cys residue at EL2.50. The Ballesteros-Weinstein indices and the AAs from bovine rhodopsin sequence used to highlight the class A key positions are in Table 12.3. Nine key positions were identified for each of the 7TMs of class B and class C GPCRs. These results are summarized in Table 1 of Fatakia *et al.*[20] The identified key positions were also found to be a clique (a completely connected subgraph). Hence a mutation in any of these positions could influence the entire cohort.

The location of the class A key positions in the crystallographic structure of bovine rhodopsin is shown in Fig. 12.11a. They all line the 7TM cavity, which has been demonstrated to harbor the ligand binding site for the small molecule-binding class A GPCRs whose 3D structure has been experimentally resolved - additional details may be found in Ref. 20. The 3D structure of class B or C GPCRs has not been experimentally resolved. Unlike many class A receptors, class B and class C receptors bind their ligands

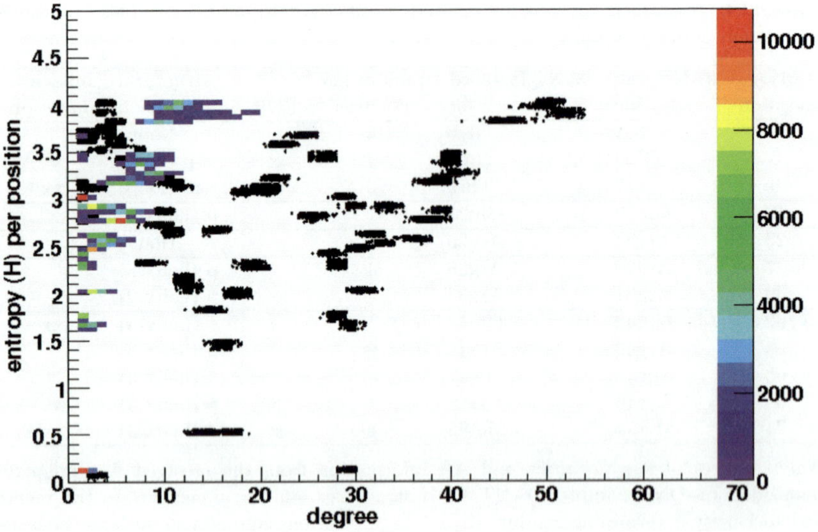

Fig. 12.14. Informational entropy ($\mathbf{H(j)}$) versus degree distribution from dataset and surrogates. The points in black represent the entropy and degree values obtained from the studies involving class A 7TMs and EL2. The simulated results from various surrogates are represented in the color coded scheme from violet to red.

through their ectodomains, not within the 7TM bundle. Consistently, the key positions for class B receptors, visualized in the crystal structure of bovine rhodopsin in Fig. 12.11b, do not form a isolated cluster within the 7TM bundle. Our analysis yielded key positions that lined the 7TM cavity proposed to bind synthetic allosteric modulators of the Ca^{2+} receptor,[32] a member of class C, as shown in Fig. 12.11c in a rhodopsin-based molecular model. In every study, key positions were found without invoking any functional specificity of the involved proteins.

The analysis was also applied to the homeodomain. Previously, a correlated mutation analysis of a 77 AA long MSA that includes the homeodomain was published by Gloor *et al.*[26] From the MSA, they listed 7 coevolving positions of greatest significance: (3, 5, 31, 42, 50, 52 and 54). Our method resulted in a set of 9 AA positions that are listed in Table 12.4. These key positions are shown in Fig. 12.12, which shows the resolved structure of Antennapedia homeodomain-DNA complex (9ANT)[33] obtained from the RCSB Protein Data Bank.[30] The key position with the highest degree (Lys46) is located in the third helix, and is shown to have a missense mutation.[34,35] DNA-DNA backbone contacts persist at five key positions

Gln6, Glu19, Arg28, Arg43 and Lys 46. Missense mutations have also been demonstrated at positions Gln6, Thr7, Glu42, Lys46 and Met54.[34,35] Arg28 and Glu42 are the leading AAs of the second and third helices.

12.4. Discussion

We have presented a method using information and graph theory to identify a group of statistically related positions in an amino acid sequence of a protein family. The identified key positions from class A and class C GPCRs were located toward the extra-cellular surface of the receptors, which formed a topologically contiguous cohort. The class B key positions were distributed across the extra-cellular and intra-cellular regions. These positions are probably not involved in ligand recognition; they may be important for the structure and/or function of these receptors. Our approach may also be useful for elucidating protein structure when the structure is unknown or partially known. Key positions share high mutual information with each other and likely indicate close spatial proximity. Such knowledge would be useful when attempting to infer the protein structure.

A major difficulty in identifying important positions is the presence of evolutionary noise, which could result in a spurious signal because sequences evolved from a common ancestor. Noivirt *et al.*[36] generated a surrogate null set by shuffling position pairs in sequences which lie within the same subfamily or which are evolutionarily close more than pairs which are evolutionarily more distant. Wollenberg and Atchley[37] generated pseudo sequences using the Jones, Taylor and Thomson (JTT) matrix,[38] ensuring that residues with high similarity were represented more often and therefore evolutionary close lineages were sampled more often. Gloor and colleagues[26,39] used a normalized MI value called $nMI(j,k)$ $(=MI(j,k)/H(j,k))$ instead of raw MI values. Subsequently, Dunn *et al.*[40] introduced a correction which involved the product of the mean of normalized MI pairs that involved the j and the k position in the MSA $(\overline{nMI(j)} \times \overline{nMI(k)})$. Little and Chen[41] calculated the least squares regression of MI, then obtained a correction term with the residuals to eliminate the stochastic bias in $nMI(j,k)$ and refined the previously posited correction term $\overline{nMI(j)} \times \overline{nMI(k)}$.

Our approach was to use the robustness of high degree in the mutual information graph to screen out such spurious correlations. We also omitted MI values from intra-TM pairs of positions in the analysis to avoid correlations within TMs . However, even if these pairs were included, the key positions remain unchanged for class A. Figure 12.13 shows the MI histogram from intra-TM pairs and inter-TM pairs on a linear and logarithmic scale. Additionally, while there was no correlation between high entropy and high degree as seen in Fig. 12.14, key positions tended to be high in both measures while data generated from surrogate sequences with high entropy had low degree and may be attributed to spurious or background MI.

The number of identified key positions may depend on the similarity of the sequences in the MSA. Using the PROTDIST program from PHYLIP,[42] the average percentage of accepted mutation among the class A receptors for each of the 7TMs were 244.5%, 226.8%, 191.8%, 245.4%, 278.4%, 194.1% and 179.6% respectively. It was not evident to us that the number of key positions identified within each TM was related to the percentage of accepted mutations among the sequences. TM3 and TM5 each have three key positions (which was the maximum for an individual TM). TM3 had the lowest percentage of accepted mutations (192%) while TM5 had the highest (278%) percentage of accepted mutations.

The frequency estimates of AA residues within columns of an MSA are sensitive to sampling errors and misalignments. However, for a given MSA the errors due to finite size effects would expectedly be random and not systematic for any given position. Therefore, instead of modifying the raw MI values themselves, our strategy was to identify positions with high correlations with many other positions. Spurious high MI pairs in such a MI graph would only lead to a small number of spurious random edges. Isolated instances of sequence misalignments may not affect the identity of the key positions because high MI position pairs associated with many other high MI position pairs were sought. Additionally, if a pair of positions have high MI, but were not associated with many other pairs then they were not selected. For example, consider MI between position 3.32 in TM3 with three positions in TM7: 7.35, 7.39 and 7.43 as shown in Fig. 12.2. Although the MI between 3.32 and 7.43 is high, 7.43 did not meet the criteria for key position.

MI has been used to identify specificity determining residues specific to subfamilies.[31,43–45] In these studies, the MI was computed between the AA residues at a position for a subfamily or subgroup and the subgroup index or label. In a later investigation, specificity determinants were extracted via Shannon entropy.[46] Both approaches identified subfamilies, and conserved residues specific to them. Our analysis identified statistical relationships that were common among AAs pairs throughout the whole family. Subfamilies were not defined in our approach. Therefore, identified key positions were not necessarily conserved within subfamilies (non-zero Shannon entropy). However, it is possible that some key positions may be conserved within subfamilies. Using the homeodomain MSA, we compared results from our method to that of specificity determining residues by Reva *et al.*[46] ("proteinkeys" online resource[47]). The leading five key positions that we identified were also identified as specificity determining residues. From a total of nine key positions that we identified, six positions were identified as specificity residues.[48] The common and unique positions identified using the homeodomain MSA are shown in Table 12.4. Seven of the nine identified positions are in the first coil and the third helix.

Five of the identified key positions were associated with the DNA-DNA backbone contacts, and one each were associated with the major groove and minor groove. We hypothesize that backbone key positions coevolved along with the major groove and minor groove key position to sustain the evolutionary conservation for homeodomain flexibility. From a bioengineering standpoint, in support of the importance of specific backbone AAs, G.D. Rose *et al.* have postulated that "Once a scaffold is established, evolution is at liberty to spawn species-adapted orthologs and paralogs that tune the sequence for optimal performance in specific microenvironments or to explore new functions".[49] From an evolutionary standpoint, Aravind *et al.* have argued that such evolutionary conservation is quintessential for the "progenitor of each superfamily" so as to give rise to a vast "diversity of sequences and multidomain architectures".[50] Jeffery W. Kelly summarized the recent work of Ranganathan and colleagues ,[51,52] who showed that "maintaining the conservation pattern in a protein family, along with a surprisingly small subset of coevolving residues, enables the generation of low-homology sequences that fold and function. These studies indicate that the number of critical interactions in protein may be smaller than previously thought".[53]

12.5. Acknowledgments

We wish to thank Professor Gregory B. Gloor at the University of Western Ontario, Canada, for the homeodomain MSA. We also wish to thank the anonymous reviewer for corrections and valuable suggestions. This work was supported by the intramural program of the NIH/NIDDK.

References

1. E. Lander, L. Linton, B. Birren, C. Nusbaum, M. Zody, J. Baldwin, K. Devon, K. Dewar, *et al.*, Initial sequencing and analysis of the human genome, *Nature*. **409**(6822), 860–921, (2001).

2. J. Craig Venter, M. Adams, E. Myers, P. Li, R. Mural, G. Sutton, H. Smith, M. Yandell, *et al.*, The sequence of the human genome, *Science.* **291**(5507), 1304–1351, (2001).
3. R. Doolittle, Similar amino acid sequences: Chance or common ancestry?, *Science.* **214**(4517), 149–159, (1981).
4. R. Doolittle, Similar amino acid sequences revisited, *Trends in Biochemical Sciences.* **14**(7), 244–245, (1989).
5. A. Lesk, Ed., *Computational molecular biology: sources and methods for sequence analysis.*
6. M. Kimura, Ed., *The neutral theory of molecular evolution.*
7. C. Chothia and A. Lesk, The relation between the divergence of sequence and structure in proteins., *The EMBO journal.* **5**(4), 823–826, (1986).
8. C. Chothia and M. Gerstein, How far can sequences diverge?, *Nature.* **385** (6617), (1997).
9. R. Russell, M. Saqi, R. Sayle, P. Bates, and M. Sternberg, Recognition of analogous and homologous protein folds: Analysis of sequence and structure conservation, *Journal of Molecular Biology.* **269**(3), 423–439, (1997).
10. B. Rost, Protein structures sustain evolutionary drift, *Folding and Design.* **2** (3), S19–S24, (1997).
11. T. Wood and W. Pearson, Evolution of protein sequences and structures, *Journal of Molecular Biology.* **291**(4), 977–995, (1999).
12. C. Chothia and J. Gough, Genomic and structural aspects of protein evolution, *Biochemical Journal.* **419**(1), 15–28, (2009).
13. R. Fredriksson, M. Lagerstrom, and H. Schioth, Expansion of the superfamily of g-protein-coupled receptors in chordates, *Annals of the New York Academy of Sciences.* **1040**, 89–94, (2005).
14. D. Perez, From plants to man: The gpcr.
15. D. Perez, Fredriksson, and Schioth, Erratum: From plants to man: The gpcr 'tree of life' (molecular pharmacology (2005) 67 (1383-1384)), *Molecular Pharmacology.* **67**(6), 2185, (2005).
16. H. Rompler, C. Staubert, D. Thor, A. Schulz, M. Hofreiter, and T. Schoneberg, G protein-coupled time travel: Evolutionary aspects of gpcr research, *Molecular Interventions.* **7**(1), 17–25, (2007).
17. W. Fitch and E. Markowitz, An improved method for determining codon variability in a gene and its application to the rate of fixation of mutations in evolution, *Biochemical Genetics.* **4**(5), 579–593, (1970).
18. S. Chakrabarti and A. Panchenko, Structural and functional roles of coevolved sites in proteins, *PLoS ONE.* **5**(1), (2010).
19. F. Pazos and A. Valencia, Protein co-evolution, co-adaptation and interactions, *EMBO Journal.* **27**(20), 2648–2655, (2008).
20. S. N. Fatakia, S. Costanzi, and C. C. Chow, Computing highly correlated positions using mutual information and graph theory for g protein-coupled receptors, *PLoS ONE.* **4**(3), (2009).
21. W. Gehring, M. Affolter, and T. Burglin, Homeodomain proteins, *Annual Review of Biochemistry.* **63**, 487–526, (1994).
22. W. Gehring, The homeobox in perspective, *Trends in Biochemical Sciences.* **17**(8), 277–280, (1992).

23. J.-S. Surgand, J. Rodrigo, E. Kellenberger, and D. Rognan, A chemogenomic analysis of the transmembrane binding cavity of human g-protein-coupled receptors, *Proteins: Structure, Function and Genetics.* **62**(2), 509–538, (2006).
24. http://lbm.niddk.nih.gov/s.fatakia/dir_msa/dir_gpcr/dir_surgand_rognan/total_alignment.fasta.
25. J. Ballesteros and H. Weinstein, Integrated methods for the construction of three-dimensional models and computational probing of structure-function relations in g protein-coupled receptors, *Methods Neurosci.* **25**, 366–428, 1995).
26. G. Gloor, L. Martin, L. Wahl, and S. Dunn, Mutual information in protein multiple sequence alignments reveals two classes of coevolving positions, *Biochemistry.* **44**(19), 7156–7165, (2005).
27. http://cjelli.biochem.fmd.uwo.ca/~ggloor/hd_new/nox_hd_90.fasta.
28. C. Shannon, A mathematical theory of communication, *Bell Syst. Tech. J.* **27**, 379–423, (1948).
29. T. Cover and J. Thomas, *Elements of Information Theory.* (Wiley, New York).
30. http://www.rcsb.org/.
31. L. Mirny and M. Gelfand, Using orthologous and paralogous proteins to identify specificity-determining residues in bacterial transcription factors.
32. J. Hu, J. Jiang, S. Costanzi, C. Thomas, W. Yang, J. Feyen, K. Jacobson, and A. Spiegel, A missense mutation in the seven-transmembrane domain of the human ca^{2+} receptor converts a negative allosteric modulator into a positive allosteric modulator, *Journal of Biological Chemistry.* **281** (30), 21558–21565, (2006).
33. E. Fraenkel and C. Pabo, Comparison of x-ray and nmr structures for the antennapedia homeodomain- dna complex, *Nature Structural Biology.* **5**(8), 692–697, (1998).
34. A. Delia, G. Tell, I. Paron, L. Pellizzari, R. Lonigro, and G. Damante, Missense mutations of human homeoboxes: A review, *Human Mutation.* **18**(5), 361–374, (2001).
35. A. D'Elia, G. Tell, I. Paron, L. Pellizzari, L. R., and G. Damante, Erratum: Missense mutations of human homeoboxes: A review (human mutation (2001) 18 (361-374)), *Human Mutation.* **19**(4), 457, (2002).
36. O. Noivirt, M. Eisenstein, and A. Horovitz, Detection and reduction of evolutionary noise in correlated mutation analysis, *Protein Engineering, Design and Selection.* **18**(5), 247–253, (2005).
37. K. Wollenberg and W. Atchley, Separation of phylogenetic and functional associations in biological sequences by using the parametric bootstrap, *Proceedings of the National Academy of Sciences of the United States of America.* **97**(7), 3288–3291, (2000).
38. D. Jones, W. Taylor, and J. Thornton, The rapid generation of mutation data matrices from protein sequences, *Computer Applications in the Biosciences.* **8**(3), 275–282, (1992).
39. L. Martin, G. Gloor, S. Dunn, and L. Wahl, Using information theory to search for co-evolving residues in proteins, *Bioinformatics.* **21**(22), 4116–4124, (2005).

Chapter 13

Statistical Modeling for Data of Positron Emission Tomography in Depression

Chung Chang

Department of Mathematical Sciences, New Jersey Institute of Technology
University Heights, Newark, New Jersey 07102-1982, USA
Currently with Department of Applied Mathematics,
National Sun Yat-sen University, Taiwan
binchung10000@gmail.com

R. Todd Ogden

Department of Biostatistics, Columbia University
722 W. 168th St., 6th floor, New York, NY 10032, USA
to166@columbia.edu

One of the main applications of Positron Emission Tomography (PET) is to estimate the density of a neuroreceptor at each location throughout the brain by measuring the concentration of a radiotracer over time and modeling its kinetics. Several statistical approaches to modeling the kinetics have been proposed in the literature. In this article, we will first briefly review the PET imaging acquisition process and then focus on reviewing several kinetic models for PET data, including compartment models, graphical models , and Basis Pursuit. We will also briefly discuss how to model the plasma function (input function) and how to use metabolites to correct the input function.

13.1. Introduction to PET Imaging and Its Application to Depression

Positron emission tomography, also called PET imaging or a PET scan, is a diagnostic examination that involves the acquisition of physiologic images based on the detection of positrons, which are tiny particles emitted from a radioactive substance administered to the patient. PET can be used to obtain neuroreceptor binding maps in the brain. To study the distribution of a target neuroreceptor, we label one of its high affinity ligands with a radioactive isotope, which serves as a tracer, and inject it into a subject's bloodstream. Then through the PET imaging modality, we measure the radioligand flow throughout the brain and therefore obtain information on how the ligand binds to receptors throughout the brain.

Statistical models have been long used to analyze PET data. Although much statistical work has been done on PET data, in this paper, we will focus mainly on what happens after the PET images have been acquired. Also, PET has been used in many areas of biomedical research, such as oncology, pharmacology, and psychiatry. In this article, we will focus on the application of PET to depression, which belongs to psychiatric application. In this paper, we will review some of the commonly used statistical models and methodologies in the literature of PET imaging. In next paragraph we will briefly introduce its application to depression. Then in Sec. 13.2, we will briefly introduce PET data and the procedure of acquisition of PET imaging. In Sec.13.3, we will review some models for plasma data and metabolites as well as for PET data.

One of the important applications of the PET scan is for the Major-depressive disorder (MDD). MDD is a mental illness characterized by a pervasive low mood, loss of interest in usual activities. A well known neuroreceptor related to MDD is the serotonin (5-hydroxytryptamine, $5 - HT_{1A}$)) neuroreceptor. To study the distribution of this receptor, several radioactivity compounds, such as 11C-DWAY, 11C-NAD-299, 18F-FCWAY, and 18F-p-MPPT, can be selected as high-affinity ligands. Using PET, we can obtain the distribution of $5 - HT_{1A}$ neuroreceptor throughout the brain.

13.2. PET Data and Acquisition of PET Imaging

13.2.1. *PET data*

PET data are collected in four dimensions, three spatial and one temporal. At a given time point, PET data consist of a digital 3-dimensional image. The image is partitioned into a grid of rectangular volume elements, called voxels, and the intensity associated to each voxel represents the concentration of the tracer at that location. At a given voxel, PET data consist of a concentration-time curve.

One common approach to modeling PET data is to fit a model to the concentration-time curve voxel by voxel. Although such approach offers high resolution information, the data contain a lot of noise. There are also potential computational difficulties in fitting hundred of thousands of voxels simultaneously.

An alternative approach to model PET data is to fit the concentration over time in each of several regions of interest (ROI). ROI-based approach is to predefine an anatomical region, which can contain thousands of voxels, and then to average the concentration of all the voxels in this region for each image. The advantage of ROI-based modeling is that we have lower noise and that we can compute efficiently. Moreover, in a voxel-based modeling we have to register first when we need to compare the data across subjects. However, the disadvantage of using ROI-based modeling is that it has low resolution and it relies upon a questionable assumption that the regions are homogeneous.

13.2.2. *Acquisition of PET imaging*

The visual PET image is actually a reconstruction of raw images (in "sinogram space") and the results can depend on the choice of reconstruction algorithm. For 2D reconstruction, the most commonly used algorithm is the analytical method called filtered back-projection (FBP). FBP is straight forward to implement but does have the property of amplifying noise in the signal. Recently, considerable interest has been shown in iterative reconstruction schema, such as the Ordered Subsets - Expectation Maximization (OSEM) algorithm (Hudson and Larkin, 1994), which possess different noise properties as compared to FBP. For 3D reconstruction, the Reprojection and filtered Back-projection (3D-RP) method of Kinahan and Roger (1990) has been the most popular, in part because of the significant computational burden of newer 3D iterative reconstruction methods. 3D-RP itself is computationally expensive, and this has led to the development of approximate 3D reconstruction algorithms. Of these, Fourier rebinning (Defrise *et al.*, 1997) which reduces the 3D problem to a series of 2D problems without significantly distorting the image and results in a significant reduction in the computational burden, has stimulated particular interest.

13.3. Modeling of Voxel-specific Time Series

13.3.1. *Compartment models*

Researchers use different models to analyze PET data. One group of commonly used models is so called compartment models. A compartment is a theoretical construct introduced for modeling purposes. We divide them in order to describe the motion of the tracer between different compartments.

In compartment models, the tracer concentration can be expressed as a sum of exponential functions convolved with the plasma input function (Gunn *et al.*, 2001). The general formula for $C_T(t)$, the total concentration in the brain at time t, is

$$C_T(t) = \sum_{i=1}^{N} \phi_i e^{-\theta_i t} \otimes C_P(t), \tag{13.1}$$

where $C_P(t)$ is the concentration in plasma at time t, N is the number of tissue compartments in the model, and $\otimes$ represents the convolution operator. The parameters $\{\phi_1, \ldots, \phi_M; \theta_1, \ldots, \theta_M\}$ are functions of the rate parameters describing movements between compartments. Later in this section we will show how to obtain the relationship between $\{\phi_1, \ldots, \phi_M; \theta_1, \ldots, \theta_M\}$ and the rate parameters.

In this paper, we only consider the most typical compartment model, the two tissue models (N=2), illustrated in Fig. 13.1. General compartment models are introduced in Gunn *et al.* (2001).

In the two tissue compartment model (see Fig. 13.1), first the tracer is injected into blood (plasma compartment) of the subject. Each tracer molecule floats around in the plasma, eventually crossing the blood brain barrier (BBB), and then it moves around ("free") trying to bind itself to one of the many receptors in the neurons. When it binds with a receptor with which it has a high affinity ("specifically Bound"), it will remain there for a while before breaking off and floating around again. When it binds with a receptor for which it has low affinity ("non-specifically bound"), it quickly detached itself. It might also re-cross the BBB and enter the plasma. After the tracer is injected into subjects blood, it goes from the blood (plasma compartment C_p) to the second compartment, which consists of both unmetabolized free ligand and non-specifically bound compartment (C_{F+NS}). Then the tracer may enter to the third, specifically bound compartment (C_{SP}) or it may cross the BBB the opposite direction and enter the plasma compartment. We assume that the tracers can change states in both directions between two neighboring compartments, so we have four kinetic constants, K_1, k_2, k_3, and k_4. K_1 is the influx constant for tracer from plasma into the free and non-specifically bound compartment. k_2, k_3, and k_4 are all fractional rate constants. A fractional rate constant is the fraction of the concentration in one compartment moving to another compartment per unit of time (min^{-1}). In this example, the second and third compartments comprise the tissue we are interested in. In general, K_1 is the influx constant for the tracer to enter the tissue. Note that more

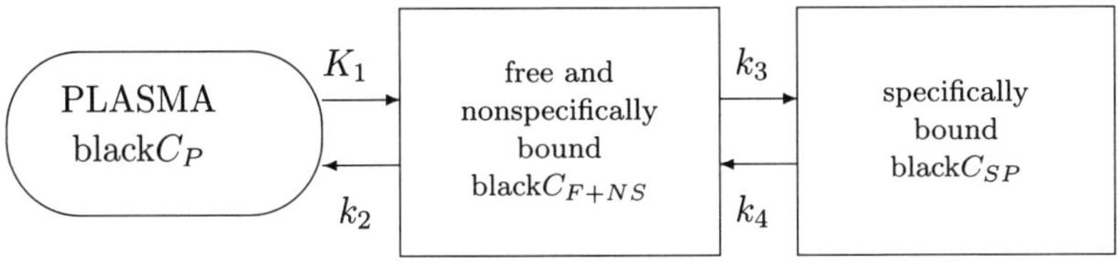

Fig. 13.1. Two tissue compartment model.

detailed description of this model can be found in Slifstein and Laruelle (2001).

The parameters for the two tissue model are often not identifiable; therefore, adding some constraints is a typical solution in practice. The two tissue model considered in this paper has only one constraint: that non-specific bindings are the same throughout the brain.

This constraint requires that the ratio $\frac{K_1}{k_2}$ to be uniform for every region. This is typically accomplished in practice by identifying a "reference" region that is thought to be devoid of receptors. As a result, for the reference region a one tissue model is appropriate, and its rate constants K_1 and k_2 can be estimated. Using the estimate of the ratio of K_1 and k_2 from the reference region, the constraint can be accomplished and applied to all the other regions.

Define C_T as the total concentration of tissue (i.e. $C_T = C_{FS}+C_{SP}$). Note that in PET study we measure the total concentration of the ligand at each point in the brain, but can not distinguish among the various states (F, NS, SP) that the ligand may be in. Assuming first-order kinetics, the model can be described in terms of the following differential equations.

$$\frac{dC_{F+NS}(t)}{dt} = -(k_2 + k_3)C_{F+NS}(t) + k_4C_{SP} + K_1C_P(t)$$
$$\frac{dC_{SP}(t)}{dt} = k_3C_{F+NS}(t) - k_4C_{SP}$$

with initial values

$$C_{F+NS}(0) = C_{SP}(0) = 0.$$

The first equation means that the changing rate of the concentration of the non-specifically bound and free compartment increases in proportion to the concentration of specifically bound compartment and plasma compartment and decreases in proportion to the concentration of itself.

To solve the differential equations above, first we define

$$K = \begin{pmatrix} -(k_2 + k_3) & k_4 \\ k_3 & -k_4 \end{pmatrix}.$$

The solution to the system of differential equations is given by

$$C_T(t) = \sum_{i=1}^{2} A_i e^{-\alpha_i t} \otimes C_P(t), \tag{13.2}$$

where

$$\begin{aligned} \alpha_{1,2} &= \{K_1 + k_2 + k_3 \pm \sqrt{(K_1 + k_2 + k_3)^2 - 4K_1k_4}\}/2, \\ A_1 &= K_1 \frac{k_3 + k_4 - \alpha_1}{\alpha_2 - \alpha_1}. \\ A_2 &= -K_1 \frac{k_3 + k_4 - \alpha_2}{\alpha_2 - \alpha_1}, \end{aligned}$$

and $-\alpha_1$ and $-\alpha_2$ are the eigenvalues of the matrix K.

From Eq. (13.1), we know that the solution is a sum of convolution of exponential functions and the input function $C_P(t)$. The data for an individual voxel can be denoted by (y_i, t_i) $i = 1, ..., F$ where y_i is the observed total tissue concentration at time point t_i of the neuroreceptor in study. The y_i values are typically assumed to be independent and $var(y_i) = \sigma^2 w_i$ (w_i may be known but σ^2 is generally unknown). If w_i are known, then we can use weighted least squares to estimate the parameters $K1$, k_1, k_3, and k_4.

$$(\hat{K}_1, \hat{k}_2, \hat{k}_3, \hat{k}_4) = \underset{K_1, k_2, k_3, k_4}{argmin} \sum_{i=1}^{F} \frac{1}{w_i}(y_i - C_T(t_i))^2 \tag{13.3}$$

Note that the important parameters expressing density of receptors like V_T, BP can be expressed as functions of K_1, k_2, k_3, and k_4. BP, binding potential, is the product of the receptor density with the affinity. V_T is the total volume of distribution (free and non-specific compartment together with specific compartment) at equilibrium, the state in which concentrations in compartments remain constant. The tissue distribution volume is the ratio of the concentration in one tissue compartment to the free tracer concentration in the plasma compartment at equilibrium.

The model shown above is a typical example of the compartment model, and called two tissue compartment model because the brain tissue is divided by two compartments (C_{F+NS} and C_{SP}). Gunn *et al.* (2001) generalized this simple model to general compartment models which allow any number of compartments.

13.3.2. *Modeling plasma model and correcting the plasma data for metabolites*

We know from equation (13.2) and (13.3) that in order to estimate the kinetic parameters in Sec. 13.1 we need the concentration of plasma ($C_P(t)$) to be estimated. We first get the blood/plasma samples and use it to fit a parametric curve (generally, the sum of three exponentials). With plasma analysis we can only measure total counts, but we are only interested in the compound that actually binds to the receptors of interest in the brain. In practice, we also need to take into account of the metabolism because the metabolites don not cross the BBB and so we have to take out the metabolism portion in estimating C_P and because the rates of metabolism of some ligands are very high in PET study (Wu *et al.*, 2006). We use the metabolite corrected arterial function to replace $C_P(t)$ to be input function in Eqs. (13.2) and (13.3). The corrected input function is calculated by multiplication of $C_P(t)$ and the function of fraction of unmetabolite. Some metabolism model is to assume that the unmetabolized fraction is a parametric curve. The parameters can be estimated by using a nonlinear regression fit for data of unmetabolized fraction. The data are (y_i, t_i) $i = 1, ..., n$ where y_i is the fraction of unmetabolite at time t_i. The model is $y_i = f(t_i) + \epsilon_i$ $i = 1, .., n$ where each ϵ_i is independent with each other and mean zero. We can also fit nonlinear mixed effects models (across subjects) for metabolite data.

13.3.3. *Basis pursuit*

In Secs. 13.1 and 13.2, we have discussed several compartment models, but how to choose a proper compartment model is still a question. In this section, we review another approach called "Basis Pursuit" (Gunn *et al.*, 2002), which is based on a general compartmental description and determines a parsimonious model consistent with the measured data.

From Eq. 13.1, we know that the total concentration C_T can be expressed as a sum of exponential functions convolved with the input function (plasma function $C_P(t)$). In general, $C_T(t)$ can be expressed as a linear combination of some basis function. Explicitly,

$$C_T(t) = \sum_{i=1}^{M} \phi_i \psi_i(t), \tag{13.4}$$

$$\psi_i(t) = \int_1^t e^{-\theta_i(t-\tau)} C_P(\tau) d\tau.$$

M is a prechosen number of basis function. The next question is to determine what the M (model complexity) and then to estimate the parameters ϕ_i. The observed data are (y_i^*, t_i) $i = 1, ..., F$ and $y_i^* = y_i + \epsilon_i$. We assume that ϵ_i are independent and $var(\epsilon_i) = \sigma^2 w_i$. If w_i is known then we can use weighted least square method to estimate parameters. However, if the number of basis function is large than or equal to the number of data point $(M \geqslant F - 1)$ then we can not estimate the parameters.

This problem can be solved by considering penalty framework. The motivation is that we want the data to be accurately described by a few compartments, such as example in Sec. 13.1. The framework will introduce a regularization parameter which controls the tradeoff between approximation error and sparseness (few compartments). The penalty function to the standard least squares metric offers the framework for this,

$$\min_{\phi} \frac{1}{2} \|\boldsymbol{W}^{1/2}(y - \boldsymbol{\Psi\phi})\|_2^2 + \mu \|\boldsymbol{\phi}\| p,$$

where μ is the penalty parameter, $\|.\|_p$ is the L_p norm, $\boldsymbol{W}$ is the inverse of covariance matrix of the ϵ_i $i = 1, ..., F$, and $\boldsymbol{\phi}$ are the basis coefficients to be determined, $\boldsymbol{y} = (y_1, \ldots, y_F)$, and $\boldsymbol{\Psi}$ is the basis matrix. Both L_1 (Basis pursuit denoising) and L_0 (atomic decomposition) promote a sparse solution which is consistent with our expectation of a compartmental model consisting of just a few tissue compartments. However, atomic decomposition is computationally demanding. Basis pursuit denoising requires the determination of the regularization parameter, μ, for penalty term. We can obtain this parameter by using cross-validation method (Shao, 1993).

13.3.4. *Graphical model*

In compartment models, we have to estimate parameters using an iteration algorithm of nonlinear regression method. Another alternative way to estimate parameters is described by (Logan *et al.* 1990), who demonstrated that there is a linear relationship between the true value of $\int_0^t C_T(t')dt'/C_T(t)$ and true value of $\int_0^t C_P(t')dt'/C_T(t)$ (where C_T and C_P are concentration in ROI and concentration in plasma, respectively) after some cut off time point. The slope of this linear relationship is V_T. The name graphical model is because the researchers plot the data of $\int_0^t C_T(t')dt'/C_T(t)$ and $\int_0^t C_P(t')dt'/C_T(t)$ and choose the cut off point by eyes.

The advantage of doing this is that this approach is model-free which means we don not need to choose a specific compartment model. Another advantage is that it only uses OLS (ordinary least square) routines to estimate the parameters. Compared to the iterative nonlinear regression method in compartment models, it is computationally convenient. Although this method is widely used, the disadvantage is that the estimator of the slope is negatively biased because both $\int_0^t C_T(t')dt'/C_T(t)$ and $\int_0^t C_P(t')dt'/C_T(t)$ are measured with error and the bias is not easily estimated. One alternative approach is likelihood estimation in graphical analysis (LEGA) described in Ogden (2003). The approach is based on the assumption of independent Gaussian error in each observation of concentration in ROI. Then by maximum likelihood method, we can get the estimator of the slope V_T. Simulation results show that this method indeed reduces the bias.

In this paper, we introduced PET data and the procedure of the acquisition of PET image, as well as plasma function and metabolites, and reviewed several commonly used models for PET data. We only focused on modelings voxel by voxel or region by region. For models incorporating with spatial information, see O'sullivan (2006) and Jiang and Ogden (2008).

Acknowledgments

The authors would like to thank the referee for the very helpful suggestions which lead to great improvement of the manuscript.

References

1. H. M. Hudson and R. S. Larkin, Accelerated Image Reconstruction Using Ordered Subsets of Projection Data. *IEEE Trans. Med. Imag.* **13**:601-609, (1994).
2. P. E. Kinahan and J. G. Rogers, Analytic 3D Image Reconstruction Using All Detected Events. *IEEE Trans. Nucl. Sci.* **36**:964-968, (1990).
3. M. Defrise, P.E. Kinahan, D. W. Townsend, C. Michel, M. Sibomana, and D. F. Newport, Exact and Approximate Rebinning Algorithms for 3d-PET Data, *IEEE Trans. Med. Imag.* **16**:145-158, (1997).
4. R. N. Gunn, S. R. Gunn, and V. J. Cunningham, Positron Emission Tomography Compartmental Models, *Journal of Cerebral Blood Flow and Metabolism*, **21**:635-652, (2001).

5. M. Slifstein and M. Laruelle, Models and Methods for Derivation of in vivo Neuroreceptor Parameters with PET and Spect Reversible Radiotracers, *Nuclear Medicine and Biology*, **28**:595-608, (2001).
6. S. Wu, R. T. Ogden, J. J. Mann, and R. V. Parsey, Optimal Metabolite Curve Fitting for Kinetic Modeling of ^{1}1C-WAY-100635, *Journal of Nuclear Medicine*, **48**:926-931, (2006).
7. J. Shao, Linear Model Selection by Cross-Validation, *J. Am. Stat. Assoc.* **13**:763-781, (1993).
8. R.N. Gunn, S. R. Gunn, F. E. Turkheimer, J. A. D. Aston, and V. J. Cunningham, Positron Emission Tomography Compartmental Models: A Basis Pursuit Strategy for Kinetic Modeling, *Journal of Cerebral Blood Flow and Metabolism*, **22**:1425-1439, (2002).
9. J. Logan, J. S. Fowler, N. D. Volkow, A. P. Wolf, S. L. Dewey, D. J. Schlyer, R. MacGregor, R. Hitzmann, B. Bendrium, S. J. Gatley, and D. R. Christman, Graphical Analysis of Reversible Radioligand Binding from Time-activity Measurements Applied to [N-^{11}C-methyl]-(-)-cocaine PET Studies in Human Studies, *Journal of Cerebral Blood Flow Metabolism*, **10**:740-747, (1990).
10. R. T. Ogden, Estimation of Kinetic Parameters in Graphical Analysis of PET Imaging Data, *Statistics in Medicine*, **22**:3557-3568, (2003).
11. F. O'Sullivan, Locally Constrained Mixture Representation of Dynamic Imaging Data from PET and MR Studies, *Biostatistics*, **7**:318-338, (2006).
12. H. Jiang and R. T. Ogden, Mixture Modeling for Dynamic PET Data, *Stat. Sin.* **18**:1341-1356, (2008).

Chapter 14

The Use of Latent Class Analysis in Medical Diagnosis

David Rindskopf

Educational Psychology
CUNY Graduate Center
New York, NY 10016, USA
drindskopf@gc.cuny.edu

Medical signs, symptoms and tests are usually imperfect, as is diagnosis (which depends on these signs, symptoms, and tests.) Latent class analysis is a statistical method that allows evaluation of these indicators, as well as diagnosis, without a gold standard. This chapter discusses latent class analysis and its extensions, and applies them in medical contexts.

The evaluation of medical signs, symptoms, and tests for the purposes of diagnosis is usually framed within the context of estimating the sensitivity and specificity of the indicator. Sensitivity is the probability that a person with the disease will be positive on the indicator; specificity is the probability that a person without the disease will be negative on the indicator. The estimation of sensitivity and specificity depends on knowing who does and does not have the disease; that is, there must be a "gold standard" for diagnosis. Rindskopf and Rindskopf (1986) applied latent class analysis to this problem, and showed that sensitivity and specificity could be estimated, under some conditions, even without a gold standard. This paper briefly reviews those findings, and then discusses extensions of latent class methods for diagnosis. These methods have attracted much attention in the medical statistics community in the past 20 years; a search of Pub Med (October, 2009) resulted in almost 1600 references for the key word term "latent class"; a selection of these is included in the Reference section.

The next section of the paper gives a brief overview of latent class analysis, followed by a simple example from the literature, and an extension to models with conditional dependence. Next, I discuss a model with a predictor of latent class, using as an example data on children's wheeze (as an indicator of asthma) measured at four ages, with asthma predicted by whether or not the mother smokes. Then I discuss a new conceptual model that adds floor and ceiling effects to logistic regression. This model is similar to models with error of measurement, but with a different interpretation. Finally, I present some implications of these models and their utility in medical diagnosis.

14.1. Latent Class Analysis: A Brief Overview

Latent class analysis hypothesizes the existence of one or more unobserved (underlying, latent) categorical variables to explain the relationships among a set of observed categorical variables. In the medical diagnosis context, the observed variables are signs, symptoms, or test results, usually dichotomized into a binary classification (positive and negative). The latent variable is true status on the disease; while this is often dichotomous (disease present or absent), it may not be (e.g., heart attack, congestive heart failure, or no heart problem). Sometimes there is more than one latent variable; each might correspond, e.g., to the presence or absence of a particular disease. In this paper, the examples are all cases in which there is only one latent variable.

The observed data are usually presented in terms of a crosstabulation of the observed variables. The statistical model has two kinds of parameters. Unconditional probabilities are the probabilities of being in each latent class (i.e., each level of the latent variable, if there is only one such variable; or each combination of levels, if there is more than one). Conditional probabilities are the probabilities of having a particular result on a test, given membership in a specific latent class. There is a set of conditional probabilities for each observed variable; these are assumed to be independent, conditional on latent class, so that the latent variable explains the relationships among the observed variables. The sensitivities and specificities of the observed measures are conditional probabilities, and thus are parameters in the model.

Models can be tested using any of a number of fit statistics appropriate for categorical data, including various statistics with chi-square distributions, and adjusted fit statistics such as Akaike's or Schwartz's Information Criteria. Care must be taken to be certain that a model is identified; that is, that all parameters are estimable. This can be accomplished through algebraic proof, or (more commonly) using numerical techniques. Sometimes more than one model is found that fits the data. Nested models can be compared to see if the extra parameters are needed, although there is some disagreement about when this is appropriate.

Once models are found that fit the data, the parameters can be interpreted. Another goal is to use Bayes' Theorem to examine how well individuals can be assigned to latent classes. In medical situations, this is the process of diagnosis. Some response patterns may be easier to classify than others. One can also determine whether it is possible to simplify the classification rules, e.g. by counting number of positive results or symptoms.

14.2. A Simple Example: Myocardial Infarction

The following example is summarized from Rindskopf and Rindskopf (1986). It illustrates the main points about the use of latent class analysis in medical diagnosis.

Data come from a study of patients admitted to an emergency room suffering from chest pain (Galen and Gambino, 1975). Each of four indicators was scored as either indicating a myocardial infarction (MI; commonly known as heart attack) or not indicating MI. The indicators included history, EKG (inverted Q-wave), and two blood tests (CPK and LDH). The data set consists of counts of the number of patients with each of the 16 possible patterns of indicators.

The data were consistent with a simple 2-class model (LR=4.29, df=6, p=.64), where the classes represented those with and without MI. The data were inconsistent with several other possible models, including the model of complete independence of indicators, and a quasi-independence model.

Table 14.1 contains the parameter estimates for the model. The unconditional probabilities of being classes 1 and 2 are about .46 and .54 respectively. To determine what these classes mean, one must examine the conditional probabilities. For each indicator, in class 1 there is a relatively high probability of the indicator being positive; for class 2, there is a relatively low probability of the indicator being positive. Therefore, class 1 represents those with MI, and class 2 represents those without MI. The indicators vary in their sensitivities and specificities; for example CPK has a high sensitivity, but low specificity, while inverted Q wave on an EKG has a high specificity but lower sensitivity. History is modest on both measures, while LDH is similar to EKG results. Sometimes a graphical presentation of results is easier to examine; Fig. 14.1 shows a plot of sensitivities by specificities, similar to ROC plots in signal detection theory. The ideal indicator would be at the point (1,1) in the upper right-hand corner. The

Table 14.1. Parameter Estimates for MI Data, 2-class unrestricted model.

		mi 0.4578	no mi 0.5422
cpk	1	1.0000	0.1956
cpk	0	0.0000	0.8044
ldh	1	0.8279	0.0269
ldh	0	0.1721	0.9731
his	1	0.7914	0.1951
his	0	0.2086	0.8049
qwa	1	0.7669	0.0000
qwa	0	0.2331	1.0000

closer an indicator is to that ideal point, the better the indicator is. The actual statistical process of diagnosis is the assignment to latent classes. The usual procedure is to use all observed variables, as shown in Table 14.2. The largest probability of error is for patients who were positive on CPK and history, but negative on LDH and Q-wave.

In medical diagnosis, unlike the typical situation in latent class analysis, one might be interested in class assignment on the basis of only a subset

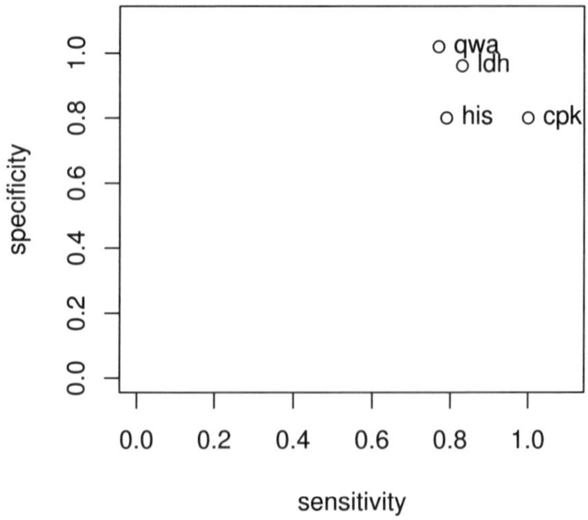

Fig. 14.1. Plot of sensitivity and specificity of indicators in MI data, similar to ROC plot.

Table 14.2. Assignment to latent class for MI data, with error probabilities for each pattern of indicators.

C	L	H	Q	F	1	2	Diag	Err
1	1	1	1	24	1.00	0.00	1	0.00
1	1	1	0	5	0.99	0.01	1	0.01
1	1	0	1	4	1.00	0.00	1	0.00
1	1	0	0	3	0.89	0.11	1	0.11
1	0	1	1	3	1.00	0.00	1	0.00
1	0	1	0	5	0.42	0.58	2	0.42
1	0	0	1	2	1.00	0.00	1	0.00
1	0	0	0	7	0.04	0.96	2	0.04
0	1	0	0	1	0.00	1.00	2	0.00
0	0	1	0	7	0.00	1.00	2	0.00
0	0	0	0	33	0.00	1.00	2	0.00

Note: C= CPK, L=LDH, H=History, Q=Q wave. Response patterns with zero observed frequency are omitted.

of the observed variables. For example, one might want to use only history and EKG results, which are available more quickly than the LDH and CPK, which are based on blood tests. Technically, of course, this is no problem; one merely applies Bayes' Theorem.

In order to see this, we will use the natural one-character names H (history) and Q (inverted Q-wave) for the observed variables we will use in this example, and X for the latent class variable. Then the model allows computation of the joint distribution of the three variables as $f(HQX) = f(X)f(H|X)f(Q|X)$ because of conditional independence of the observed variables given latent class. Then the marginal distribution f(HQ) of H and Q is obtained by summing f(HQX) over the over X. The conditional distribution of X given the observed variables H and Q is $f(X|HQ) = f(HQX)/f(HQ)$.

The results of applying these calculations are shown in Table 14.3. If Q wave is positive, history is irrelevant; the patient had an MI. If both are negative, the patient probably did not have an MI. If Q wave is negative but history is positive, there is quite a bit of uncertainty.

14.3. Conditional Dependence Models

In certain situations, two or more indicators might be more strongly related than is allowed by the conditional independence model. For example, out of a series of tests, two might be based on similar chemical principles, and

Table 14.3. Using a subset of indicators to make a diagnosis.

Q-wave	History	$P(MI\|Q, H)$	Prop.
0	0	.04853	.47
0	1	.44394	.18
1	0	1.00000	.06
1	1	1.00000	.29

thus give more similar results than would otherwise be the case. The usual latent class model can be extended to allow for these relationships.

Qu, Tan and Kutner (1996) proposed a more elaborate model to account for conditional dependence, but it is not clear that this is necessary in most cases. Their first example was a dataset with four tests of HIV status, and two of the four were expected to be related due to similar procedures and principles involved. The usual two-class latent class model did not fit the data, but a simple model with the two tests allowed to correlate did fit the data well, so the more elaborate random-effects model was not needed. A model with conditional dependence between a pair of items can be fit using common programs for latent class analysis. Suppose we call the observed items A, B, C, and D, and the latent variable X. Using notation borrowed from hierarchical loglinear models, the usual latent class model would have the relationships [AX] [BX] [CX] [DX], showing that each observed variable was related directly only to the latent variable X. To specify conditional dependence between A and B, one could either add [AB] alone or [AB] and [ABX], which would use 1 or 2 (respectively) additional degrees of freedom.

In a second example, with 7 pathologists diagnosing 118 slides for cancer, Qu *et al.* claimed that their model was needed, but in fact reported incorrect results (they inadvertently reported the same fit statistics for this data set as for their first example). For the cancer data, usual fit statistics are not easy to interpret because of the sparsity of the data: 118 cases are spread over 128 cells of a contingency table. But the AIC and BIC indicate that the usual two-class model fits reasonably well, as both are well below zero. Qu *et al.* proposed a plot to examine possible conditional dependence, and the plot indicates possible residual correlations between pathologists 2 and 5, and also between pathologists 4 and 6. A simple latent class model with these correlations included fits much better than the usual model. These models can be compared even with sparse data, just as nested models in logistic regression can be compared, by subtracting

Table 14.4. Fit of models to cancer data.

Model[a]	L^2	df	AIC	BIC
Usual 2-class model	62.37	112	-161	-472
+ (2,5) and (4,6) correlated	36.19	110	-183	-489
+ (2,6) and (4,7) correlated	18.67	108	-197	-497

[a]First model is usual 2-class model; second has additional terms allowing conditional dependence between two pairs of raters; third is like second but with additional terms allowing conditional dependence between two more pairs.

likelihood ratio statistics. Including two other correlations indicated in the plot in Qu *et al.* produced a further improvement in the fit. Table 14.4 contains the results of fitting these models.

14.4. Latent Class Analysis with a Categorical Predictor of Class: Wheeze as an Indicator of Asthma in Children

In some analyses one or more predictors of the latent class is available. Consider data on wheeze in children from the Six Cities study (see, e.g., Cunningham *et al.*, 1994), for which data from one city are widely used as an example data set. Children were assessed for presence or absence of wheeze each year from age 7 through 10 years. If no other information were available, a latent class analysis similar to that for the MI data would be suitable. In this case, one such additional variable was whether or not the mother smoked. Theoretically, passive smoke might affect asthma in children.

Using X to represent the latent class variable and M to represent whether the mother smoked, one can write the part of the model that involves these two variables as $[M][X|M]$ to show that the distribution of X depends on M. This is one way to factor the joint distribution of X and M; it is equivalent to factoring as $[X][M|X]$, which would fit the same but would provide estimates of the proportion in each latent class, and the conditional probability of the mother smoking given latent class. The latter results are displayed below for this model.

The results of fitting a latent class model with mother's smoking as a predictor of class are contained in Table 14.5. First, note that at each age at which wheeze was measured, the probability of wheeze was somewhat high in class 1, and very low in class 2; this makes class 1 the class with probable asthma. This class contains about 16 percent of the children in the study;

Table 14.5. Parameter estimates for wheeze data.

	class:	1	2
	p(class):	0.1607	0.8393
smo	yes	0.4487	0.3290
age7	yes	0.5924	0.0796
age8	yes	0.7222	0.0636
age9	yes	0.7019	0.0542
age10	yes	0.5119	0.0418

Note: LR = 13.6793, df=20, p=0.8464.
Data are available online at `http://www.statsci.org/data/general/wheeze.html`

the class without asthma contains about 84 percent of the children. While specificity of wheeze was good (relatively low) at each age, sensitivity was not as high as might be desired, being generally above .50 but less than .75.

The probability of the mother smoking was about 45 percent among children with asthma, and 33 percent among children without asthma. To test whether this effect was significant, the model was run again, constraining the conditional probabilities of smoking to be equal in both classes. This model fit the data very well.

For the model with equal rates of smoking in the two classes, Bayes' Theorem was used to assign children to class based on the four observed measures of wheeze. The results are shown in Table 14.6. In general, as would be expected, the most difficult response patterns to assign were those where wheeze occurred on two occasions and did not occur on the other two. Even in these cases, the probability was relatively high (about .6 to .8) that the child actually had asthma. In fact, because of the approximately equal sensitivities and specificities across ages, it is possible to deduce a simple rule for assignment to classes: If wheeze occurred not at all or only once, the child probably does not have asthma; if wheeze occurred two or more times, the child probably does have asthma. Such a simple rule will not always suffice, even for a two-class model.

Other contributors to the literature who have proposed similar models include Clogg (1981); Dayton and Macready (1988, 2002); Hagenaars (1990); Vermunt (1997); and Vermunt and Magidson (2002).

Formally, this model is equivalent to a two-group latent class model, with equality of all conditional probabilities across the two groups (mothers who do and do not smoke). There are several advantages of the current formulation over the two-group form of the model. First, it is easier to extend this approach to the case with multiple predictors. Second, this

Table 14.6. Assignment to latent classes, wheeze data, for model with no effect of mom's smoking; sorted by modal class.

A7	A8	A9	A10	P(1)	P(2)	Class	P(err)
1	1	1	1	0.9961	0.0039	1	0.0039
1	1	1	2	0.9131	0.0869	1	0.0869
1	1	2	1	0.8597	0.1403	1	0.1403
1	2	1	1	0.8693	0.1307	1	0.1307
2	1	1	1	0.9373	0.0627	1	0.0627
1	1	2	2	0.2030	0.7970	2	0.2030
1	2	1	2	0.2165	0.7835	2	0.2165
1	2	2	1	0.1388	0.8612	2	0.1388
1	2	2	2	0.0067	0.9933	2	0.0067
2	1	1	2	0.3832	0.6168	2	0.3832
2	1	2	1	0.2660	0.7340	2	0.2660
2	1	2	2	0.0148	0.9852	2	0.0148
2	2	1	1	0.2822	0.7178	2	0.2822
2	2	1	2	0.0161	0.9839	2	0.0161
2	2	2	1	0.0094	0.9906	2	0.0094
2	2	2	2	0.0004	0.9996	2	0.0004

Note: If positive 0 or 1 time, class = 1; If positive 2 or more times, class = 2.

method gives a direct estimate of the relationship between the predictor and the latent variable; this is even more important with multiple predictors. Third, it is easy to extend this model to the case where there are continuous or quasi-continuous predictors (e.g. several levels of a quantitative variable). The following example illustrates a variation on this model.

14.5. Logistic Regression with Floor and Ceiling Effects

Logistic regression models assume that for a low enough value of the predictor(s), the probability of a response is zero, and that for a high enough value of the predictor(s), the probability of a response is one. In some applications, one or both of these assumptions is likely to be false. For example, in predicting graduation from college using SAT scores, there will undoubtedly be a proportion of students who graduate in spite of low SAT scores, and another group who will not graduate in spite of high SAT scores.

In this section we consider models in which the asymptotes of the logistic regression equation are not constrained to be zero and one as they are in traditional models. The idea is presented graphically in Fig. 14.2. The predicted probabilities for the model can be expressed as

$$f + (c - f)\left\{\frac{\exp(b_0 + b_1X)}{1 + \exp(b_0 + b_1X)}\right\}$$

where f is the floor, c is the ceiling, and the expression in braces is the usual logistic regression model. Because the expression in braces is a proportion, and will therefore always be between 0 and 1, the observed probabilities will always be predicted to be between f and c.

This idea has some precedent in related areas. In item response theory (IRT) models, guessing parameters allow items to have a probability greater than zero of being responded to correctly, primarily to adapt to guessing on true-false or multiple choice tests. Finney proposed a probit model with a floor effect in toxicity studies; he called it a model for toxicity with natural mortality. He never seems to have had a ceiling effect for situations in which a certain proportion of animals were not affected by a poison or drug, and his floor effect never found its way into standard computer programs for logistic regression or probit models.

Technically, the model is identical to a model for errors in variables proposed by Ekholm and Palmgren (1982). The interpretation is different,

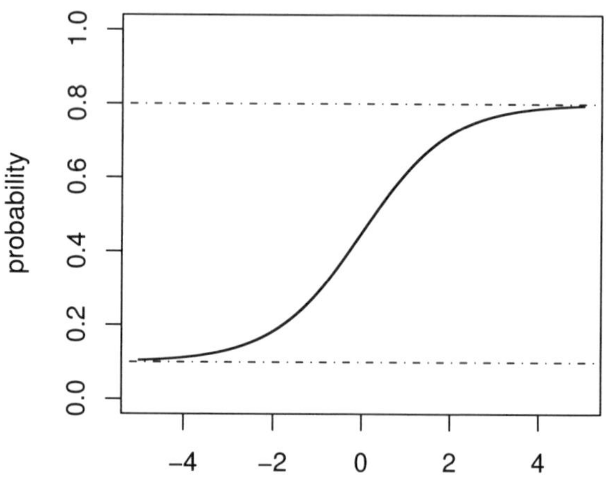

Fig. 14.2. Logistic regression with floor and ceiling effect.

Table 14.7. Coal miner data, wheeze, model with floor and ceiling effects.

	Latent Class 1	Latent Class 2
Marginal prob	0.7052	0.2948
Wheeze status		
yes	0.0000	0.6792
no	1.0000	0.3208
Age group		
1	0.9224	0.0776
2	0.8904	0.1096
3	0.8474	0.1526
4	0.7915	0.2085
5	0.7219	0.2781
6	0.6395	0.3605
7	0.5480	0.4520
8	0.4532	0.5468
9	0.3616	0.6384

even though the fit of the model is identical. In this case, Ekhlolm and Palmgren would interpret the floor and ceiling effect as solely due to errors of measurement.

The model with floor and ceiling effects (or, equivalently, the logistic regression model with errors of measurement) is also equivalent to a special kind of latent class model. This model has only one (binary) observed indicator, and one or more predictors of the latent variable. With a single continuous predictor having five or more levels, the model is identified, even though traditional latent class models would require four or more indicators. The interpretation of the latent class model would be similar to that of the Ekholm and Palmgren model, rather than the model with floor and ceiling effects.

Table 14.7 presents the results of an analysis of data on wheeze in coalminers, originally from Ashford and Sowden (1970). These data were analyzed by Ekholm and Palmgren using their errors-in-variables model. Here we should get the same results, but the interpretation is different. The floor effect could also be due to some miners contracting asthma (or wheeze alone) from other causes, and the ceiling effect could be miners whose lungs are particularly resistant to developing wheeze, and would not develop it no matter how long they were miners. Only replicate measurements at the same time period would allow an analysis that would separate these possibilities from errors of measurement.

In keeping with the different interpretation of this model, the entries in Table 14.7 are not all the same as in the usual latent class output. The marginal probabilities are the unconditional probabilities of being in each latent class. Wheeze status contains the conditional probability of having wheeze or not, given class membership. The numbers for age group, however, are the reverse of the usual interpretation; they are the conditional probability of latent class membership given age, and must be read in rows instead of columns. For example, miners in age group 1 have a probability of about .92 of being in latent class 1, which represents no wheeze.

The results are consistent with those of Ekholm and Palmgren. The floor is 0; that is, extrapolating to younger ages would give negligible probability of wheeze. The ceiling is about .68, so extrapolating to older miners would give a probability of .68 of wheeze; we would predict that about 32 percent of miners would never get wheeze.

14.6. Discussion

Latent class analysis has many advantages over traditional methods of (i) estimating sensitivity and specificity, and (ii) developing diagnostic rules on the basis of medical tests and indicators. The lack of need for a gold standard is the most important reason for preferring latent class models. In addition, one can determine whether certain simple rules for diagnosis (e.g. symptom counts) are reasonable.

Traditional latent class models can be extended in several ways that are useful in medical statistics. First, one can construct and test prediction models for the true (latent) status on the disease. Second, these extensions can allow the estimation of sensitivity and specificity with fewer indicators than a traditional latent class model. Third, with appropriate data, it is possible to separate floor and ceiling effects from measurement error. This requires repeat observations at the same time, and either several times of measurement, or a set of predictors.

Acknowledgments

I would like to thank the anonymous referee for comments that were most helpful in revising this chapter, Thethach Chuaprapaisilp for his assistance with LaTeX, and the editors for their gracious help (and infinite patience) with various editorial issues that arose in preparing this chapter for publication.

References

Albert, P. S., McShane, L. M. and Shih, J. H. (2001). Latent class modeling approaches for assessing diagnostic error without a gold standard: with applications to p53 immunohistochemical assays in bladder tumors, *Biometrics* **57**, 2, pp. 610–619.

Ashford, J. R. and Sowden, R. R. (1970). Multi-variate probit analysis, *Biometrics* **26**, 3, pp. 535–546.

Clogg, C. C. (1981). New developments in latent structure analysis, in D. J. Jackson and E. F. Borgatta (eds.), *Factor Analysis and Measurement in Sociological Research: A Multi-Dimensional Perspective* (Sage Publications, London), pp. 215–246.

Cunningham, J., Dockery, D. W. and Speizer, F. E. (1994). Maternal smoking during pregnancy as a predictor of lung function in children, *American Journal of Epidemiology* **139**, 12, pp. 1139–1152.

Dayton, C. M. and Macready, G. B. (1988). Concomitant-variable latent-class models, *Journal of the American Statistical Association* **83**, 401, pp. 173–178.

Dayton, C. M. and Macready, G. B. (2002). Use of categorical and continuous covariates in latent class analysis, in A. McCutcheon and J. Hagenaars (eds.), *Advances in latent class modeling* (Cambridge University Press).

Ekholm, A. and Palmgren, J. (1982). A model for a binary response with misclassifications, in R. Gilchrist (ed.), *GLIM 82: Proceedings of the International Conference on Generalised Linear Models* (Springer-Verlag, New York), pp. 128–143.

Espeland, M. A. and Handelman, S. L. (1989). Using latent class models to characterize and assess relative error in discrete measurements, *Biometrics* **45**, 2, pp. 587–599.

Finney, D. J. (1952). *Probit analysis*, 2nd edn. (Cambridge University Press, Cambridge).

Galen, R. S. and Gambino, S. R. (1975). *Beyond normality* (Wiley, New York).

Goetghebeur, E., Liinev, J., Boelaert, M. and Van der Stuyft, P. (2000). Diagnostic test analyses in search of their gold standard: latent class analyses with random effects, *Statistical Methods in Medical Research* **9**, 3, pp. 231–248.

Goldstein, J. M., Santangelo, S. L., Simpson, J. C. and Tsuang, M. T. (1990). The role of gender in identifying subtypes of schizophrenia—A latent class analytic approach, *Schizophrenia Bulletin* **16**, 2, pp. 263–275.

Hagenaars, J. A. (1990). *Categorical longitudinal data: Log-linear panel, trend, and cohort analysis* (Sage, Newbury Park, CA).

Hagenaars, J. A. and McCutcheon, A. L. (eds.) (2002). *Applied latent class analysis* (Cambridge University Press, Cambridge, UK).

Jørgensen, P. and Jensen, J. (1990). Latent class analysis of deluded patients, *Psychopathology* **23**, 1, pp. 46–51.

Qu, Y., Tan, M. and Kutner, M. H. (1996). Random effects models in latent class analysis for evaluating accuracy of diagnostic tests, *Biometrics* **52**, 3, pp. 797–810.

Rindskopf, D. and Rindskopf, W. (1986). The value of latent class analysis in medical diagnosis, *Statistics in Medicine* **5**, 1, pp. 21–27.

Sullivan, P. F., Kessler, R. C. and Kendler, K. S. (1998). Latent class analysis of lifetime depressive symptoms in the national comorbidity survey, *American Journal of Psychiatry* **155**, 10, pp. 1398–1406.

Uebersax, J. S. (1993). Statistical modeling of expert ratings on medical treatment appropriateness, *Journal of the American Statistical Association* **88**, pp. 421–427.

Uebersax, J. S. (1994). Latent class analysis of substance abuse patterns, in L. Collins and L. Seitz (eds.), *Advances in data analysis for prevention intervention research*, NIDA research monograph, No. 142 (National Institute on Drug Abuse, Rockville, MD), URL `http://www.drugabuse.gov/pdf/monographs/142.pdf`.

Uebersax, J. S. and Grove, W. M. (1990). Latent class analysis of diagnostic agreement, *Statistics in Medicine* **9**, 5, pp. 559–572.

Vermunt, J. K. (1997). *Log-linear models for event histories* (Sage Publications, Thousand Oakes, CA).

Vermunt, J. K. and Magidson, J. (2002). Latent class cluster analysis, in J. A. Hagenaars and A. L. McCutcheon (eds.), *Advances in latent class models* (Cambridge University Press, Cambridge, UK).

Walter, S. D. and Irwig, L. M. (1988). Estimation of test error rates, disease prevalence and relative risk from misclassified data: A review, *Journal of Clinical Epidemiology* **41**, 9, pp. 923–937.

Young, M. A. (1983). Evaluating diagnostic criteria: A latent class paradigm, *Journal of Psychiatric Research* **17**, pp. 285–296.

Young, M. A. and Tanner, M. A. (1983). Recent advances in the analysis of qualitative data with applications to diagnostic classification, in R. D. Gibbons and M. Dysken (eds.), *Statistical and Methodological Advances in Psychiatric Research* (Spectrum Publications, New York).

Chapter 15

Subset Selection in Comparative Selection Trials

Cheng-Shiun Leu*, Ying Kuen Cheung† and Bruce Levin‡

Department of Biostatistics
Columbia University, New York, USA
**cl94@columbia.edu*
†yc632@columbia.edu
‡bl6@columbia.edu

When several treatment regimens are possible candidates for a large phase III study, but too few resources are available to evaluate each relative to a standard, conducting a multi-arm randomized selection trial is a useful strategy to remove inferior treatments from further consideration. When the study has a relatively quick endpoint, frequent interim monitoring of the trial becomes ethically and practically appealing. In this paper, we propose a class of sequential procedures designed to select a subset of treatments that offer clinically meaningful improvements over the control group, or to declare that no such subset exists. The proposed procedures are easy to implement, allow sequential elimination of inferior treatments and sequential recruitment of promising treatments while preserving the type I error rate, and can be applied on a flexible monitoring schedule in terms of calendar time.

15.1. Introduction

We address the problem of selecting a subset of treatments in multi-arm randomized clinical trials with several experimental regimens and control arms of standard treatments and/or placebo. The primary objective of these trials is to identify a subset of experimental regimens superior to the control(s), if such a subset of treatments exists. We call such trials *comparative selection* trials. From an ethical and economic viewpoint, it is beneficial to execute interim decisions so that treatments that are unlikely to emerge as successful may be dropped before the end of the trial. Several authors such as Thall, Simon, and Ellenberg (1988), Schaid, Wieand,

and Therneau (1990), and Bischoff and Miller (2005) have proposed two-stage designs to deal with the special case of selecting one treatment arm to compare with one control arm. A typical strategy of these designs is to eliminate the empirically inferior regimens at the end of a selection stage, and further randomize patients to the remaining superior arm and the control arm for final comparison. However, investigators often desire to monitor the trial data more frequently so as to act on any clear early evidence; this is especially so when a new biomarker or technology is used to ascertain the study endpoint. For example, Stallard and Todd (2003) allow multiple looks in the comparison stage between the one selected regimen and the control, and Cheung (2008) proposes a selection method with a sequential elimination of inferior treatments again for the special case of selecting one active treatment. In a somewhat different setting, Follmann, Proschan, and Geller (1994) extend the α-spending approach (Lan and DeMets, 1983) to multi-arm trial settings so that treatments may be selected based on pairwise comparisons with a prespecified overall type I error rate. Follman, Proschan, and Geller's method does allow a selection of several treatments, although such group-sequential approaches generally impose additional computational difficulties in the construction of stopping boundaries. At this time there is intense interest in adaptive trial designs with a selection component, and some actual trials have been designed along these lines; see, e.g., Levy *et al.* (2006) and Haley *et al.* (2010).

In this paper, we propose a class of sequential subset selection procedures designed to identify a subset of a pre-specified size $b \geq 1$ of active treatments that do better than control treatments, if possible. The procedures meet various requirements for multi-arm clinical trials: they are easy to implement, provide flexibility in scheduling interim analyses, and allow sequential elimination of inferior treatments and sequential recruitment of promising treatments while preserving the type I error rate. We explain these terms below. We treat the general case of comparing one or more experimental treatments with one or more control treatments, which could include standard active treatments and/or placebos. In our approach, selecting subsets is as easy to handle as selecting a single treatment. We treat only the case of fixed subset size, however, rather than the random subset size approach of Gupta (1956, 1965). The fixed subset size design will be most practical when budgetary constraints predetermine the number of selected treatments that can be studied further. For fixed b, the experimenter should have cogent reasons for requiring a set of no fewer than b treatments.

15.2. Terminology and Procedures

15.2.1. *Notation*

While patient outcomes may follow any distribution, we focus here on the case of binary outcomes primarily for their simplicity and clarity, in which case we may talk about tossing a set of c coins and the reader will understand that we mean observing binary patient outcomes on c treatments. Suppose then that we have $c \geq 2$ coins, with labels in the set $C = \{1, \cdots, c\}$. Among these c coins, there are c_0 "control" coins in the set C_0 and c_1 "active" coins in the set C_1 such that $C_0 \cup C_1 = C$ and $c_0 + c_1 = c$. For $i = 0$ or 1 representing controls and actives, respectively, let $p_j^{(i)}$ be the probability of heads (or success) on a single toss and $w_j^{(i)} = p_j^{(i)}/(1-p_j^{(i)})$ be the odds (of success) for the jth coin in C_i. Without loss of generality, we assume that $p_1^{(i)} \geq p_2^{(i)} \geq \cdots \geq p_{c_i}^{(i)}$ for $i = 0, 1$ and let $p_1 \geq \cdots \geq p_c$ be the ordered pooled success probabilities with the technical convention that whenever coins have tied probabilities, coins in C_0 are assigned the smallest subscripts. Of course, these true parameters and their ordering are not known in practice. For any integer b with $1 \leq b \leq c$, let (b) denote an ordered integer b-tuple of the form $(b) = (j_1, ..., j_b)$ with $1 \leq j_1 < \cdots < j_b \leq c$.

For any given b with $1 \leq b < c$, an interesting and reasonable goal for a comparative selection trial is to decide whether or not there exists a subset of b treatments every one of which is better than any control treatment, and if so, to identify such a subset. We call such a subset a "better-than-control b-tuple" and abbreviate this as a BTC(b). Alternate terms which may be useful depending on the context are a "better-than-placebo" or "better-than-standard" or "better-than-symptomatic"[a] subset of therapies. Formally, a BTC(b) is a subset $(b) \subseteq C_1$ such that $j \in (b) \Rightarrow p_j > p_1^{(0)}$.

[a] "Better-than-symptomatic" refers to a problem in the search for treatments for neurodegenerative diseases like Parkinson's disease (PD). Some medications alleviate the motor dysfunction symptoms of PD patients, but do not actually alter the progression of their disease. An important challenge is to distinguish new treatments that might truly modify the disease progression by retarding the neurodegeneration from those that "merely" treat the symptoms but do not change the course of the disease. This challenge was the topic of a workshop entitled "Demonstrating Disease-modifying Effects for the Treatment of Parkinson's Disease: Drug Development and Regulatory Issues" sponsored by the American Association of Pharmaceutical Scientists, FDA, the Michael J. Fox Foundation, and the Parkinson Study Group, in April, 2008.

We may formalize our goal by defining two hypotheses:

H_0: There does not exist a BTC(b), i.e., $p_b^{(1)} \leq p_1^{(0)}$,

H_1: There does exist a BTC(b), i.e., $p_b^{(1)} > p_1^{(0)}$.

If we accept H_0 then we will declare that no (b) is better than control, whereas if we reject H_0 in favor of H_1, then we will identify a b-tuple of active treatments which we will declare as a BTC(b). In fact, we will let our sequential selection procedure provide the following simple rejection criterion: If the selected subset, call it $(b)^*$, contains only active treatments, i.e., $(b)^* \subseteq C_1$, then we reject H_0 and declare that $(b)^*$ is a BTC(b). If, however, $(b)^*$ contains one or more control treatments, i.e., $(b)^* \cap C_0 \neq \emptyset$, then we do not reject H_0 and we declare that no BTC(b) exists. Therefore, a *correct declaration* may refer either to the event $(b)^* \cap C_0 \neq \emptyset$ (when H_0 is true) or to the event $(b)^* \subseteq C_1$ (when H_1 is true and $(b)^*$ is truly a BTC(b).) We denote either such event by $[cd]$.

If $c_1 > b$ there are several ways to choose $(b)^* \subseteq C_1$, and several of these may be BTC(b), but not all may be BTC(b) because $p_1^{(0)} \geq p_j$ for some $j \in (b)^*$. Under H_1, if $(b)^* \subseteq C_1$ but is not truly a BTC(b), we call that event a *false declaration* even though it leads us correctly to reject H_0. For example, with $b = 1$, $c = 3$, $c_1 = 2$, $c_0 = 1$, the parameters $p_1^{(1)} = 0.20$, $p_2^{(1)} = 0.10$, $p_1^{(0)} = 0.15$ are in H_1. But if we select the worst coin, which is active, we would correctly reject H_0 though our declaration that the selected coin is better than control would be false. There can be other false declaration events, e.g., when $(b)^* \cap C_0 \neq \emptyset$ under H_1, leading to both a type II error and a declaration of no BTC(b); or when $(b)^* \subseteq C_1$ under H_0, leading to both a type I error and a declaration that $(b)^*$ is a BTC(b).

When H_1 is true, the b-tuple of best coins, $(1, 2, ..., b)$, is a BTC(b), but as noted above, there may be others. Thus $P[(b)^* \text{ is a BTC}(b)] \geq P[(b)^* = (1, ..., b) = \text{best}(b)]$. Also, if $(b)^*$ is truly a BTC(b) it contains only active treatments, which results in a correct declaration and causes us correctly to reject H_0, though as noted above, H_0 can be rejected without a correct declaration. Thus $P[\text{Reject } H_0] \geq P[cd]$. It follows that for any vector of odds, $\underline{w} = (w_1, ..., w_c) = (p_1/q_1, ..., p_c/q_c)$ where $q_j = 1 - p_j$, if $\underline{w}$ is contained in H_1 we have

$$P_{\underline{w}}[\text{Reject } H_0] \geq P_{\underline{w}}[cd] = P_{\underline{w}}[(b)^* \text{ is a BTC}(b)] \geq P_{\underline{w}}[(b)^* = \text{best}(b)].$$

Thus if we design the procedure to guarantee that either of the two right-most terms is at least some pre-specified probability, say P^*, the procedure will have good probability of correct declaration (i.e., $P_{\underline{w}}[cd] \geq P^*$), and its power (i.e., $P_{\underline{w}}[\text{Reject } H_0]$) will be at least as good if not better.

In terms of controlling the probabilities of type I and II errors and the probabilities of correct declaration, we consider several requirements as follows.

(R1) If H_0 is true, we require that $P_{\underline{w}}[\text{Reject } H_0] \leq \alpha$, or equivalently, $P_{\underline{w}}[cd] \geq 1 - \alpha$, for a pre-specified α $(0 < \alpha < 1)$.

For the sake of efficiency, however, there may be occasions in which we may wish to relax the control of the type I error rate unless the control success probabilities are bounded by a pre-specified constant p_0 $(0 < p_0 \leq 1)$. Here p_0 represents a reasonable upper bound for the best control (placebo, standard, etc.) success probability. In such cases we may demand that (R1) hold true whenever $w_1^{(0)} \leq w_0$ for $w_0 = p_0/(1 - p_0)$. If the constraint is violated, we allow $P_{\underline{w}}[\text{Reject } H_0]$ to exceed α. This gives us (R2):

(R2) For given $w_0 > 0$, we require that $P_{\underline{w}}[\text{Reject } H_0] \leq \alpha$ whenever $\underline{w} \in H_0$ and $j \in C_0 \Rightarrow w_j \leq w_0$.

If we are further willing to assume that for given constants $\Psi_L \geq 1$ and $\Psi_U \geq 1$, $w_j \geq w_0/\Psi_L$ for all $j \in C_0$ and $w_j \leq w_0 \Psi_U$ for all $j \in C_1$, then we may require that (R1) hold true only under those assumptions. This gives us (R3):

(R3) For given $w_0 > 0$, $\Psi_L \geq 1$, and $\Psi_U \geq 1$, we require that $P_{\underline{w}}[\text{Reject } H_0] \leq \alpha$ whenever $\underline{w} \in H_0$, $w_0/\Psi_L \leq w_{c_0}^{(0)} \leq w_1^{(0)} \leq w_0$ and $w_1^{(1)} \leq w_0 \Psi_U$.

Requirement (R1) is a special case of (R2) with $p_0 = 1$ and $w_0 = \infty$, and requirement (R2) is a special case of (R3) with $\Psi_L = \Psi_U = \infty$.

To control the probability of correct declaration under the alternative hypothesis we consider two requirements.

(R4) If H_1 is true with $w_b^{(1)}/w_1^{(0)} \geq \theta > 1$ for a pre-specified "separation odds ratio" design constant θ between the bth best active treatment and the best control, then we require that $P_{\underline{w}}[(b)^* = \text{BTC}(b)] \geq P^*$ for a pre-specified P^* $(1/c < P^* < 1)$.

(R5) If, in fact, $w_b^{(1)} > w_{b+1}^{(1)} \geq w_1^{(0)}$ with $w_b^{(1)}/w_{b+1}^{(1)} \geq \theta$, then we require that $P_{\underline{w}}[(b)^* = \text{best}(b)] \geq P^*$.

The design constant θ represents a minimal clinically meaningful effect size for which we require a high probability P^* of correct declaration. We discuss how to design the procedure for each probability requirement (R1)–(R5) in Sec. 15.2.2.

To illustrate the above requirements, consider the following example. Suppose we have $c_1 = 3$ active coins and $c_0 = 2$ control coins and we wish to select a BTC pair of coins ($b = 2$). Assume that $\alpha = 0.05$, $P^* = 0.80$, and $\theta = 3$. If, in truth, $p_1^{(1)} = p_2^{(1)} = p_3^{(1)} = p_1^{(0)} = p_2^{(0)} = p$, then (R1) requires the selection procedure to identify a 2-tuple that includes at least one control coin, in which case we will declare "there is no better-than-control 2-tuple" and will not reject H_0, with probability no less than 0.95 ($= 1 - \alpha$). However, if $p \leq p_0 = 0.3$ and we design the procedure under that assumption, then the selection procedure will only need to satisfy (R2) rather than the more stringent (R1), i.e., we will require that $P_{\underline{w}}[\text{Reject } H_0]$ be less than or equal to 0.05 only if $p \leq 0.3$. If, in truth, $p > 0.3$, we allow the type I error rate to be greater than 0.05. Moreover, suppose we further restrict the parameter space to limit all the control odds to lie between w_0/Ψ_L and w_0 and all active odds no bigger than $w_0\Psi_U$ for given constants Ψ_L and Ψ_U. Then we will require the selection procedure to satisfy (R3) rather than (R1) or (R2). On the other hand, if the true p's are such that $w_2^{(1)} \geq 3w_1^{(0)}$, then we require the procedure to select 2 active coins, in which case we will declare "there is a better-than-control 2-tuple" (the selected one), with probability greater than or equal to 0.80 ($= P^*$) so that the procedure fulfills requirement (R4). If in fact $w_2^{(1)} \geq 3w_3^{(1)} \geq 3w_1^{(0)}$ then we require selection of the *best* two coins with probability at least 0.80 to meet (R5).

To accomplish these goals we use the Levin-Robbins-Leu (LRL) procedure (Leu and Levin, 2008) with *data augmentation* for control coins. The LRL procedure is actually a family of subset selection procedures which allow sequential elimination of inferior coins, sequential recruitment of superior coins, both, or neither. Any of the procedures guarantees the probability of correctly selecting the best b-tuple at a pre-specified level, P^*, when there is adequate separation, θ, between the odds of the bth and $(b+1)^{st}$ best coins. Data augmentation refers to a systematic enhancement of the treatment success tallies of given coins. We show below that with data augmentation for control coins, the modified procedure not only preserves the probability of correct declaration under (R4) or correct selection of the best b-tuple under (R5) when H_1 is true with adequate separation, it also controls the probability of type I error under H_0. We describe the procedure and its properties in detail with examples below. We indicate that there are two ways to augment the control data, stochastically or deterministically. We mainly discuss the stochastic method as it preserves the binary nature of the outcomes after augmentation and therefore allows the useful theoretical properties developed in our previous work to be applied directly. The deterministic method has certain other advantages to be discussed briefly in Sec. 15.2.3.

15.2.2. *Procedures and properties*

The original LRL procedure with elimination and recruitment (LRL-$\mathcal{E}/\mathcal{R}$, see Leu and Levin, 2008) proceeds as follows. A criterion or "reference" integer $r \geq 1$ is chosen in advance (see below for how to choose r). Begin tossing the c coins with vector-at-a-time sampling. We eliminate any "inferior" coin as soon as it falls r heads behind the coin or coins with the currently held b^{th} largest tally. We also recruit any "superior" coin as soon as it pulls r heads ahead of the coin or coins with the currently held $(b+1)^{st}$ largest tally. By *eliminate* we mean that a coin is withdrawn from the competition with no further tosses and is viewed as lying outside the set of b best coins. By *recruit* we mean that a coin is withdrawn from the competition with no further tosses and is selected to be among the set of b best coins. After each elimination and/or recruitment, we iterate the procedure with the remaining coins until b coins are recruited and/or $c-b$ coins are eliminated.

To specify the procedure more precisely, let $\mathbf{X}^{(n)}=(X_1^{(n)}, X_2^{(n)}, \cdots, X_c^{(n)})$ be the vector of tallies that reports the cumulative number of successes observed for each coin after n tosses, and let $\mathbf{X}^{[n]} = (X_1^{[n]}, X_2^{[n]}, \cdots, X_c^{[n]})$ be the ordered $\mathbf{X}^{(n)}$ vector with $X_1^{[n]} \geq X_2^{[n]} \geq \cdots \geq X_c^{[n]}$. Let $N_r^{(b,C)}$ be the *time of first elimination or recruitment* in a c-coin game with coins C:

$$N_r^{(b,C)} = \inf\{n \geq 1 : X_b^{[n]} - X_c^{[n]} = r\} \wedge \inf\{n \geq 1 : X_1^{[n]} - X_{b+1}^{[n]} = r\}.$$

At time $N_r^{(b,C)}$ we eliminate any coin j with $X_j^{(n)} = X_c^{[n]}$ if $X_b^{[n]} - X_c^{[n]} = r$ and/or recruit any coin j with $X_j^{(n)} = X_1^{[n]}$ if $X_1^{[n]} - X_{b+1}^{[n]} = r$. If fewer than b coins are recruited and/or fewer than $c-b$ coins are eliminated, the procedure continues, starting from the current tallies of the remaining subset of coins $C' \subset C$, and iterating with $N_r^{(b',C')}$, wherein $c' = |C'|$ replaces c and b' is b minus the total number of coins recruited at time $N_r^{(b,C)}$. Continuing in this way, if b'' coins remain to be recruited out of c'' coins still competing, we stop whenever there is a simultaneous recruitment of b'' coins and elimination of $c'' - b''$ coins, at which point a total of b coins will have been recruited and $c - b$ coins eliminated. Upon stopping we declare the subset of recruited coins as the b best. The procedure always identifies a well-defined subset of b coins with no ties.

Let $(b)^*$ denote the random b-tuple selected by the LRL-$\mathcal{E}/\mathcal{R}$ procedure. Define random variables $Y_1, ..., Y_{c-1}$ based on $(b)^*$ as follows. For $a = 1, ..., c-1$, let Y_a count the number of coins among the a truly best coins that are contained in $(b)^*$:

$$Y_a = \sum_{j=1}^{a} I[j \in (b)^*].$$

Many events of interest can be expressed in terms of the random variables $Y_1, ..., Y_{c-1}$. For example, selecting at least one of the two best coins is the event $[Y_2 \geq 1]$. A selection of the b best coins is the event $[Y_b = b]$ which, following tradition, we call a *correct selection* (cs). Note that $[cs]$ and $[cd]$ are not identical events.[b] For purposes of applying the following theorem, we may write the event of a correct selection equivalently as $[cs] = [Y_b \geq b]$ since Y_b can be no greater than b.

Theorem 15.1. *it Let $0 \leq y_1 \leq \cdots \leq y_{c-1}$ be nonnegative integers and let $A[\underline{y}]$ denote the event $A[\underline{y}] = [Y_1 \geq y_1, ..., Y_{c-1} \geq y_{c-1}]$. Then for any odds parameter vector $\underline{w} = (w_1, ..., w_c)$ with $w_1 \geq \cdots \geq w_c$, each member of the LRL family of procedures satisfies*

$$P_{\underline{w}}\{A[\underline{y}]\} \geq \sum_{(b) \in A[\underline{y}]} w^r_{(b)} \Big/ \sum_{(b)} w^r_{(b)},$$

where the sum in the numerator is over all b-tuples that would lead to event $A[\underline{y}]$, the sum in the denominator is over all $\binom{c}{b}$ possible b-tuples, and $w^r_{(b)} = w^r_{j_1} \cdots w^r_{j_b}$ for $(b) = (j_1, ..., j_b)$.

For example, if $b = 1$, Theorem 15.1 provides a lower bound to the probability of selecting the *best* coin: $P[cs] \geq w_1^r/(w_1^r + \cdots + w_c^r)$. This useful result allows us to determine r to guarantee $P[cs] \geq P^*$ if $w_1/w_2 \geq \theta$.

Leu and Levin (2008) prove Theorem 15.1 for the non-adaptive LRL procedure without elimination or recruitment (just stop when $X_b^{[n]} - X_{b+1}^{[n]} = r$). We conjecture that, remarkably, Theorem 15.1 also holds true for the procedure with elimination only, recruitment only, or both elimination and

[b] As noted above, $[cs] \Rightarrow [cd]$ but not conversely if $c_1 > 1$ under H_1 or if $c_0 > 1$ under H_0.

recruitment. The conjecture has been established by rigorous proof in certain special cases for each member of the family ($b = 1$ or $b = c - 1$; $b = 2$, $c = 4$; see Levin and Robbins, 1981, and Leu and Levin, 1999a and b, 2004, 2006, 2008) and by algebraic computation in many other cases. The conjecture is completely supported by extensive simulations. Although the proof of the completely general case is still an open problem, in what follows we shall assume Theorem 15.1 holds in all cases.

For the problem at hand, we apply the LRL-$\mathcal{E}/\mathcal{R}$ procedure using *augmented* control outcomes. The idea is to enhance the probability of selecting a control coin under H_0, thereby limiting the type I error rate. Any other LRL procedure may also be used.

First consider controlling the type I error rate for H_0. The stochastic augmentation procedure for control coins goes as follows. Let Z_j denote a binary outcome for coin j, so that $Z_j \sim Bin(1, p_j)$. For any control coin $j \in C_0$, after each toss, we preserve the original outcome if $Z_j = 1$; however, if $Z_j = 0$, we augment the data by adding an independent Bernoulli random variable $U_j \sim Bin(1, \pi)$, where π is given below. Thus the augmented outcome is $Z'_j = Z_j + U_j(1 - Z_j)$. Notice that the augmented outcome Z'_j still has a binary distribution, $Bin(1, p_j + \pi q_j)$.

Under H_0, the *least favorable configuration* (LFC) for the success odds is of the form $w_2^{(0)} = \cdots = w_{c_0}^{(0)} = 0$, $w_1^{(1)} = \cdots = w_{b-1}^{(1)} = \infty$, and $w_1^{(0)} = w_b^{(1)} = w_{b+1}^{(1)} = \cdots = w_{c_1}^{(1)} = w$ for some w. More precisely, this configuration minimizes the *lower bound* provided by Theorem 15.1. This suffices for our needs; we need not prove this configuration minimizes the exact $P[cd]$. Under the LFC, if we aim to amplify the odds of the best control coin by a factor of Ω, we must choose π such that

$$\frac{p_1^{(0)} + \pi q_1^{(0)}}{1 - (p_1^{(0)} + \pi q_1^{(0)})} = \frac{p_1^{(0)} + \pi q_1^{(0)}}{q_1^{(0)}(1 - \pi)} \geq \Omega \frac{p_1^{(0)}}{q_1^{(0)}}.$$

Assume for the moment that $p_1^{(0)}$ is known. Then we must choose π satisfying

$$\pi \geq \frac{(\Omega - 1)p_1^{(0)}}{1 + (\Omega - 1)p_1^{(0)}}. \tag{15.1}$$

To fulfill (R1), we need $P_{\underline{w}'}[\text{Reject } H_0] \leq \alpha$ or equivalently $P_{\underline{w}'}[cd] \geq 1 - \alpha$. We show below using Theorem 15.1 that

$$P_{\underline{w}'}[cd] = P_{\underline{w}'}[\text{Select the best control}] \geq \frac{\Omega^r}{\Omega^r + c - b}, \tag{15.2}$$

where $\underline{w}'$ denotes the augmented odds based on the LFC. Thus we may choose Ω^r such that

$$\frac{\Omega^r}{\Omega^r + c - b} \geq 1 - \alpha. \tag{15.3}$$

Therefore, once r is chosen, we set $\Omega = \{(c-b)(1-\alpha)/\alpha\}^{1/r}$ and specify π using (15.1), assuming $p_1^{(0)}$ is known. However, as (R1) assumes nothing about $p_1^{(0)}$, without making further assumptions on $\underline{p}$, we can only choose π conservatively as

$$\pi = \frac{(\Omega - 1)}{\Omega}, \text{ i.e., we set } p_0 = 1,$$

in order to bound the probability of type I error under α. This would clearly be extremely conservative for most practical purposes.

On the other hand, if we are willing to assume an upper bound for $p_1^{(0)}$, say p_0, and relax the requirement from (R1) to (R2) (i.e., to control the type I error rate less than or equal to α whenever $p_1^{(0)} \leq p_0$ but not necessarily if $p_1^{(0)} > p_0$), then we may use a less conservative π by replacing $p_1^{(0)}$ by p_0 in (15.1), i.e., to use

$$\pi = \frac{(\Omega - 1)p_0}{1 + (\Omega - 1)p_0}.$$

Note that the actual $P[cd]$ will exceed $\Omega^r/(\Omega^r + c - b)$ if the actual $p_1^{(0)} < p_0$. This is because the augmentation using $\pi = (\Omega - 1)p_0/(1 + (\Omega - 1)p_0)$ will give an an actual control coin with success probability of $p_1^{(0)}$ an odds of

$$\frac{p_1^{(0)} + \frac{(\Omega-1)p_0}{1+(\Omega-1)p_0} q_1^{(0)}}{1 - \{p_1^{(0)} + \frac{(\Omega-1)p_0}{1+(\Omega-1)p_0} q_1^{(0)}\}} = \frac{p_1^{(0)} + (\Omega - 1)p_0}{q_1^{(0)}},$$

which, compared with $p_1^{(0)}/q_1^{(0)}$, gives an odds ratio of $\Omega' = 1 + (\Omega - 1)p_0/p_1^{(0)}$, which is $\geq \Omega$ if and only if $p_1^{(0)} \leq p_0$. Thus we see that $\pi = (\Omega - 1)/\Omega$ will generally yield a conservative type I error rate since rarely would $p_0 = 1$ be a sharp or even practical upper bound for $p_1^{(0)}$. (R2) or (R3) will therefore be the requirement of greatest interest under H_0.

To prove (15.2), we consider two cases. Case 1 of the LFC is that the common odds $w_1^{(0)} = w_b^{(1)} = \cdots = w_{c_1}^{(1)} = w \geq (\Omega - 1)p_0$. After augmentation, the control coins with $w_2^{(0)} = \cdots = w_{c_0}^{(0)} = 0$ have odds of $\pi/(1-\pi) = (\Omega - 1)p_0$. Then a correct declaration is at least as likely as a correct selection of the best b-tuple (consisting of all $b - 1$ active coins with $w_j^{(1)} = \infty$ and the best control with augmented success probability $p_1^{(0)} + \pi q_1^{(0)}$). Since the $b-1$ best coins will be selected with probability 1, it suffices to bound $P[cs]$ in a $c-b+1$ coin game with best coin odds $\Omega' w$ and selecting the best $b' = 1$ coin. The lower bound provided by Theorem 15.1 for this game is

$$P[cd] \geq P[cs] \geq \frac{(\Omega' w)^r}{(\Omega' w)^r + (c_1 - b + 1)w^r + (c_0 - 1)\{(\Omega - 1)p_0\}^r}$$
$$= \frac{\Omega'^r}{\Omega'^r + (c_1 - b + 1) + (c_0 - 1)\{(\Omega - 1)p_0/w\}^r}.$$

But in the present case $(\Omega - 1)p_0/w \leq 1$, hence $P[cd] \geq \Omega'^r/(\Omega'^r + c - b)$. As mentioned above, $\Omega' \geq \Omega$ for $p_1^{(0)} \leq p_0$, whence $P[cd] \geq \Omega^r/(\Omega^r + c - b)$.

Case 2 of the LFC is that the common odds $w_1^{(0)} = w_b^{(1)} = \cdots = w_{c_1}^{(1)} = w < (\Omega - 1)p_0$. In this case a correct declaration is the event $[Y_{c_0} \geq 1]$, i.e., the $b' = 1$ selected coin is one of the c_0 best coins after augmentation, with best odds $\Omega' w$ and the remaining control coins with augmented odds $(\Omega - 1)p_0$ (again where we have reduced the game by excluding the $b - 1$ perfect active coins). Theorem 15.1 provides the lower bound

$$P[cd] \geq P[Y_{c_0} \geq 1] \geq \frac{(\Omega' w)^r + (c_0 - 1)\{(\Omega - 1)p_0\}^r}{(\Omega' w)^r + (c_0 - 1)\{(\Omega - 1)p_0\}^r + (c_1 - b + 1)w^r}$$
$$= \frac{\Omega'^r + (c_0 - 1)\{(\Omega - 1)p_0/w\}^r}{\Omega'^r + (c_0 - 1)\{(\Omega - 1)p_0/w\}^r + (c_1 - b + 1)}.$$

Now in this case, $(\Omega - 1)p_0/w \geq 1$, and since the bound is monotonic increasing in $(c_0 - 1)\{(\Omega - 1)p_0/w\}^r$ we have

$$P[cd] \geq \frac{\Omega'^r + (c_0 - 1)}{\Omega'^r + (c_0 - 1) + (c_1 - b + 1)}$$
$$= \frac{\Omega'^r + c_0 - 1}{\Omega'^r + c - b} \geq \frac{\Omega'^r}{\Omega'^r + c - b} \geq \frac{\Omega^r}{\Omega^r + c - b}.$$

This completes the proof of (15.2).

Under (R3) we assume further that all the control odds lie between w_0/Ψ_L and w_0, and that all active odds are bounded above by $w_0\Psi_U$ for given constants Ψ_L and Ψ_U. In such cases the LFC is $w_2^{(0)} = \cdots = w_{c_0}^{(0)} = w_0/\Psi_L$, $w_1^{(1)} = \cdots = w_{b-1}^{(1)} = \Psi_U w_0$, and $w_1^{(0)} = w_b^{(1)} = w_{b+1}^{(1)} = \cdots = w_{c_1}^{(1)} = w_0$ under H_0. Then (R3) requires (after augmentation)

$$
\begin{aligned}
P_{\underline{w}'}[cd] &= P_{\underline{w}'}[\text{Select at least one control coin}] \\
&\geq P_{\underline{w}'}[\text{Select the best control coin in } (b)^*] \\
&\geq P_{\underline{w}'}[\text{Select the best control coin with the best } (b-1) \text{ active coins}] \\
&= P_{\underline{w}'}[cs] \;\geq\; \frac{\Psi_U^{(b-1)r}\Omega^r}{D}
\end{aligned}
\tag{15.4}
$$

where

$$
\begin{aligned}
D =& \sum_{i=0}^{(b-1)} \sum_{j=0\vee(b-c_0-i)}^{(c_1-b+1)\wedge(b-i-1)} \binom{b-1}{i}\binom{c_1-b+1}{j}\binom{c_0-1}{b-i-j-1}\Psi_U^{ir}\Psi_L^{-(b-i-j-1)r}\Omega^{(b-i-j)r} \\
&+\sum_{i=0}^{(b-1)} \sum_{j=0\vee(b-c_0-i+1)}^{(c_1-b+1)\wedge(b-i)} \binom{b-1}{i}\binom{c_1-b+1}{j}\binom{c_0-1}{b-i-j}\Psi_U^{ir}\Psi_L^{-(b-i-j)r}\Omega^{(b-i-j)r}.
\end{aligned}
$$

To satisfy (R3), we solve for Ω^r in

$$
\frac{\Psi_U^{(b-1)r}\Omega^r}{D} = 1-\alpha. \tag{15.5}
$$

Once r is determined, Ω is too, and finally π is chosen as above.

We note that in the special case $\Psi_L = \Psi_U = 1$, which envisions only one point in H_0, viz., $p_1 = \cdots = p_c = p_0$, (15.5) is overly conservative. In this case, in (15.4) we can bound the first equality directly using Theorem 15.1:

$$
\begin{aligned}
P_{\underline{w}'}[cd] &= P_{\underline{w}'}[\text{Select at least one control coin}] \\
&= P[Y_{c_0} \geq 1] \geq \frac{\sum_{i=1}^{b\wedge c_0}\binom{c_0}{i}\binom{c_1}{b-i}\Omega^{ri}}{\sum_{i=0}^{b\wedge c_0}\binom{c_0}{i}\binom{c_1}{b-i}\Omega^{ri}} = 1-\alpha,
\end{aligned}
\tag{15.6}
$$

and (15.6) can be used to solve for Ω^r. This special case may not be of especially great interest, however, since if $\underline{p} \in H_0$ but not all p_j are equal, (15.6) will not control the type I error rate.

The integer r is determined to guarantee the probability of correct declaration under H_1. Under (R4), the LFC is $w_1^{(0)} = \cdots = w_{c_0}^{(0)} = w_{b+1}^{(1)} = \cdots = w_{c_1}^{(1)} = w$, say, and $w_1^{(1)} = \cdots = w_b^{(1)} = \theta w$. Assuming for the moment that $\theta > \Omega$, the augmented odds are $\underline{w}' = (\theta w, \cdots, \theta w, \Omega w, \cdots, \Omega w, w, \cdots, w)$ with b, c_0, and $c_1 - b$ equal terms, respectively. Using the augmented data, Theorem 15.1 implies

$$\begin{aligned}P_{w'}[cd] &= P_{w'}[\text{Select } b \text{ active coins}]\\ &\geq P_{w'}[\text{Select the best } (b)]\\ &\geq \frac{\theta^{br}}{\sum_{i=0}^{b}\sum_{j=0\vee 2b-c_1-i}^{c_0\wedge b-i}\binom{b}{i}\binom{c_0}{j}\binom{c_1-b}{b-i-j}\theta^{ri}(\Omega^r)^j} \geq P^*. \qquad (15.7)\end{aligned}$$

Notice that Ω appears in (15.7) only as Ω^r which has already been determined by the type I error constraints (R1), (R2), or (R3), leaving r as the only unknown. To guarantee $P_{w'}[cd] \geq P^*$ under (R4), our choice of r must satisfy (15.7). The LFC for selecting the best b-tuple under (R5) is the same as for (R4). Thus we use (15.7) to choose r under either (R4) or (R5). Once r is found, set $\Omega = (\Omega^r)^{1/r}$, and set $\pi = (\Omega - 1)/\Omega$ under (R1) and $\pi = (\Omega - 1)p_0/(1 + (\Omega - 1)p_0)$ under (R2) or (R3). It can be shown that for any α, P^*, and θ the value of r satisfying (15.7) will always yield $\Omega < \theta$ (proof omitted).

This completes the description of the stochastic augmentation procedure, choice of r and Ω, and the selection procedure to achieve control of the type I error rate for H_0 and $P[cd] \geq P^*$ under requirements (R1)–(R5).

15.2.3. *Example*

Consider the design of a dose selection trial in which there are two active treatments ($c_1 = 2$) and one control ($c_0 = 1$). The trial objective is to select one better-than-control treatment ($b = 1$) for a binary outcome determined, e.g., by magnetic resonance imaging (Fisher *et al.*, 2006). We require the following: if there is an active coin with odds at least twice as great as the control coin (i.e., the odds ratio $\geq \theta = 2$), we want the probability of correct selection to be at least $P^* = 0.80$, but if $w_1^{(1)} \leq w_1^{(0)}$, we want to reject H_0 with probability no greater than $\alpha = 0.05$. Then under (R1) or (R2) we must augment the control coin odds by a factor of Ω such that $\Omega^r/(\Omega^r + 2) \geq 1 - \alpha$ by (15.3). This determines $\Omega^r \geq 38$. As we also require $P[cd] \geq 0.80$ under (R4), the criterion integer r must satisfy (15.7), i.e., $P[cd] = P[cs] \geq \theta^r/(1 + \theta^r + \Omega^r) \geq 0.80$. So we choose $r = 8$ and $\Omega = 38^{1/r} \sim 1.6$. At termination, if either active coin is selected, then we declare "there exists a better-than-control treatment" and we select that one as the best; on the other hand, if the control coin wins, then we declare "there is no better-than-control treatment". By choosing $\Omega^r = 38$ (or greater) and $r = 8$ (or greater), the procedure will allow us to make a correct declaration with probability at least 0.95 if the control coin is at least as good as the best active coin, whereas the probability of correct selection of a better-than-control active coin will be at least 0.80 if there indeed exists an active coin with odds twice as large as that of the control coin. If $p_0 = 0.2$ under (R2), then we augment the control coin with $U \sim Bin(1, \pi)$ with $\pi = (0.6)(0.2)/1.12 \sim 0.107$.

Another way to augment the control data is deterministic. The deterministic method produces the augmented outcome $Z_j^{''} = Z_j + \pi(1 - Z_j)$ for all control coins. Note however that $Z_j^{''}$ is no longer a binary variable. The LRL procedure is modified to eliminate when $X_b^{''[n]} - X_c^{''[n]} \geq r$ and/or recruit when $X_1^{''[n]} - X_{b+1}^{''[n]} \geq r$. Theorem 15.1 is not immediately applicable for the deterministic augmentation procedure, but simulations show that with the previously determined Ω and r for stochastic augmentation, the probabilities of correct declaration are even greater when using deterministic augmentation, although larger sample sizes are required on average. This is illustrated below.

15.3. Simulation Studies

In the following illustration, we present some operating characteristics obtained by simulation under three scenarios. Suppose $b = 2$, $c = 4$, $c_1 = c_0 = 2$, $\alpha = 0.10$, $p_0 = 0.5$, $\theta = 3$, and $P^* = 0.90$. Under these design constants, we need $\Omega^r \geq 18$ from (15.3) and $r \geq 6$ from (15.7) to guarantee the probability of correct declaration at least 0.90 with type I error under 0.10. Thus $\Omega = 18^{1/6} \simeq 1.6189$ and $\pi = 0.2363$ from (15.1) with $p_1^{(0)} = p_0 = 0.5$. In Table 15.1 we report the probability of correct declaration; the expected number of rounds (i.e., sampled vectors), N; the expected total number of tosses (patients), T; and the expected total number of failures, F. In the first row we simulate under the design alternative at $\underline{p} = (0.75, 0.75, 0.5, 0.5)$ for the two active and two control coins, respectively. In the second row we simulate under H_0 at the LFC $(p_1^{(1)}, p_1^{(0)}, p_2^{(1)}, p_2^{(0)}) = (1, 0.5, 0.5, 0)$. The third row simulates under the special null hypothesis $p_1 = \cdots = p_c = 0.5$, i.e., (R3) with $\Psi_L = \Psi_U = 1$.

We confirm that with stochastic augmentation, the type I error is controlled under the null hypothesis that the pair of active treatments is not a BTC pair, and that the probability of correct declaration is no less than P^* under the design alternative. Note that under the special null case H_{00}: $p_1 = \cdots = p_c = p$ at $p = 0.5$ the P[Type I error] is well below $\alpha = 0.10$.

The following example compares the stochastic augmentation procedure

Table 15.1. Simulation results for the null hypothesis and design alternative. $\Omega = 18^{1/6}$, $\pi = 0.2363$, designed under (R2) with $p_0 = 0.5$, $r = 6$.

Success probabilities	*Lower bound*	$P[cd]$	$E[N]$	$E[T]$	$E[F]$
$(\mathbf{0.75}, \mathbf{0.75}, 0.5, 0.5)^a$	0.910	0.926	68.0	211.8	78.9
$(\mathbf{1}, 0.5, \mathbf{0.5}, 0)^b$	0.947	0.948	45.5	116.6	59.9
$(0.5, 0.5, \mathbf{0.5}, \mathbf{0.5})^c$	0.998	0.999	72.6	221.6	110.8

[a]Design alternative. [b]Null hypothesis LFC. [c]Equiprobable coins.
Active coin success probabilities are in boldface. All simulations used 100,000 replications.

with the deterministic augmentation procedure. In this comparison, we assume the same design parameters as in the previous illustration. We used 100,000 replications of coordinated binary sequences, i.e., we used the same set of sequences of binary c-vectors for the deterministic method and for the stochastic method under each scenario. As shown in Table 15.2, the deterministic method has a somewhat higher probability of correct declaration, though the stochastic method has smaller $E[N]$, $E[T]$, and $E[F]$ than the deterministic method.

Table 15.2. Comparisons between stochastic and deterministic augmentation. $\Omega = 18^{1/6}$, $\pi = 0.2363$, designed under (R2) with $p_0 = 0.5$, $r = 6$.

Success probabilities	*Method**	*LB***	$P[cd]$	$E[N]$	$E[T]$	$E[F]$
(**0.75**, **0.75**, 0.5, 0.5)	S	0.910	0.926	68.0	211.8	78.9
(**0.75**, **0.75**, 0.5, 0.5)	D	0.910	0.973	71.5	228.8	86.4
(**1**, 0.5, **0.5**, 0)	S	0.947	0.948	45.5	116.6	59.9
(**1**, 0.5, **0.5**, 0)	D	0.947	0.978	51.7	129.9	66.8
(0.5, 0.5, **0.5**, **0.5**)	S	0.998	0.999	72.6	221.6	110.8
(0.5, 0.5, **0.5**, **0.5**)	D	0.998	0.999	81.4	253.0	126.5

*S = Stochastic D = Deterministic.
**Lower bound.
Active coin success probabilities are in boldface. All simulations used 100,000 coordinated replications.

15.4. Discussion

The proposed procedure accomplishes the goal of guaranteeing a high probability of correctly selecting a better-than-control subset of fixed size with simultaneous control of the type I error rate. Compared with non-adaptive selection procedures, the feature of allowing early elimination of inferior treatments and recruitment of superior treatments not only reduces the cost of the study but also decreases the average number of failures, which has ethical appeal.

How do these procedures compare to a classical fixed sample size hypothesis test of equal success probabilities? For a classical hypothesis test of H_{00} at level $\alpha = 0.10$ with fixed sample sizes of n patients per group, to achieve 90% power to reject H_0 at (0.75, 0.75, 0.50, 0.50) with a chi-squared test of homogeneity would require $T = 4n = 4 \times 45 = 180$ patients with $E[F] = 90$ under H_0 and $E[F] = 67.5$ under the design H_1. This comparison is not entirely fair, because the new procedure achieves more goals than merely testing H_{00}: it selects a BTC(b) with good $P[cd]$ under a wider null, H_0.

A fairer comparison would be to compare the classical hypothesis test with the less conservative selection procedure given by (6) for testing H_{00}. This is illustrated in Table 15.3. Here $r = 4$, $\Omega = 1.1257$, $\pi = 0.0591$. For the classical test, if we reject H_0 we select the two coins with best tallies (breaking ties if necessary).

Table 15.3. Comparisons with the fixed sample size design. $\Omega = 1.1257$, $\pi = 0.0591$, designed under (R3) with $\Psi_L = \Psi_U = 1$, $p_0 = 0.5$, and $r = 4$.

	Lower bound	$P[cd]$	$E[N]$	$E[T]$	$E[F]$
Design alternative					
$\underset{\sim}{p} = (\mathbf{0.75}, \mathbf{0.75}, 0.5, 0.5)$					
without curtailment	0.926	0.941	27.0	84.9	31.6
with curtailment	0.926	0.941	26.0	82.6	30.8
fixed	0.90*	< 0.90*	45.0	180.0	67.5
Equiprobable coins					
$\underset{\sim}{p} = (0.5, 0.5, \mathbf{0.5}, \mathbf{0.5})$					
without curtailment	0.900	0.912	48.7	143.4	71.7
with curtailment	0.900	0.912	26.2	91.4	45.7
fixed	0.90**	0.900	45.0	180.0	90.0

*This is statistical power. Not every rejection will result in a correct declaration of a BTC(b), so $P[cd] < 0.90$.
**This is $1 - \alpha$.
Active coin success probabilities are in boldface. All simulations used 100,000 replications.

We note finally that if all one wanted to do is to test H_0, the procedure does not need to go to the end. Curtailment (without complete selection) is possible to test H_0 under the following conditions:

(i) If any control coin is recruited, then we stop and accept H_0.

(ii) If the number of active coins still in the game is less than the number of coins still to be selected, then we stop and accept H_0.

(iii) If all control coins have been eliminated (and all recruited coins are actives), then we stop and reject H_0.

Although curtailment under conditions (i) or (ii) makes sense, in a comparative selection trial one would generally not want to curtail under (iii). One would rather continue to the end to select a BTC(b). This would be necessary in order to guarantee the required probability of correct selection. Table 15.3 shows the effect of curtailment under conditions (i) or (ii) only.

Finally, we note that it may seem unusual to augment control treatment outcomes stochastically in an actual clinical trial. How could the procedure be replicated or audited? Could the system be subverted? In fact, modern clinical trials almost always use replicable randomization lists, e.g., randomized permuted block schemes, that can be prepared in advance, replicated, and audited at any time. It is no more difficult to construct a sequence of binary outcomes following $Bin(1, \pi)$ that can be prepared in advance and audited at any time. By such a device the "unreproducibility" of a truly stochastic scheme can be avoided. If deterministic augmentation is used this issue does not arise, although other steps must be taken to preserve the blinding, since control arms will possess non-integer tallies. See Cheung (2008) for further discussion of this point.

Acknowledgment

We thank the anonymous referee for carefully reading our paper and for providing very helpful comments.

References

Bischoff, W. and Miller, F. (2005). Adaptive two-stage test procedures to find the best treatment in clinical trials. *Biometrika* **92,** 197-212.

Cheung, Y. K. (2008). Simple sequential boundaries for treatment selection in multi-armed randomized clinical trials with a control. *Biometrics* **64,** 940-949.

Fisher M, Cheung K, Howard G, and Warach S (2006). New pathways for evaluating potential acute stroke therapies. *International Journal of Stroke* **1,** 52-58.

Follmann, D. A., Proschan, M. A., and Geller, N. L. (1994). Monitoring pairwise comparisons in multi-armed clinical trials. *Biometrics* **50,** 325-336.

Gupta, S. S. (1956). On a decision rule for a problem in ranking means, Mimeograph Series 150, Institute of Statistics, Chapel Hill: University of North Carolina.

Gupta, S. S. (1965). On Some Multiple Decision (Selection and Ranking) Rules, *Technometrics* **7,** 225-245.

Haley, E.C., Thompson, J.L.P., Grotta, J.C., Lyden, P.D., Hemmen, T.G., Brown, D.L., Fanale, C., Libman, R., Kwiatkowski, T.G., Llinas, R.H., Levine, S.R., Johnston, K.C., Buchsbaum, R., Levy, G., and Levin, B., for the Tenecteplase in Stroke Investigators (2010). Phase IIB/III Trial of Tenecteplase in Acute Ischemic Stroke: Results of a Prematurely Terminated Randomized Clinical Trial. *Stroke* **41,** 707-711.

Lan, K. K. G. and DeMets, D. L. (1983). Discrete sequential boundaries for clinical trials. *Biometrika* **70,** 659-663.

Leu, C.-S. and Levin, B. (1999a). On the probability of correct selection in the Levin-Robbins sequential elimination procedure. *Statistica Sinica* **9,** 879-891.

Leu, C.-S. and Levin, B. (1999b). Proof of a lower bound formula for the expected reward in the Levin-Robbins sequential elimination procedure. *Sequential Analysis* **18,** 81-105.

Leu, C.-S. and Levin, B. (2004). Selecting the Best Subset of b out of c coins with the Levin-Robbins sequential elimination procedure: proof of the lower bound formula for the probability of correct selection in the case $b = 2$, $c = 4$, Technical Report #B-91, Department of Biostatistics, Columbia University, July 1, 2004.
Available at http://biostats.bepress.com/columbiabiostat/.

Leu, C.-S. and Levin, B. (2006). Proof of the lower bound formula for the probability of correct binomial subset selection with the Levin-Robbins-Leu sequential elimination and recruitment procedure in the case $b = 2$, $c = 4$. Technical Report #B-98, Department of Biostatistics, Columbia University, December 15, 2006.
Available at http://biostats.bepress.com/columbiabiostat/.

Leu, C.-S. and Levin, B. (2008). On a conjecture of Bechhofer, Kiefer, and Sobel for the Levin-Robbins-Leu Binomial subset selection procedures. *Sequential Analysis* **27,** 106-125.

Levin, B. and Robbins, H. (1981). Selecting the highest probability in binomial or multinomial trials. *Proceedings of the National Academy of Sciences of the United States of America* **78,** 4663-4666.

Levy, G., Kaufmann, P., Buchsbaum, R., Montes, J., Barsdorf, A., Arbing, R., Battista, V., Zhou, X., Mitsumoto, H., Levin, B., Thompson, J.L.P. (2006). A two-stage design for a phase II clinical trial of coenzyme Q10 in ALS. *Neurology* **66**, 660-663.

Schaid, D. J., Wieand, S., and Therneau, T. M. (1990). Optimal two-stage screening designs for survival comparisons. *Biometrika* **77,** 507-513.

Stallard, N. and Todd, S. (2003). Sequential designs for phase III clinical trials incorporating treatment selection. *Statistics in Medicine* **22,** 689-703.

Thall, P. F., Simon, R., and Ellenberg, S. S. (1988). Two-stage selection and testing designs for comparative clinical trials. *Biometrika* **75,** 303-310.

Index